AF761209

Analytical Chemistry and Microchemistry

nova
science publishers
www.novapublishers.com

Analytical Chemistry and Microchemistry

FTIR Spectroscopy: Advances in Research and Applications
Adriana S. França, PhD (Editor)
Leandro S. Oliveira, PhD (Editor)
2022. ISBN: 978-1-68507-571-2 (Softcover)
2022. ISBN: 978-1-68507-939-0 (eBook)

Trace Metals: Sources, Applications and Environmental Implications
Oscar M. Thygesen (Editor)
2022. ISBN: 978-1-68507-797-6 (Softcover)
2022. ISBN: 978-1-68507-827-0 (eBook)

Mass Spectrometry: Theory and Applications
William O. Nichols (Editor)
2021. ISBN: 978-1-53619-790-7 (Hardcover)
2021. ISBN: 978-1-53619-874-4 (eBook)

Aflatoxins: Biochemistry, Toxicology, Public Health, Policies and Modern Methods of Analysis
Spyridon Kintzios (Editor)
Sofia Mavrikou (Editor)
2020. ISBN: 978-1-53616-785-6 (Hardcover)
2019. ISBN: 978-1-53616-786-3 (eBook)

Gas Chromatography: History, Methods and Applications
Percy Henrichon (Editor)
2020. ISBN: 978-1-53617-350-5 (Softcover)
2020. ISBN: 978-1-53617-352-9 (eBook)

More information about this series can be found at
https://novapublishers.com/product-category/series/analytical-chemistry-and-microchemistry/

Larry D. Crenshaw
Editor

Chemometrics

Advances in Applications and Research

nova
science publishers

www.novapublishers.com

Copyright © 2022 by Nova Science Publishers, Inc.

All rights reserved. No part of this book may be reproduced, stored in a retrieval system or transmitted in any form or by any means: electronic, electrostatic, magnetic, tape, mechanical photocopying, recording or otherwise without the written permission of the Publisher.

We have partnered with Copyright Clearance Center to make it easy for you to obtain permissions to reuse content from this publication. Simply navigate to this publication's page on Nova's website and locate the "Get Permission" button below the title description. This button is linked directly to the title's permission page on copyright.com. Alternatively, you can visit copyright.com and search by title, ISBN, or ISSN.

For further questions about using the service on copyright.com, please contact:
Copyright Clearance Center
Phone: +1-(978) 750-8400 Fax: +1-(978) 750-4470 E-mail: info@copyright.com.

NOTICE TO THE READER

The Publisher has taken reasonable care in the preparation of this book, but makes no expressed or implied warranty of any kind and assumes no responsibility for any errors or omissions. No liability is assumed for incidental or consequential damages in connection with or arising out of information contained in this book. The Publisher shall not be liable for any special, consequential, or exemplary damages resulting, in whole or in part, from the readers' use of, or reliance upon, this material. Any parts of this book based on government reports are so indicated and copyright is claimed for those parts to the extent applicable to compilations of such works.

Independent verification should be sought for any data, advice or recommendations contained in this book. In addition, no responsibility is assumed by the Publisher for any injury and/or damage to persons or property arising from any methods, products, instructions, ideas or otherwise contained in this publication.

This publication is designed to provide accurate and authoritative information with regard to the subject matter covered herein. It is sold with the clear understanding that the Publisher is not engaged in rendering legal or any other professional services. If legal or any other expert assistance is required, the services of a competent person should be sought. FROM A DECLARATION OF PARTICIPANTS JOINTLY ADOPTED BY A COMMITTEE OF THE AMERICAN BAR ASSOCIATION AND A COMMITTEE OF PUBLISHERS.

Additional color graphics may be available in the e-book version of this book.

Library of Congress Cataloging-in-Publication Data

ISBN: 979-8-88697-274-0

Published by Nova Science Publishers, Inc. † New York

Contents

Preface

Chemometrics is a discipline of chemistry that finds correlation between specific data using mathematical and statistical methods. During any thorough research, the scientists are handling vast amounts of data related to the samples which are being researched. In this type of research, finding the correlation (similarities or differences) between analyzed samples and data is of great importance. In the first chapter, commonly used chemometrics for spectral modeling transfer is examined. The second chapter provides an analytical tool to detect fraud when olive oil is illegally blended with VOs or a 'legal' blend is falsely labelled with respect to the botanical nature of the oils mixed and/or the percentage of each oil in the declared mixture. H-NMR spectral data of olive and virgin olive oils and their mixtures with the VOs most commonly used to make blends was analysed by pattern recognition techniques to develop multivariate classification and regression models, which were organised in a decision tree to afford a stepwise strategy for the aimed purposes. The next chapter focuses on a metabolomics approach based on H-NMR fingerprinting and multivariate data analysis for virgin olive oil stability assessments. In the fourth chapter, the authors review unsupervised methods using both principal component analysis (PCA) and hierarchical cluster analysis (HCA). Using these methods, they were able to spot the correlation between the samples and underlying data structures without the potential bias of scientists about the previous knowledge of data samples.

Chapter 1 - A situation of the model cannot give correct prediction results due to the difference in spectra measured by different instruments is often encountered in the process of spectral analysis, especially when the model established on one spectrometer is used on another spectrometer. The first way to address this problem, is to improve the standardization of instrument hardware processing, improve the level of processing technology, and reduce the differences between the host and the secondary in terms of devices. There have been many reports on the standardization of spectroscopic instrument hardware such as calibrating the wavelength accuracy, absorbance accuracy, resolution, spectral response line type and symmetry of the spectrometer through sharp-line emission light source, standard materials, etc. As well, there have been developments on instruments by optimizing optical components, assembly processes, and control strategies with performance indicators within the allowable range of variation.

In recent years, some portable instruments can also transfer spectra between the same type. Surely, different applications have various requirements for the consistency of instruments. Models established by different methods and different systems also have different tolerances for divergence between instruments. Due to the sharp peaks of spectrum, Raman spectrometry can better solve the problem of consistency between instruments via hardware. Besides spectroscopic instruments, other instruments like mass spectrometers have also encountered similar problems, but most of which can be solved by simple peak calibration.

Although there has been a development for decades, differences between different brands of instruments still exist, e. g. the difference between a grating type and a Fourier transform type spectrometers. Due to these differences, the inapplicability of the multivariate models would produce unacceptable systematic prediction bias. Thus, the solution to this prediction bias is called Model Transfer or Instrument Transfer. In the area of machine learning, the relevant keywords are Transfer Learning, Domain Adaptation, and Multi-Task Learning, etc.

Transfer between the spectra establishes the functional relationship between the spectra measured by the host and secondary through a mathematical method. There are two ways of spectral conversion. One is to convert the host's calibration spectra, and then re-establish a calibration model suitable for the secondary spectra, that is called Reverse Standardization in the literatures. The other is to convert the secondary spectra, and directly use the host model to predict the result. Algorithm of two ways is essentially the same and needs to be selected according to different applications. A robust calibration is established through the preprocessing, wavelength variable screening algorithms, etc. Or the host calibration set is expanded by adding spectra under different test conditions, spectra from secondary, etc., to establish a global calibration, or hybrid calibration. Thus, this chapter mainly introduces the model transfer methods.

Chapter 2 - The high price of olive oil, its distinctive sensory profile and its reputation as a healthy source of dietary fats make olive oil a target for fraud. The most common types of olive oil fraud are illegal blending with other vegetable oils (VOs) or low-quality olive oils, deliberate mislabelling of less expensive classes of olive oils, other VOs or their blends with olive oils, and mislabelling of geographical origin or Protected Designation of Origin (PDO) declaration. Olive oil adulteration, being one of the biggest financial frauds in the agricultural sector, evidenced the need to update and harmonize analytical methods for quality and authenticity control of olive oil. A novel stepwise strategy based on the 1H-NMR fingerprint of edible oils and multivariate data analysis has been developed in order to assure the authenticity and traceability of olive oils and their declared blends with VOs. This approach provides an analytical tool to detect fraud when olive oil is illegally blended with VOs or a 'legal' blend is falsely labelled respect to the botanical nature of the oils mixed and/or the percentage of each oil in the declared mixture. 1H-NMR spectral data of olive and virgin olive oils and their mixtures with the VOs most commonly used to make blends, i.e., sunflower, high oleic sunflower, corn, virgin and refined avocado, virgin and refined hazelnut, soybean oil, refined palm olein and desterolised high oleic sunflower oils, was analysed by pattern recognition techniques to develop multivariate classification and regression models, which are organised in a decision tree to afford a stepwise strategy for the aimed purposes. Partial least squares discriminant analysis (PLS-DA) provided satisfactory, stable and robust binary classification models with recognition and prediction abilities of 90| 100% of correct hits for most of the models to distinguish the type of olive oil and identify the VO in the blend. Partial least squares regression (PLS-R) afforded regression models with excellent precisions and acceptable accuracies to determine the percentage of VO in the mixture. The effectiveness of the proposed strategy was tested with blind samples, the results of which were satisfactory and confirmed its potential to support regulations and control bodies.

Chapter 3 - The quality of virgin olive oil (VOO) is related to its oxidative stability, storage, and sensory and nutritional properties. VOO oxidation leads to the formation of off-flavour substances, degradation of its bioactive antioxidants and accumulation of degradation compounds, causing the loss of its sensory and health-promoting qualities, its commercial value and consumer acceptance. VOO resistance to oxidation depends on its chemical composition and its exposure to pro-oxidant factors such as oxygen, light and temperature. VOO stability during storage has been studied by an untargeted metabolomics approach based on 1H-NMR fingerprinting and pattern recognition. In a preliminary study, the stability of a VOO was studied under normal storage conditions, i.e., in dark and at room temperature, over a period of 43 months to evaluate the approach for virgin olive oil stability assessment. Principal component analysis (PCA) of the 1H-NMR fingerprints allowed to visualize the evolution of VOO with time. In a further study, shelf life conditions during commercialization of VOOs were simulated, and a representative set of VOOs covering the full range of possible chemical compositions were exposed to these conditions: light (500 lux for 12h/day) at 25°C for 12 months or stored in the dark at 25 °C, 30°C and 35°C for 24 months. Supervised multivariate data analysis of the 1H-NMR spectra of VOOs provided classification models to evaluate VOO freshness and verify the light exposure of the VOO during storage, and regression models to determine VOO storage time and tentatively the best before date of a fresh VOO. The different pattern recognition techniques disclosed VOO compositional changes due to hydrolytic and oxidative degradation taking place during storage, and confirmed that light and increasing temperature enhance these processes. The presence of characteristic resonances of hydroperoxides (primary oxidation products) and the decrease of 1H signals assigned to phenolic compounds, fatty acids, squalene and native (E)-2-hexenal present in fresh VOO revealed its oxidative degradation. The emergence of low intensity 1H signals of saturated aldehydes with increasing temperature in darkness, as well as in light exposure, meant that the secondary oxidation process has started at a low rate and yield in the stored oils. The decrease of the 1H signals of triacylglycerides and sn-1,2-diacylglycerides, and increase of sn-1,3-diacylglycerides indicated that VOO hydrolytic degradation and diacylglyceride isomerisation was occurring. 1H-NMR fingerprinting of VOO together with pattern recognition techniques afford relevant information to assess VOO quality regarding legal, sensory and health-promoting aspects.

Chapter 4 - Chemometrics is a discipline of chemistry that finds correlation between specific data using mathematical and statistical methods. During any thorough research, the scientists are handling vast amounts of data related to the samples which are being researched. In this type of research, finding the correlation (similarities or differences) between analyzed samples and data is of great importance. The contribution of chemometrics is in its way to objectively analyze the results and groups them based on their common characteristics and not per assumptions of scientists. This prevents subjective analysis of given results and shows correlation between the results which cannot be always identified easily in matrix that can often consist of a lot of data. For this type of analysis, many different methods can be used, which are divided in unsupervised and supervised.

In this chapter, the authors will be talking about unsupervised methods where principal component analysis (PCA) and hierarchical cluster analysis (HCA) are part of. Using these methods, the authors can spot the correlation between the samples and underlying data structures without the potential bias of scientists about the previous knowledge of data of the samples. Theoretical basics and practical contribution of these methods can be better understood in a specific example. The PCA and HCA methods were applied to a set of data on element accumulation in a great number of mushrooms. Both review and the authors' own data were used regarding the wild edible mushrooms gathered in the forests of European countries (Serbia, Poland, and Turkey) and China. The essential elements were used as variables, and depending on their content in mushrooms, the groups of mushrooms were formed. The graphical representation of HCA method gives the ability of determining the number of clusters, but not the reason for clustering of mushrooms. Unlike HCA, the graphical representation of PCA gives more information and explanation about grouping of certain mushroom species. The PCA method shows a certain clustering of samples which can be explained by minimal or maximum content of certain elements which are characteristic for that cluster.

Chapter 1

Commonly Used Chemometrics for Spectral Modeling Transfer

Yue Huang[1,*] **and Xiaoli Chu**[2]
[1]College of Food Science and Nutritional Engineering,
China Agricultural University, Beijing, China
[2]Department of Analytical Research,
Sinopec Research Institute of Petroleum Processing, Beijing, China

Abstract

A situation of the model cannot give correct prediction results due to the difference in spectra measured by different instruments is often encountered in the process of spectral analysis, especially when the model established on one spectrometer is used on another spectrometer. The first way to address this problem, is to improve the standardization of instrument hardware processing, improve the level of processing technology, and reduce the differences between the host and the secondary in terms of devices. There have been many reports on the standardization of spectroscopic instrument hardware such as calibrating the wavelength accuracy, absorbance accuracy, resolution, spectral response line type and symmetry of the spectrometer through sharp-line emission light source, standard materials, etc. As well, there have been developments on instruments by optimizing optical components, assembly processes, and control strategies with performance indicators within the allowable range of variation.

In recent years, some portable instruments can also transfer spectra between the same type. Surely, different applications have various requirements for the consistency of instruments. Models established by different methods and different systems also have different tolerances for

* Corresponding Author's Email: huangyue@cau.edu.cn.

In: Chemometrics
Editor: Larry D. Crenshaw
ISBN: 979-8-88697-274-0
© 2022 Nova Science Publishers, Inc.

divergence between instruments. Due to the sharp peaks of spectrum, Raman spectrometry can better solve the problem of consistency between instruments via hardware. Besides spectroscopic instruments, other instruments like mass spectrometers have also encountered similar problems, but most of which can be solved by simple peak calibration.

Although there has been a development for decades, differences between different brands of instruments still exist, e. g. the difference between a grating type and a Fourier transform type spectrometers. Due to these differences, the inapplicability of the multivariate models would produce unacceptable systematic prediction bias. Thus, the solution to this prediction bias is called Model Transfer or Instrument Transfer. In the area of machine learning, the relevant keywords are Transfer Learning, Domain Adaptation, and Multi-Task Learning, etc.

Transfer between the spectra establishes the functional relationship between the spectra measured by the host and secondary through a mathematical method. There are two ways of spectral conversion. One is to convert the host's calibration spectra, and then re-establish a calibration model suitable for the secondary spectra, that is called Reverse Standardization in the literatures. The other is to convert the secondary spectra, and directly use the host model to predict the result. Algorithm of two ways is essentially the same and needs to be selected according to different applications. A robust calibration is established through the preprocessing, wavelength variable screening algorithms, etc. Or the host calibration set is expanded by adding spectra under different test conditions, spectra from secondary, etc., to establish a global calibration, or hybrid calibration. Thus, this chapter mainly introduces the model transfer methods.

Keywords: model transfer, spectral analysis, chemometrics

Challenges and Solutions of Spectral Analysis

A situation is often encountered in the process of spectral analysis, when the model established on one spectrometer (host, master, primary, parent instrument) is used on another spectrometer (secondary, child instrument), the model cannot give correct prediction results due to the difference in spectra measured by different instruments (Shenk, Westerhaus and Templeton, 1985; Vogt and Booksh, 2004). The first thing to solve this problem, is to improve the standardization of instrument hardware processing, improve the level of processing technology, and reduce the differences between the host and the secondary in terms of devices. Instrument standardization makes the spectra from the same samples measured by different instruments as consistent as possible. There have been many reports on the standardization of spectroscopic instrument hardware such as calibrating the wavelength accuracy, absorbance accuracy, resolution, spectral response line type and symmetry of the spectrometer through sharp-line emission light source, standard materials, etc (Blanco et al., 1995; Fearn et al., 1996; Ray and McCreery, 1997; Mann and Vickers, 1999; Busch et al., 2000; Martinsen et al., 2008; Martinsen et al., 2010). As well, there have been developments on instruments by optimizing optical components, assembly processes, and control strategies with performance indicators within the allowable range of variation (Isaksson et al., 2003; Yang et al., 2003). This is called the First Principle of instrument calibration, which is the most fundamental basis of modern spectroscopic analysis technology (Soyemi et al., 2001; Workman, 2016).

For Fourier instruments of the same or even different types, it is basically possible to directly transfer the spectra through the instrument calibration method (Cinier and Guilment, 1998; Wang et al., 1998; Ridder et al., 2014; Ridder, Steeg and Price, 2014; Terrell, 2015; Xu et al., 2020). In recent years, some portable instruments can also transfer spectra between the same type (Hacisalihoglu et al., 2016; Sun, Hsiung and Smith, 2019). Surely, different applications have various requirements for the consistency of instruments (Aldridge et al., 1996; Barnes et al., 2010; Isabelle et al., 2016). Models established by different methods and different systems also have different tolerances for divergence between instruments (Rodgers et al., 2001; Bakeev and Kurtyka, 2005; Via et al., 2005; Sun et al., 2016; Pissarda et al., 2021). Due to the sharp peaks of spectrum, Raman spectrometries can better solve the problem of consistency between instruments via hardwares (Hutsebaut, Vandenabeele and Moens, 2005; Choquette et al., 2007; Rodriguez et al., 2011; Chen et al., 2015). Besides spectroscopic instruments, other instruments like mass spectrometers have also encountered similar problems, but most of which can be solved by simple peak calibration (Pavón et al., 2003; Coleman et al., 2007).

Although there has been a development for decades, differences between different brands of instruments still exist, e. g. the difference between a grating type and a Fourier transform type spectrometers. Due to these differences, the inapplicability of the multivariate models would produce unacceptable systematic prediction bias (Bergman et al., 2003; Drennen, 2003; Sohn et al., 2009). Thus, the solution to this prediction bias is called Model Transfer or Instrument Transfer (Noord, 1994; Chu, Yuan and Lu, 2001; Fearn, 2001;

Zhang, Cai and Shao, 2017). In the area of machine learning, the relevant keywords are Transfer Learning, Domain Adaptation, and Multi-Task Learning, etc (Malli et al., 2017).

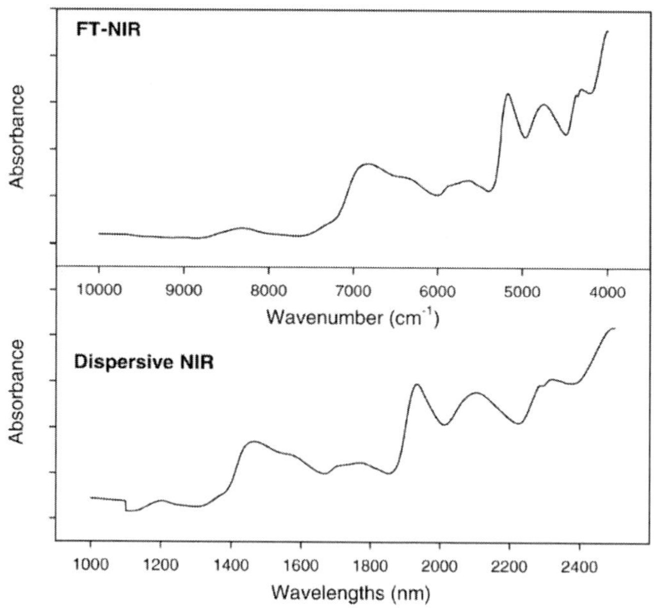

Figure 1. Near-infrared spectra of barley samples on dispersive and Fourier transform instruments.

Model transfer mainly includes the following three types of solutions (Feudale et al., 2002; Zhang et al., 2009; Shi, Li and Chu, 2019):

(1) Transfer in spectra
Transfer between the spectra establishes the functional relationship between the spectra measured by the host and secondary through a mathematical method. There are two ways of spectral conversion. One is to convert the host's calibration spectra, and then re-establish a calibration model suitable for the secondary spectra, that is called Reverse Standardization in the literatures (Lima and Borge, 2002; Leion et al., 2005; Rukundo et al., 2020). The other is to convert the secondary spectra, and directly use the host model to predict the result. Algorithm of two ways is essentially the same and needs to be selected according to different applications.

(2) Regression coefficient conversion

Regression coefficient conversion is conducted to the host model so as to make it suitable for the secondary spectra (Forina et al., 1995). Also, prediction results of the host model can be corrected, such as the Slope/Bias Correction (SBC) method, etc., to eliminate the systematic deviation of the prediction results (Dardenne, 2002; Hopkins, 2003).

(3) Robust calibration

A robust calibration is established through the preprocessing, wavelength variable screening algorithms, etc. Or the host calibration set is expanded by adding spectra under different test conditions, spectra from secondary, etc., to establish a global calibartion, or hybrid calibration (Ozdemir, Mosley and Williams, 1998; Dardenne and Welle, 1998). In some cases, this method is also called Model Updating or Calibration Maintenance. For instance, with the aging of the electronic components, detectors, and optical components of the instrument, as well as changes in the instrument hardware caused by the environment, these inevitable changes in the instrument will lead to changes in the spectra (Kramer and Small, 2007). In practices, the most commonly used methods are the first and third ones (Workman, 2018), the second is rarely used. This chapter mainly introduces the first and third types of methods.

Classic Algorithms

Commonly used methods for converting spectra among different instruments (or under different conditions) include: Spectral Substration Correction (SSC), Shenk's Algorithm, Direct Standardization (DS), and Piecewise Direct Standardization (PDS), etc. These methods usually require a representative set of standard samples (15-30 samples), which is called the standard model transfer method.

Spectral Substration Correction

Calculate the average spectrum $\bar{x}_{ms}$ and $\bar{x}_{ss}$ of the host standard and secondary sample spectral matrix X_{ms} and X_{ss} respectively, then calculate the difference between the average spectrum of them $\Delta \bar{x} = \bar{x}_{ms} - \bar{x}_{ss}$.

For the unknown spectrum $\boldsymbol{x}_{s,un}$ from the secondary spectroscopy, convert with formula $\boldsymbol{x}_{s,un}{}^{P}=\boldsymbol{x}_{s,un}+\Delta\bar{x}$, acquire the spectrum $\boldsymbol{x}_{s,un}{}^{P}$ consistent with $\boldsymbol{x}_{m,un}$, the final result is calculated by the calibration model established by the host (Smith and Jee, 2004).

Shenk's Algorithm

Shenk's algorithm consists of two main steps as wavelength correction and absorbance correction (Shenk and Westerhaus, 1991; Bouveresse, Massart and Dardenne, 1994; Hoffmann and Zanier-Szydlowski, 1999). The following introduces this algorithm by taking the conversion of spectrum from the secondary to host (secondary→host) as an example.

(1) Wavelength correction
  a. Corresponding to $\boldsymbol{X}_{ms,l}$, choose the spectral part $\boldsymbol{X}_{ss,j+k+1}$ with window size of $(k+j+1)$ from the matrix $\boldsymbol{X}_{ss}$ in the secondary instrument, calculate the correlation coefficients of $\boldsymbol{X}_{ms,i}$ and $\boldsymbol{X}_{ss,i-j}$, $R_{ss,i-j+1}$, …, $\boldsymbol{X}_{ss,i+k-1}$, $\boldsymbol{X}_{ss,i+k}$, respectively. If the coefficient r_l of $\boldsymbol{X}_{ms,l}$ $(i-j \leq l \leq i+k)$ and $\boldsymbol{X}_{ms,i}$ is max, it is indicated that the $\boldsymbol{l}$ wavelength from the secondary instrument corresponds to the $\boldsymbol{i}$ wavelength from the host. To obtain more accurate results, wavelengths $\boldsymbol{l}$-1, $\boldsymbol{l}$, $\boldsymbol{l}$+1 and their corresponding correlation coefficients $\boldsymbol{r}_{l-1}$, $\boldsymbol{r}_l$, $\boldsymbol{r}_{l+1}$ are selected to establish a univariate quadratic parabola model as $r=a_i+b_i i+c_i i^2$. The secondary wavelength $\boldsymbol{i}$* corresponding to the host wavelength $\boldsymbol{i}$ would be obtained from this fitted parabola.
  b. Loop $\boldsymbol{i}$ to find all corresponding $\boldsymbol{i}$*.
  c. Establish the one-dimensional quadratic parabola wavelength calibration model $\boldsymbol{i}^*=A+B\boldsymbol{i}+C\boldsymbol{i}^2$ using the obtained $\boldsymbol{i}$* and $\boldsymbol{i}$.

(2) Absorbance correction

(3) After wavelength correction, calculate the absorbance matrix $\boldsymbol{X}_{ss,i}{}^*$ of the secondary wavelength $\boldsymbol{i}$* by the interpolation method, then find the regression coefficients sa_i and sb_i using the linear equation $\boldsymbol{X}_{ms,i}=sa_i+sb_i\boldsymbol{X}_{ss,i}{}^*$.

(4) For the unknown spectrum $\boldsymbol{x}_{s,un}$ from the secondary instrument, wavelength fitting calibration curve $\boldsymbol{i}^*=A+B\boldsymbol{i}+C\boldsymbol{i}^2$ is used to calibrate the wavelength, calculate $\boldsymbol{x}_{s,un}{}^*$ by the interpolation method, and

finally the spectral conversion result conisistent with the host is obtained by $\boldsymbol{x}_{s,un}{}^{p}=sa_i + sb_i\, \boldsymbol{x}_{s,un}{}^{*}$.

Direct Standardization

DS algorithm uses the conversion matrix $\boldsymbol{F}$ to convert the unknown sample spectrum $\boldsymbol{x}_{s,un}$ measured from the secondary machine to $\boldsymbol{x}_{s,un}{}^{p}$. Conversion matrix $\boldsymbol{F}$ is calculated by $\boldsymbol{X}_{ms}=\boldsymbol{X}_{ss}\boldsymbol{F}$ through the least square as $\boldsymbol{F}=\boldsymbol{X}_{ss}{}^{+}\boldsymbol{X}_{ms}$ (Wang, Veltkamp and Kowalski, 1991; Dreassi et al., 1998), where $\boldsymbol{X}_{ss}{}^{+}$ is the generalized inverse matrix of $\boldsymbol{X}_{ss}$, $\boldsymbol{F}$ is a $m \times m$ -dimensional matrix (m is the number of wavelength variables).

For the spectrum $\boldsymbol{x}_{s,un}$ measured from the secondary machine, conversion is conducted by $\boldsymbol{x}_{s,un}{}^{p}=\boldsymbol{x}_{s,un}\boldsymbol{F}$, the final result is calculated by the calibration model established by the host.

Piecewise Direct Standardization

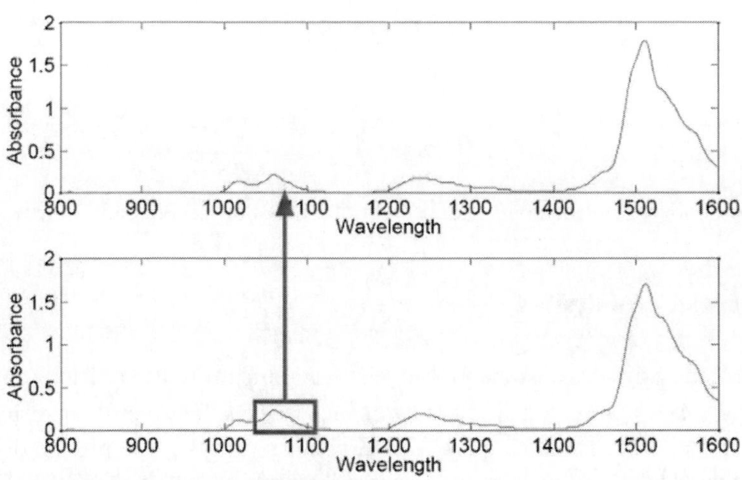

Figure 2. PDS algorithm scheme.

In PDS algorithm, as shown in Figure 2, standard sample spectral matrix $\boldsymbol{X}_{ss,j+k+1}$ with the window width (j+k+1) on both sides of the i-th wavelength point from the secondary instrument ((i-j)th to (i+k)th points) and standard sample spectral maxtrix $\boldsymbol{X}_{ms,i}$ from the host i-th wavelength point are used to calculate the conversion coefficient F_i of the i-th wavelength point. Then, the conversion matrix $\boldsymbol{F}$ of all wavelengths is obtained by moving i point by point (Wang and Kowalski, 1992; 1993; Wang, Lysaght and Kowalski, 1992; Bouveresse et al., 1996).

When calculating the conversion matrix $\boldsymbol{F}$, besides PLS method, as shown in Figure 3, methods such as artificial neural networks can also be used (Despagne, Walczak and Massart, 1998; Duponche et al., 1999). Moreover, spectra can also be converted and transmitted in the Fourier transform domain or the wavelet domain, etc (Greensill et al., 2001; Igne and Hurburgh, 2010).

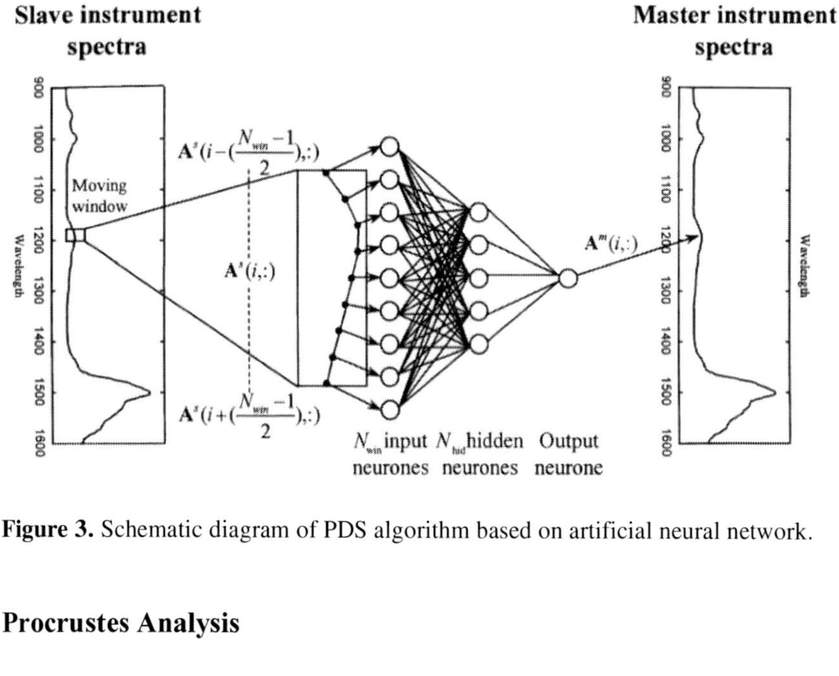

Figure 3. Schematic diagram of PDS algorithm based on artificial neural network.

Procrustes Analysis

Statistically, procrustes analysis is used to compare two matrices $X_1(m \times p_1)$ and $X_2(m \times p_2)$, and to find the conversion matrix $\boldsymbol{F}$ between the matrices $\boldsymbol{X}_1$ and X_2, where m is the samples numbe, p_1 and p_2 are the variables number.

Detailed algorithm is as follows (Anderson and Kalivas, 1999; Chu, Yuan and Lu, 2002).

(1) Perform singular value decomposition on the matrices $\boldsymbol{X}_1$ and $\boldsymbol{X}_2$ respectively, $\boldsymbol{X}_1=\boldsymbol{U}_1\boldsymbol{S}_1\boldsymbol{V}_1^t$, $\boldsymbol{X}_2=\boldsymbol{U}_2\boldsymbol{S}_2\boldsymbol{V}_2^t$, where $\boldsymbol{X}_1$ and $\boldsymbol{X}_2$ represent the spectral matrix measured by the host and slave (after averaging or standardization pretreatment), $\boldsymbol{U}$ is score matrix, $\boldsymbol{V}$ is loading matrix,

and $\boldsymbol{U}$ and $\boldsymbol{V}$ matrices contain the rotation information between the spectral matrices, $\boldsymbol{S}$ matrix contains the stretching information between the spectral matrices.

(2) Calculate $\boldsymbol{Z}_1 = \boldsymbol{U}_{1g}\boldsymbol{S}_{1g}$ and $\boldsymbol{Z}_2 = \boldsymbol{U}_{2h}\boldsymbol{S}_{2h}$, separately, where g and h represent the number of PCs used to find $\boldsymbol{Z}_1$ and $\boldsymbol{Z}_2$ Respectively.

(3) Calculated conversion matrix $\boldsymbol{F}$ between $\boldsymbol{Z}_1$ and $\boldsymbol{Z}_2$ by $\boldsymbol{F} = \boldsymbol{Z}_2^{+}\boldsymbol{Z}_1$, where $\boldsymbol{Z}_2^{+}$ is the generalized inverse matrix calculated by $\boldsymbol{Z}_2^{+} = \boldsymbol{S}_{2g}^{-1}\boldsymbol{U}_{2g}^{t}$.

(4) A spectrum $\boldsymbol{x}_{un}$ measured from a slave can be converted into spectrum $\boldsymbol{x}_{un}^{p}$ consistent with the master through the conversion matrix $\boldsymbol{F}$ and the loading matrix $\boldsymbol{V}$.

Target Transformation Factor Analysis

Target transformation factor analysis (TTFA) is also a transfer method based on PCA. Its core concept is to use the target transfer method to make the principal component score (virtual component concentration) of slave equal to that of host. Main steps are as follows (Wang, Yuan and Lu, 2005):

(1) Perform PCA on the standard sample spectral matrix of the host to obtain the load and score matrix:

$$\boldsymbol{X}_m = \boldsymbol{T}_m\boldsymbol{P}_m^t$$

(2) Perform PCA on the standard sample spectral matrix of the slave to obtain the load and score matrix:

$$\boldsymbol{X}_s = \boldsymbol{T}_s\boldsymbol{P}_s^t$$

(3) Establish the mathematical relationship between host and slave:

$$\boldsymbol{T}_m = \boldsymbol{T}\boldsymbol{T}_s$$

(4) Solve the transformation matrix by generalized inverse operation:

$$\boldsymbol{T} = \boldsymbol{T}_m\boldsymbol{T}_s^t(\boldsymbol{T}_s\boldsymbol{T}_s^t)^{-1}$$

(5) Conversion of spectrum from host to slave can be expressed as:

$$\boldsymbol{X}_s^{\mathrm{P}} = \boldsymbol{X}_m P_m \boldsymbol{T}^+ P_s^{\,t}$$

where, the conversion matrix, $\boldsymbol{F} = \boldsymbol{P}_m \boldsymbol{T}^+ \boldsymbol{P}_s^t$.

Maximum Likelihood Principal Component Analysis

Maximum likelihood principal component analysis (MLPCA) treats model transfer as an issue of missing data (Andrews and Wentzell, 1997).

Combine the standard sample spectra of host and slave:

$$\boldsymbol{x}_{i,comb} = [\boldsymbol{x}_{i,m}, \boldsymbol{x}_{i,s}],\ i=1, 2, \ldots, n$$

where n is number of standard samples.

Besides, other samples of the calibration set on the host that have no corresponding spectra on the slave can be expressed as:

$$\boldsymbol{x}_{i,comb}^{\#} = [\boldsymbol{x}_{i,m}, \boldsymbol{x}_{i,s}^{\#}],\ i=1, 2, \ldots, m$$

where m denotes the number of calibration set samples minus standard set samples, $\boldsymbol{x}_{i,s}^{\#}$ denotes the missing spectra from the slave.

$\boldsymbol{x}_{i,comb}$ and $\boldsymbol{x}_{i,comb}^{\#}$ can be expressed as a matrix:

$$\boldsymbol{X}_{comb} = \begin{bmatrix} \boldsymbol{X}^* \\ \boldsymbol{X}^{\#} \end{bmatrix}$$

where $\boldsymbol{X}^*$ represents combined spectral matrix without missing data in the standard sample set, $\boldsymbol{X}^{\#}$ represents combined spectral matrix of other samples in calibration set with missing data in the slave.

Perform MLPCA on:

$$\boldsymbol{X}_{comb}\colon \begin{bmatrix} \boldsymbol{X}^* \\ \boldsymbol{X}^{\#} \end{bmatrix} = \begin{bmatrix} \boldsymbol{U}^* \\ \boldsymbol{U}^{\#} \end{bmatrix} \boldsymbol{D}\boldsymbol{P}^{\mathrm{t}}$$

Then, the spectrum $\boldsymbol{x}_{i,m}$ collected on the host can be converted into the spectrum on the slave according to the following formula:

$$\widehat{\boldsymbol{x}}_{i,s} = \boldsymbol{U}^{\#}(\boldsymbol{U}^{*\mathrm{t}}\boldsymbol{U}^{*})^{-1}\boldsymbol{U}^{*\mathrm{t}}\boldsymbol{x}_{i,m}$$

Afterward, based on the MLPCA algorithm, Folch-Fortuny et al., (2017) used the trimmed scores regression (TSR) method to deal with the issue of missing data, and proposed the TSR method for model transfer.

Slope/Bias Correction

Different from the above methods that are based on the conversion between spectra, there is a method based on the conversion between the prediction results, that is, the functional relationship between the prediction results obtained by the host and slave is established, which is named Slope/Bias correction (SBC) algorithm (Bouveresse and Massart, 1996; Sales, Callao and Rius, 1999; Tillmann, Reinhardt and Paul, 2000).

Calibration model established on the host is used to predict the analysis results $\boldsymbol{y}_{mp}$ and $\boldsymbol{y}_{sp}$ of the host and slave standard sample spectral arrays $\boldsymbol{X}_{ms}$ and $\boldsymbol{X}_{ss}$, respectively. Assume that $\boldsymbol{y}_{mp}$ and $\boldsymbol{y}_{sp}$ have a relationship of $\boldsymbol{y}_{mp}=a\times\boldsymbol{y}_{sp}+b$, the least squares method can be used to obtain a and b.

As for the unknown spectrum $\boldsymbol{x}_{s,un}$ measured by the slave, first the calibration model established by the host is used to calculate $\boldsymbol{y}_{sp,\ un}$, and then formula $\boldsymbol{y}_{sp,un}{}^{P}=b+a\times\boldsymbol{y}_{p,un}$ is used to calculate the corrected analysis result $\boldsymbol{y}_{sp,\ un}{}^{P}$.

Normally, the SBC algorithm is not recommended. Because if the spectral difference between instruments is significant, it would be difficult or even impossible to identify the out-of-bounds samples of the model.

Improvement of Classic Algorithms

In the process of model transfer, the selection of standard samples is particularly important (Hong et al., 1994). For the model update problem, Capron et al., (2005) compared the effects of the weighting of calibration samples and the selection of representative samples, and the results showed that the selection effect of representative samples is better. Siano et al., (2007) compared the influence of different transfer standard selection methods on the spectrum conversion effect of PDS, etc., and the results showed that the optimal K-dissimilarity selection (OptiSim) method is better than the K-S method. Li H. et al., (2011) replaced the Euclidean distance in the K-S algorithm with the Mahalanobis distance, and selected the transfer standard sample through the improved K-S algorithm. In the PDS algorithm, the samples selected by the Mahalanobis distance are more representative, because the combination of concentration differences and spectral differences can better represent the differences between samples. Zhou Z. et al., (2020) proposed a transfer standard set selection method based on Markov chain, and the result is better than the K-S method.

From the perspective of optimizing the selection method of the transfer standard set, Liang et al., (2016) optimized the transfer matrix of the PDS algorithm and proposed the Rank-KS-PDS model transfer algorithm. When Rank-KS selects the transfer standard sample set, it comprehensively considers the influence of sample spectral space and sample concentration space, overcomes the shortcomings of K-S algorithm insensitive to low concentration areas, and improves the accuracy of model transfer. Based on the idea of Backward Selection variables, Zheng et al., (2020) proposed a backward selection iterative method to transfer the selection of the standard sample set. Through this method, the transfer effect of the standard sample set is better than the standard sample set selected by the KS algorithm. Aiming at the problem of difficulty in obtaining and storing standard samples of natural plant models, Ni et al., (2018) proposed a method for preparing standard samples, this kind of standard sample is stable under normal temperature and pressure, similar to the color of various natural plants, with constant spectrum and good reproducibility, and can be used for a long time in the transfer process of the near-infrared model of a variety of natural plants.

Aiming at the limitations of the SBC algorithm in solving nonlinear problems, Xin X. et al., (2017) established a linear function relationship between the prediction results of the host and the slave by introducing high powers, both Lagrange interpolation and Newton interpolation are used to find the parameters, and the nonlinear fitting of the two sets of data is realized. Cao Y. et al., (2018) proposed a method for selecting PDS algorithm parameters (number of standard samples, number of PLS main factors, window width) based on the angle of the spectral space. Zhang et al., (2017) used sampling error profile analysis (SEPA) to optimize the PDS algorithm's window width and PLS main factor number and other parameters, and proposed the SEPA-PDS method.

Blanco et al., (2010; 2011) used the spectral difference of a set of standard sample sets under different conditions combined with weighting to calculate a mutation matrix, and randomly added it to the calibration set spectrum, which solved the problem of the laboratory-prepared drug calibration samples used for production process analysis. On this basis, Wang J. et al., (2014) improved the SSC algorithm, the spectrum compensation between the master and the slave does not use the average spectral difference of the standard sample set spectrum array uniformly, but for each sample in the calibration set, select the most similar standard sample set to compensate for the spectral difference between the master and slave, and the compensation spectrum is weighted by the concentration ratio between the calibration set samples and the standard set samples. On the basis of the SSC algorithm, Li et al., (2019) respectively compensated different correction vectors for different types of correction sets, and then through continuous model updates, the influence of different measurement environments on the identification of corn haploid grains by NIRS can be eliminated. Based on the traditional SBC algorithm, Wang et al., (2016) proposed a dual-domain model delivery strategy. In this method, the master model is used to predict the spectra of the master and slave standard sample sets respectively, and the transfer model is established by using the ratio of the predicted value and the spectrum of the slave standard sample set. For the spectra collected by the slave, first use the host model to calculate the initial value, then use the transfer model to calculate the ratio, and finally get the final prediction result through the ratio. Li X. et al., (2011) established a PLS model between the spectral difference and the predicted concentration difference between two NIRS instruments to achieve the correction of the predicted concentration from the slave machine. Tan et al., (2002) improved the unstandardized finite impulse response transfer algorithm (Finite Impulse Response, FIR), which eliminated the peak problem caused by the FIR algorithm and improved the transfer effect of the FIR method (Sum and Brown, 1998). Bouveresse et al., (1995) used the local weight regression algorithm for the correction between spectral absorbance and improved Shenk's algorithm.

Before performing DS and PDS operations, Wang Z. et al., (1995) used additive background correction to improve the effect of spectral transmission. In response to the nonlinear conversion between spectra in the classic PDS algorithm, and the appearance of discontinuities and even abnormal peaks (Gemperline et al., 1996), Wang H. et al., (2016) replaced the PLS regression in the PDS algorithm with a radial basis function neural network, and obtained better results. Chen X. et al., (2019) believe that the appearance of the peak in the PDS algorithm originates from the coefficient of the larger norm in the PLS model, and its essence is an over-fitting problem, they used a linear regression method with coefficient norm penalty to establish a spectral transfer model, and the transferred spectra were smoother and more robust.

Univariate correction is a special case where the PDS method takes a window width of 1. Yang H. et al., (2014) proposed the Simple Linear Regression Direct Standardization (SLRDS) method, this univariate correction method is more suitable for the case where the spectrum has a small linear difference. Norgaard, (1995) also used this method in fluorescence spectroscopy, which is called the single wavelength standardization (SWS). Galvao et al., (2015) calculated the covariance matrix based on the univariate spectral correction using the spectral residuals transferred from the master and slave of the standard sample set, and then established the model through a robust regression method. This method has certain advantages in the case of fewer wavelength variables. Lu H. et al., (2019) used the minimum angle regression to first select the characteristic wavelength variable, and then used the unary linear direct correction method to transmit the spectrum, which further improved the transmission effect. Wang W. et al., (2019) used the dynamic time warping algorithm (DTW) in speech analysis to correct the spectral wavelengths on the two instruments, and then use the unary regression or multiple regression algorithm to correct the absorbance and obtain satisfactory results. In order to prevent excessive warping, Zou et al., (2019) proposed a variable penalty dynamic time warping (VPdtw), which has a better transmission effect on the NIRS than the PDS algorithm.

Yan and Zhang, (2015) gave different weights to the wavelength variables in the moving window of the PDS algorithm, and proposed a windowed PDS algorithm (WPDS) based on ridge regression and penalty terms. The PDS algorithm can be regarded as all wavelength variables assigned the same weight. A special case of the WPDS algorithm. The double window PDS algorithm (DWPDS) is an extension of the classic PDS algorithm, that is, a window of a certain width is taken from the master and slave spectra, and the spectrum conversion matrix is established window by window. Oliveri et al., (2013) used the idea of the dual-window PDS algorithm to calculate the transfer coefficient between the average spectrum of the standard sample spectral matrix transferred from the host and the slave, and the transfer coefficient was calculated using the classic least square method (Figure 4). Greensill et al., (2001; 2002) also used the dual-window PDS algorithm to transfer different array types of near-infrared spectra, but the best result is the wavelet transform combined with DS algorithm. Ottaway and Kalivas, (2015) improved the DS and PDS algorithms by adding higher-order terms and derivative terms, which can solve the problem of nonlinear differences between different spectroscopic instruments to a certain extent.

On the basis of the DS algorithm, Chen W et al., (2016) used Extreme Learning Machine Auto-encoder to establish the relationship between the master and slave standard sample spectral matrixes, and obtained stable spectral transmission results through an integrated strategy. Laref et al., (2018) used SVM for DS algorithm to obtain the signal conversion of different electronic nose instruments, and the standard set samples were obtained by the SPXY method. The superposition partial least squares (VIP-PLS) method improved by the variable projection importance coefficient proposed by Li X. et al., (2018) first rearranges the wavelengths and divides them into a series of non-overlapping spectral intervals, and then transmits them through the DS algorithm.

Tan et al., (2001) used wavelet transform spectroscopy to denoise and compress the signal, then reconstructed spectral signals of different scales through inverse wavelet transform, and then carried out transfer operations on the spectral signals of different scales respectively, and proposed a wavelet hybrid direct standardization (WHDS). Chen D. et al., (2017) also proposed a similar model transfer method. Yoon et al., (2002) first perform wavelet transform on the spectrum to obtain wavelet coefficients, and then use the DS algorithm to transform. For large spectral matrixes, this method can reduce the time for spectral conversion and modeling, which is called wavelet transform direct standardization (WTDS). Tan and Li, (2007) use wavelet packets to decompose the spectrum, and realize the transmission of spectrum between different instruments through wavelet packet coefficient transformation, which is called wavelet packet transform standardization (WPTS).

After Ni et al., (2010) divide the spectrum by wavelet, use the PDS algorithm to transfer each sub-spectrum, and then use the consensus modeling strategy to establish a PLS model for each converted sub-spectrum one by one, which is called stacked dual-domain piecewise direct standardization (SDDPDS). Poerio and Brown, (2017) combined the dual-domain wavelet transform with orthogonal projection, and proposed the dual-domain transfer using orthogonal projection (DDTOP).

Based on the direct orthogonal signal correction (DOSC) algorithm, Lin et al., (2013) used the regression of the virtual standard average spectrum to eliminate the background differences between sample batches, and proposed the Orthogonal Space Regression (OSR) model.

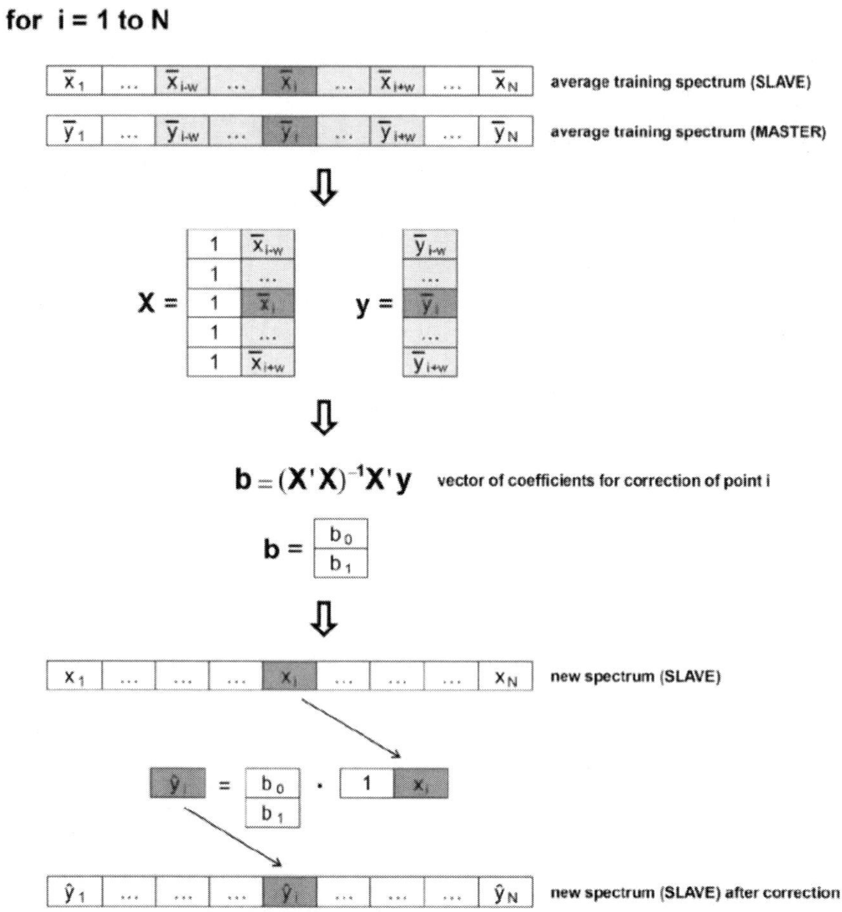

Figure 4. Spectral transfer method based on the concept of double-window PDS.

Delivery method, it can correct the systematic errors between the spectra of multiple batches of preparations, and realize the model transfer of the chlorogenic acid quantitative model between batches during the water extraction process of the honeysuckle pilot test. On this basis, Yang and Wang et al., (2019; 2019) proposed a guided orthogonal projection technology combined with a model transfer method of SBC, which realized the transfer of the near-infrared quantitative model of the small test moisture to the pilot test. Wang Q. et al., (2020) used random forest to select the wavelength of the NIRS, and then used the DOSC algorithm to preprocess the spectrum to realize the sharing of models between different instruments.

For the spectra collected at different temperature points, the conversion can be realized through the PDS algorithm. However, the traditional PDS algorithm cannot transmit the spectrum at any temperature between different temperature points (Lin, 1998). Based on the PDS algorithm, Wulfert et al., (2000) and Barring et al., (2001) proposed a continuous piecewise direct standardization (CPDS) algorithm. The conversion matrix $\boldsymbol{F}_{\Delta T}$ between two different temperatures and the temperature difference ΔT were subjected to polynomial regression to obtain two conversion matrix $\boldsymbol{F}_T$ at any temperature between two temperature points.

In order to eliminate the influence of temperature on the online determination of electroplating bath composition content by AC voltammetry analyzer, Jaworski et al., (2017; 2018) proposed continuous direct standardization (CDS), and used temperature as a variable to participate in the calibration model to eliminate the influence of temperature on the prediction results.

New Developments in Algorithms

Canonical Correlation Analysis

Canonical correlation analysis (CCA) is a multivariate analytical method to study the correlation between two sets of variables, and which can reveal the linear dependence of them. CCA algorithm considers that the information of the measured object between two sets of spectra from the host and slave is consistent, and should be linearly related to each other, while the noise and interference information are random and independent. This method first performs canonical correlation analysis on the host and slave spectral matrix, and then converts the obtained canonical correlation variables. Conversion of canonical correlation variables can extract spectral conversion information from the overall spectra, and can filter out noise and interference. Specific calculation is as follows (Wei et al., 2008):

(1) Execute CCA on the standard sample spectral matrix of the host and slave respectively:

$$\boldsymbol{L}_m=\boldsymbol{X}_m\boldsymbol{W}_m$$
$$\boldsymbol{L}_s=\boldsymbol{X}_s\boldsymbol{W}_s$$

$\boldsymbol{L}_m$ and $\boldsymbol{L}_s$ are the score matrix of the host standard sample spectral matrix $\boldsymbol{X}_m$ and the slave standard sample spectral array $\boldsymbol{X}_s$ respectively, $\boldsymbol{W}_m$ and $\boldsymbol{W}_s$ are the loading matrix of $\boldsymbol{X}_m$ and $\boldsymbol{X}_s$ respectively.

(2) Calculate the conversion matrix $\boldsymbol{F}$:

$$\boldsymbol{F}_1 = \boldsymbol{L}_s^+ \boldsymbol{L}_m$$
$$\boldsymbol{F}_2 = \boldsymbol{L}_m^+ \boldsymbol{X}_m$$
$$\boldsymbol{F} = \boldsymbol{W}_s \boldsymbol{F}_1 \boldsymbol{F}_2$$

(3) Conversion of the slave spectrum $\boldsymbol{x}_{un}$ to the host spectrum can be expressed as $\boldsymbol{x}_{\mathrm{un}}{}^{\mathrm{P}} = \boldsymbol{x}_{\mathrm{un}} \boldsymbol{F}$.

CCA algorithm only considers extraction of the maximum correlation of typical variables, which may introduce redundant information that has nothing to do with the target, thereby complicating the model transfer function. On this basis, Zheng et al., (2014) proposed to use PLS to extract the factors that are related to the target value with the largest variance, and then used CCA to perform spectral conversion, which improves the pertinence of spectral conversion to a certain extent.

Before CCA conversion, Bin et al., (2017) used wavelet transform to preprocess the original spectra, and performed CCA by wavelet coefficients (WTCCA), which achieved a better transfer effect. Fan et al., (2018) calculated the principal components of the host's standard sample spectral matrix $\boldsymbol{X}_m$ by the latent variables of host PLS model, and then performed spectral conversion (PC-CCA). The validation effect was better than the single CCA method.

Similar with CCA, based on spectral regression (SR), Peng et al., (2011) decomposed on $\boldsymbol{X}_m$ and $\boldsymbol{X}_s$, and proposed the spectral regression transfer algorithm. It puts the issue of solving the characteristic function in the regression model, avoids the frequent eigenvalue decomposition process of dense matrix, and improves the calculation efficiency. Zhang J. et al., (2019) decomposed the spectra based on multi-level simultaneous component analysis (MSCA), and proposed a two-level strategy for spectral transfer algorithm.

Spectral Space Transformation

Spectral space transformation (SST) obtained the combined spectrum matrix $\boldsymbol{X}_{comb} = [\boldsymbol{X}_m, \boldsymbol{X}_s]$ by combining the standard set spectra $\boldsymbol{X}_m$ and $\boldsymbol{X}_s$ measured by the host and slave. PCA on it, then obtain the loading of the combined spectrum matrix, and calculate the spectrum conversion matrix (Du et al., 2011). Caculation is as follows:

(1) Combine the standard set spectra $\boldsymbol{X}_m$ and $\boldsymbol{X}_s$ measured by the host and slave to obtain a combined spectrum matrix:

$$\boldsymbol{X}_{comb} = [\boldsymbol{X}_m, \boldsymbol{X}_s]$$

(2) PCA on the combined spectral matrix $\boldsymbol{X}_{comb}$.

$$\boldsymbol{X}_{comb} = \boldsymbol{T}\left[\boldsymbol{P}_m{}^{t}, \boldsymbol{P}_s{}^{t}\right] + \boldsymbol{E}$$

where, $\boldsymbol{P}_m{}^{t}$ and $\boldsymbol{P}_s{}^{t}$ are loading of the host and slave matrix.

(3) Calculate the conversion matrix $\boldsymbol{F}$.

$$\boldsymbol{F} = \boldsymbol{I} + \left(\boldsymbol{P}_s{}^{t}\right)^{+}\left(\boldsymbol{P}_m{}^{t} - \boldsymbol{P}_s{}^{t}\right)$$

where, $\boldsymbol{I}$ is the identity matrix.

(4) Conversion of the slave spectrum $\boldsymbol{x}_{un}$ to the host spectrum can be expressed as:

$$\boldsymbol{x}_{un}{}^{P} = \boldsymbol{x}_{un}\boldsymbol{F}$$

The structure of the SST algorithm is relatively simple to be calibrated, and it can still maintain good prediction results under a low standard sample number. Similarly, Liu Y et al., (2018) used ICA to decompose the combined spectral matrix obtained from multiple instruments, and the expression of the conversion matrix is consistent with the SST algorithm.

Alternating Trilinear Decomposition

Alternating trilinear decomposition (ATLD) is an algorithm commonly used to decompose three-dimensional data. For a group of standard samples with collected spectra from different instruments, a three-dimensional matrix $\underline{\boldsymbol{X}}$ can be obtained, the dimension of which is $I \times J \times K$, where I is the number of standard samples, J is the number of spectral points, K is the number of instruments. ATLD algorithm can decompose $\underline{\boldsymbol{X}}$ into three matrices as $\boldsymbol{A}(I \times N)$, $\boldsymbol{B}(J \times N)$, $\boldsymbol{C}(K \times N)$, where N is the number of contributing factors, $\boldsymbol{A}$ represents the relative concentration matrix of the standard sample, $\boldsymbol{B}$ represents the relative spectral intensity matrix of the standard sample, and $\boldsymbol{C}$ represents the instrument information matrix. Algorithm is as follows (Liu, Cai and Shao, 2014).

(1) Decompose $\underline{\boldsymbol{X}}$ by ATLD:

$$x_{ijk} = \sum_{n=1}^{N} a_{in} b_{jn} c_{kn} + e_{ijk}, \quad (i=1,2,\ldots, I, \; j=1,2,\ldots,J, \; k=1,2,\ldots,K)$$

(2) Calculate the conversion matrix $\boldsymbol{F}$:

$$\boldsymbol{F}_k = \text{diag}(\boldsymbol{c}_k)\boldsymbol{B}^{t}$$

where $\boldsymbol{c}_k$ is the kth row of matrix $\boldsymbol{C}$.

(3) For the spectrum $\boldsymbol{x}_{k1,new}$, new collected from the k1 instrument, it can be converted to the spectrum $\boldsymbol{x}_{k2,trans}$ on the k2 instrument by the following formula:

$$\boldsymbol{x}_{k2,\ trans} = \boldsymbol{x}_{k1,\ new}\boldsymbol{F}$$

where, $\boldsymbol{F} = \boldsymbol{I} + \boldsymbol{F}_{k1}^{+} (\boldsymbol{F}_{k2} - \boldsymbol{F}_{k1})$.

ATLD and SST algorithm have the same spectral conversion matrix formula, except that before the factor analysis, the SST algorithm expresses the standard sample spectral matrix collected from multiple instruments in an unfolded manner, while ATLD algorithm is a cube matrix. As for cube data decomposition, besides ATLD, there are also PARAFAC and Tucker 3 that can also be used. Kompany-Zareh and Berg, (2010) used Tucker3 algorithm to treat the spectral conversion between different instruments as the missing data filling problem of the tensor array, and realized the model transfer from FT-Raman instrument to CCD-Raman spectrometer.

Multi-Task Learning

Multi-Task learning (MTL) calculating the transfer matrix can be summarized as solving a convex optimization problem with a regular trace norm. This regular optimization method can extend the linear conversion between the spectra of different instruments to the non-linear spectral conversion relationship. Compared with other conversion methods like neural networks, this method solves the final transfer matrix through a convex optimization problem. So it can efficiently and quickly obtain the global optimal solution, while requiring fewer preset parameters. Calculation is as follows (Yu, Ji and Kang, 2016).

(1) Calculate the Gram matrix of the spectral matrix $\boldsymbol{X}_s$ of the slave standard sample:

$$\boldsymbol{K}=\boldsymbol{X}_s\,\boldsymbol{X}_s^{\mathrm{t}}$$

(2) Perform eigenvalue decomposition on $n\times n$ order $\boldsymbol{K}$ matrix:

$$\boldsymbol{K}=\boldsymbol{U}\boldsymbol{D}\boldsymbol{U}^{\mathrm{t}}$$

where $\boldsymbol{D}$ is a diagonal matrix containing all eigenvalues, and each column of $\boldsymbol{U}$ is a corresponding eigenvector.

(3) Solve a regular problem about the trace norm of matrix $\boldsymbol{B}$ of order $n\times p$ (n is the number of standard samples, p is the number of wavelengths):

$$\min_{\boldsymbol{B}}||\boldsymbol{X}_m - \boldsymbol{U}\boldsymbol{D}^{\frac{1}{2}}\boldsymbol{B}||_{\mathrm{F}}^{2} + \rho||\boldsymbol{B}||_{\mathrm{tr}}$$

where, $||\bullet||_{\mathrm{F}}$ represents the Frobenius norm of the matrix, $||\bullet||_{\mathrm{tr}}$ represents the trace norm of the matrix, and ρ is the regular term coefficient. Method of Accelerated Proximal Gradient can be used to solve the convex optimization problem of trace norm regularization.

(4) Calculate the conversion matrix $\boldsymbol{F}$:

$$\boldsymbol{F} = \boldsymbol{X}_S^{\mathrm{t}}\boldsymbol{U}\boldsymbol{D}^{-\frac{1}{2}}\boldsymbol{B}$$

(5) Conversion of the slave spectrum $\boldsymbol{x}_{\mathrm{un}}$ to the host spectrum can be expressed as:

$$\boldsymbol{x}_{\mathrm{un}}{}^{\mathrm{P}} = \boldsymbol{F}^{\mathrm{t}}\boldsymbol{x}_{\mathrm{un}}$$

Boucher et al., (2017) also proposed a near-end regularization method for spectral conversion between different instruments based on the regularization framework, which is called Proximal Methods, which used alternating direction method of multipliers (ADMM) to solve convex optimization problems. It has a good effect on the conversion from a narrow band range spectrum to a wide band range spectrum.

As for the transfer of multiple qualitative models, Hu et al., (2012) proposed an optimization framework using the maximum margin criterion (MMC) criterion:

$$\mathrm{argmin}||\mathbf{X}_m - \mathbf{X}_{\mathrm{s}}\mathbf{F}||_{\mathrm{F}}^2 + \rho||\boldsymbol{F}^{\mathrm{t}}(\boldsymbol{S}_w - \boldsymbol{S}_b)\boldsymbol{F}||_{\mathrm{tr}}$$

where, $\boldsymbol{S}_{\mathrm{w}}$ is the intra-class scatter matrix, $\boldsymbol{S}_{\mathrm{b}}$ is the inter-class scatter matrix, $\boldsymbol{F}$ is the spectral conversion matrix. MMC algorithm has certain advantages in transfering spectra for qualitative analysis.

Generalized Least Squares

Generalized least squares (GLS) takes the difference matrix of the host and slave standard samples as a reference to establish a weighted filtering model to eliminate the influence of the difference between the instruments on the spectra. Main steps of algorithm are as follows (Cogdill, Anderson and Drennen, 2002; Martens et al., 2003).

(1) Calculate the mean-centered difference matrix of the host and slave standard sample spectra:

$$\boldsymbol{X}_d = (\boldsymbol{X}_m - \overline{\boldsymbol{X}_m}) - (\boldsymbol{X}_s - \overline{\boldsymbol{X}_s})$$

(2) Calculate the covariance matrix of $\boldsymbol{X}_d$:

$$\boldsymbol{C}_d = \frac{\boldsymbol{X}_d^t \boldsymbol{X}_d}{n-1} + \alpha \boldsymbol{I}$$

where, n is the number of standard samples, α is the coefficient (usually $1\times e^{-6}$), and $\boldsymbol{I}$ is the identity matrix.

(3) Perform singular value decomposition on the covariance matrix $\boldsymbol{C}_d$:

$$\boldsymbol{C}_d = \boldsymbol{U}\boldsymbol{S}\boldsymbol{V}^t$$

(4) Calculate the weighted filter matrix $\boldsymbol{W}$:

$$\boldsymbol{W} = \boldsymbol{V}\boldsymbol{S}_{adj}^{+}\boldsymbol{V}^t$$

where, $\boldsymbol{S}_{adj} = \textbf{sqrt}(\frac{S \times m}{\textbf{trace}(\boldsymbol{S})})$, m is the number of wavelength variables.

(5) Transform the spectral matrix of the host calibration set and that of the slave prediction set separately:

$$\boldsymbol{X}_{\text{trans}} = (\boldsymbol{X} - \overline{\boldsymbol{X}})\boldsymbol{W}$$

Other Algorithms

Based on the principle of orthogonal projection, Andrew and Fearn (2004) obtained the average spectrum of a set of standard samples on multiple instruments, and performed PCA on the average spectra. The first few representive projection space formed by the different spectral loadings between the instruments can perform transfer by orthogonal projection (TOP) on the spectra on multiple instruments. From the concept of TOP, Zhu Y. et al., (2008) replaced the average spectral matrix with a change matrix of a set of sample repetitive spectra, performed PCA to obtain the projection space, and proposed an error removal by orthogonal subtraction (EROS). Subsequently, in view of EROS, Zeaiter et al., (2006) calculated the virtual spectrum of the slave through the kernel function. According to the difference matrix between the measured and virtual spectrum, the PCA was performed to obtain the projection space, and the dynamic orthogonal projection (DOP) was proposed. Dabros et al., (2007) used the DOP algorithm for the maintenance of the online infrared calibration model, and achieved good application results. Igne et al., (2009) reviewed and compared the effects of the above-mentioned orthogonal projection algorithms for the transfer of NIRS models. Siska and Hurburgh (2001) used wiener filter (Wiener Filter) method to process the spectra for the transfer of fixed filter-type NIRS instruments, and proposed a method of optimizing the master instrument.

Chen Z. et al., (2005) proposed a loading space standardization (LSS) algorithm for the influence of temperature on the spectrum. PCA was performed on the spectra collected from the standard set samples at different temperatures, and then the quadratic function relationship between the spectral loading and the temperature at each temperature was established. Spectrum measured at a certain temperature can be standardized in the loading space to obtain corresponding spectrum at standard temperature. Afterward, Chen and Morris (2008) combined the LSS algorithm with the optical path length estimation and correction (OPLEC) and proposed the extended loading space standardization (ELSS), which was used to correct the effects of temperature and composition changes on the spectra. Similarly, Shi X. et al., (2014) effectively eliminated the influence of temperature changes on the NIRS during the production of sugar and flavoring based on LSS algorithm.

Zhang L. et al., (2011) regarded model transfer as a global affine transformation problem:

$$\hat{\boldsymbol{x}}_{i,n} = \boldsymbol{a}_i x_{i,n} + \boldsymbol{b}_i, \quad i = 1, \dots, k, \quad n = 1, \dots, N$$

where, $\boldsymbol{x}_{i,n}$ represents the spectrum of the nth standard sample on the ith slave instrument, k represents the number of slave instruments, and N represents the number of standard samples, $\hat{\boldsymbol{x}}_{i,n}$ represents the spectrum of the nth standard sample sample on the ith slave instrument transferred to the host. The transfer coefficients $\boldsymbol{a}_i$ and $\boldsymbol{b}_i$ are calculated by the robust weighted least square algorithm (RWLSA). Deshmukh et al., (2014) also used a robust regression method to achieve inter-station signal transfer among the electronic nose systems for emission detection in paper mills.

Zhao Y. et al., (2019) proposed an algorithm of calibration transfer based on affine invariance (CTAI) for model transfer based on affine invariance.

This method first establishes PLS model to obtain the score matrix and prediction vector of the host spectra, as well as, the pseudo score matrix and pseudo prediction vector of the slave spectra. Secondly, the regression coefficients of host and slave are obtained by least squares respectively, and then the angle and deviation between host and slave are calculated by the regression coefficients. Finally, the prediction result of the slave spectra is obtained based on the affine transformation.

Munoz et al., (2005) and Folch-Fortuny et al., (2017) proposed a new spectral conversion method using Joint-Y partial least squares regression (JYPLS):

$$\boldsymbol{Y}_J = \begin{bmatrix} \boldsymbol{Y}_m \\ \boldsymbol{Y}_s \end{bmatrix} = \begin{bmatrix} \boldsymbol{T}_m \\ \boldsymbol{T}_s \end{bmatrix} \boldsymbol{Q}_J^{\mathrm{t}} + \boldsymbol{E}$$
$$\boldsymbol{X}_m = \boldsymbol{T}_m \boldsymbol{P}_m^{\mathrm{t}} + \boldsymbol{E}$$
$$\boldsymbol{X}_s = \boldsymbol{T}_s \boldsymbol{P}_s^{\mathrm{t}} + \boldsymbol{E}$$
$$\boldsymbol{T}_m = \boldsymbol{X}_m \boldsymbol{W}_m$$
$$\boldsymbol{T}_s = \boldsymbol{X}_s \boldsymbol{W}_s$$

where, $\boldsymbol{Y}_m$ and $\boldsymbol{Y}_s$ are the concentration matrixes of the calibration set and standard set samples of the host. $\boldsymbol{T}_m$ and $\boldsymbol{T}_s$ are score matrixes of $\boldsymbol{X}_m$ and $\boldsymbol{X}_s$, respectively. $\boldsymbol{P}_m$ and $\boldsymbol{P}_s$ are loading matrixes of $\boldsymbol{X}_m$ and $\boldsymbol{X}_s$, respectively. $\boldsymbol{W}_m$ and $\boldsymbol{W}_s$ are weighting matrixes of $\boldsymbol{X}_m$ and $\boldsymbol{X}_s$, respectively. $\boldsymbol{Q}_J^{\mathrm{t}}$ is the loading matrix of the combined $\boldsymbol{Y}$ concentration matrix.

Spectrum $\boldsymbol{x}_{i,m}$ in the host calibration set can be converted into the spectrum $\hat{\boldsymbol{x}}_{i,s}$ of the slave by the following formula:

$$\hat{\boldsymbol{x}}_{i,s} = \left(\boldsymbol{Q}_J \, \boldsymbol{Q}_J^{\mathrm{t}}\right)^{-1} \boldsymbol{Q}_J \, \boldsymbol{x}_{i,m}$$

Shan et al., (2020) proposed a joint spectral subspace spectral transfer method (JPCA) based on principal component analysis and kernel principal component analysis. The method combines the spectral matrixes of the host and slave standard sample sets for PCA or kernel PCA, so as to obtain the conversion matrix through least squares in the low-dimensional feature space.

Khaydukova et al., (2017) proposed a standardization method with regularization coefficients (SRC) based on Tikhonov regularization (TR), which uses the following formula to obtain the conversion matrix:

$$\boldsymbol{F} = (\boldsymbol{X}_s^{\mathrm{t}} \boldsymbol{X}_s + a)^{-1} (\boldsymbol{X}_s^{\mathrm{t}} \boldsymbol{X}_m)$$

In the formula, a is the regularization coefficient, usually from 1 to 30000.

Spectrum $\boldsymbol{x}_{i,m}$ in the host calibration set can be converted into the spectrum $\hat{\boldsymbol{x}}_{i,s}$ of the slave by the following formula:

$$\hat{\boldsymbol{x}}_{i,m}=\boldsymbol{x}_{i,s}\boldsymbol{F}$$

Zhao Y. et al., (2019) used the PLS model of the host to project the spectra of host and slave standard sample sets respectively, and then established the conversion between two instruments in the PLS projection space. For the spectrum collected from the slave, the projection vector on the host was obtained by conversion relationship in the projection space, and then the PLS model of the host produced the final prediction result. Zhang F. et al., (2018) used $\boldsymbol{X}$ weight matrix of PLS model built by the host to project the host standard sample spectrum to further obtain the matrix $\boldsymbol{L}$, and then to obtain the conversion matrix $\boldsymbol{F}$ between $\boldsymbol{L}$ and the slave standard sample spectral matrix. It was a PLS-based weight matrix transfer method. Chen Z. et al., (2011) used the host model to predict the deviation of the concentration value of the slave standard sample set from the actual concentration, and established a deviation prediction model for systematic prediction error correction (SPEC).

Mou et al., (2016) aimed the minimization of the robust Cauchy Estimator function of subspace learning, and proposed a robust spectral conversion method. It calculated the shared base matrix of the master and slave spectra and its corresponding expression coefficients in an iterative manner, and then established a conversion matrix based on the expression coefficients, which can reduce the impact of out-of-bounds samples and spectral noise on the spectral transfer. Seichte et al., (2017) proposed a Bayesian model transfer method based on Lagrangian multiplier method and hierarchical model, that used the Markov chain Monte Carlo method to estimate the error bounds. It was successfully performed in model transfer of oxygen sensors. Subsequently, they used similar methods to eliminate the influence of oxygen on the determination of carbon dioxide content by mid-infrared spectroscopy (Seichte et al., 2017). Based on the multiple block orthogonal projections to latent structures (OnPLS), Skotare Seichte et al., (2019) proposed a joint and unique multiblock analysis (JUMBA) for spectral transfer of multiple instruments.

Andries, (2017) used the penalty matrix decomposition algorithm of domain adaptation in transfer learning method, such as transfer component analysis (TCA) and scatter component analysis (SCA) for spectral matrix conversion and model maintenance. Similarly, Liu C. et al., (2019) used the TCA to realize the NIRS transfer of edible oil on different instruments. In terms of soil researches, TCA was successfully applied to transfering the NIRS among arsenic and available phosphorus models (Tao et al., 2019; Zheng W. et al., 2020). Specically, TCA addresses the problem of different data distribution in the source domain and the target domain, and maps the data from two domains to a high-dimensional regenerative kernel Hilbert space, where, the data distance between the source and target are minimized, while their respective internal attributes to the greatest extent retains.

Shi G. et al., (2019) transfered the NIRS of the two types of wood by adaptive component analysis (ACA) and established a deep migration model of oak wood defect classification with the color wood data as the source domain and oak wood data as the target domain. Shan et al., (2020) proposed a joint spectral subspace method for model transfer based on PCA and KPCA. Based on the principle of domain adaptation, Nikzad-Langerodi et al., (2018) and Mishra, (2020) proposed a domain-invariant PLS (di-PLS) method, which can be used for unsupervised, semi-supervised and supervised spectral model maintenance and model transfer. Huang G. et al., (2020) also proposed a partial least squares method for domain adaptation. A transfer sample based coupled task learning (TCTL) method was proposed based on transfer learning and multi-task learning, which could be used for transfer between electronic nose instruments and compensation for drift over time (Yan and Zhang, 2016). Based on the active learning algorithm from machine learning, Hu and Li, (2019) solved the problem of multivariate quantitative correction for hyperspectral imaging of different types of blueberries through iterative screening of correction samples.

Li Q. et al., (2019) proposed a double digital projection slit algorithm for Raman spectrometers with different resolutions to solve the problem of spectral consistency. Gradient descent method was adopted to obtain the optimal solution of the conversion matrix, and a better transfer effect was achieved. In the study of Liu et al., the deep autoencoder (DAE) method was employed to establish a nonlinear mapping between spectra of different NIR instruments. An error function penalty term based on conditional probability and parameter maximum likelihood method was designed, and the network parameters of deep self-encoding was optimized by combining with gradient back propagation algorithm (Liu, Zhu and Chen, 2020).

In addition to the conversion of the calibration spectra matrix, some new methods have also been applied in order to realize the transfer of the host calibration model. Liu, Cai and Shao, (2016) proposed a linear model correction (LMC) which can realize the transfer of PLS model regression coefficients. Subsequently, the method was further improved to obtain the globally optimized regression coefficients (Zhang et al., 2018). Kauppinen et al., (2014) realized the transfer of the online NIR model of the moisture content of drug freeze-drying process by converting the PLS regression coefficients. With regard to the evaluation issue of the transferred model, Eskildsen et al., (2016) proposed to use the prediction results of the model instead of basic reference data to evaluate the effect of the model transfer.

Global Calibration, Robust Calibration and Model Update

Three keywords of global calibration, robust calibration and model update (or model maintenance) are often refered confusingly in spectral multivariate calibration analysis, and actually they have a lot in common. The global model calibartion, also known as the augmented hybrid calibration, hybrid calibration, spiking method in the literature, usually refers to the expansion of the calibration set of the host by adding spectra under different test conditions, spectra measured from different instruments, etc. to build a model, so as to achieve model sharing under different instruments, different measurement environments or different sample types. Robust calibration often refers to the establishment of a model that is not sensitive to external influence factors through spectral preprocessing algorithms, wavelength variable screening algorithms, etc (Adhihetty et al., 1991; Fearn, 1997). Therefore, it can be considered that the establishment of a global calibration is a means to achieve the robust model, and both two methods can also be combined to establish a robust global model. While, the model updating or calibration maintenance covers a broader range. In the traditional concept, when encountering samples outside of the model (chemical composition or physical state of the sample changes) or the instrument is aging over time, what we need is model update or model maintenance. In general concept, the process of establishing a global model or a robust calibration actually also belongs to the category of model update or model maintenance (Zeaiter et al., 2004, 2005). Thus, global calibration, robust calibration and model updates (or model maintenance) are usually employed to solve model transfer problems (Hong et al., 2018).

Koehler et al., (2000, 2000) introduced a self-designed FIR matrix to filter the MIR interference data, combined with model update, accurately classifying the data obtained by two MIR remote sensing spectrometers based on piecewise linear discriminant analysis (PLDA). FIR algorithm was used to eliminate the changes in the spectra from the same instrument at different times and under different environmental conditions, and a more robust NIRS model for predicting soil organic matter content was established (Small et al., 1990; Song and Qin, 2015). By means of spectral error analysis, Wang Y. et al., (2019) proved that the combination of the first derivative and SNV can significantly improve the model transfer effect between Fourier NIR spectrometers with integrating sphere diffuse reflectance measurement.

Milanez et al., (2017) involved successive projections algorithm (SPA) to select characteristic wavelength variables for establishing NIRS and UV-vis models. The discriminant models can be used among different instruments, and their recognition accuracy was equivalent to DS and PDS methods. Similar study was reported that, SPA and competitive adaptive reweighted sampling (CARS) were combined to select special wavelengths, and together with SBC algorithm, the established NRIS model predicted the soluble solid content of apples picked for consecutive years (Fan et al., 2019). A double CARS strategy was proposed by Zheng K. et al., (2019) for NIRS global modeling. Based on the concept of standard deviation, Ni et al., (2018, 2019) and Zhang L. et al., (2020) proposed a stable and consistent wavelength variable selectiion method between spectrometers, which can establish a robust calibration model and realize the sharing of models on multiple instruments. Hong S. et al., (2020) used the scale invariant features transform (SIFT) algorithm to screen the stable characteristic wavelengths for establishing NIRS model of the total alkaloids of tobacco leaves, which can realize the standard-free transfer of the model. Xu Z. et al., (2020) proposed a correlation-analysis-based wavelength selection (CAWS) based on the correlation coefficient between the host and slave spectra, which was used to establish a robust calibration model and obtained good results. Based on the multi-model consensus strategy, Zhang L. et al., (2002, 2003) selected wavelength variables to filter and weight by PLS regression coefficients, and screened calibration set samples through a method similar to stepwise multiple linear regression to select variables. Guided model reoptimization (GMR) was proposed to solve the problem of model update and model transfer.

When establishing the global model between host and slave, it is necessary to collect the spectra of a certain number of representative samples from the slave, and the corresponding concentration value is granted to get better results (Stork and Kowalski, 1999). To address this problem, Stout and Kalivas, (2006), Kalivas et al., (2009) proposed a strategy of models update and transfer based on the weighting method of slave samples and the Tikhonov regularization framework.

The common weighting modeling can be expressed as:

$$\begin{pmatrix} \boldsymbol{y} \\ \lambda \boldsymbol{y}_{\mathrm{L}} \end{pmatrix} = \begin{pmatrix} \boldsymbol{X} \\ \lambda \boldsymbol{L} \end{pmatrix} \boldsymbol{b}$$

where, $\boldsymbol{y}$ is the concentration vector of the host calibration set, $\boldsymbol{X}$ is the spectra matrix of the host calibration set, $\boldsymbol{y}_{\mathrm{L}}$ is the concentration vector of the slave calibration, $\boldsymbol{L}$ is the spectra matrix of the slave calibration, λ is the weighted value, and $\boldsymbol{b}$ is the regression coefficient.

Under Tikhonov regularization (L2 regularization), the optimization framework is:

$$\min\left(||\boldsymbol{Xb} - \boldsymbol{y}||_2^2 + \lambda^2||\boldsymbol{Lb} - \boldsymbol{y}_L||_2^2\right)$$

The solution of this equation is:

$$\widehat{\boldsymbol{b}} = (\boldsymbol{X}^{\mathrm{t}}\boldsymbol{X} + \lambda^2\boldsymbol{L}^{\mathrm{t}}\boldsymbol{L})^{-1}(\boldsymbol{X}^{\mathrm{t}}\boldsymbol{y} + \lambda^2\boldsymbol{L}^{\mathrm{t}}\boldsymbol{y}_{\mathrm{L}})$$

If $\boldsymbol{L}$ represents the difference spectra matrix of a set of standard sample sets measured on the host and slave respectively, or a set of spectral baseline background matrix under different test conditions, or a set of spectra matrix of blank samples, etc., its corresponding $\boldsymbol{y}_{\mathrm{L}}$ becomes a zero concentration vector. Then the optimization framework can be simplified as:

$$\min\left(||\boldsymbol{Xb} - \boldsymbol{y}||_2^2 + \lambda^2||\boldsymbol{Lb}||_2^2\right)$$

The solution of this equation is:

$$\widehat{\boldsymbol{b}} = (\boldsymbol{X}^{\mathrm{t}}\boldsymbol{X} + \lambda^2\boldsymbol{L}^{\mathrm{t}}\boldsymbol{L})^{-1}\boldsymbol{X}^{\mathrm{t}}\boldsymbol{y}$$

In order to obtain more stable regression coefficients, based on the idea of ridge regression, the weighting method modeling can be improved as:

$$\begin{pmatrix} \boldsymbol{y} \\ \boldsymbol{0} \\ \lambda \boldsymbol{y}_{\mathrm{L}} \end{pmatrix} = \begin{pmatrix} \boldsymbol{X} \\ \tau \boldsymbol{I} \\ \lambda \boldsymbol{L} \end{pmatrix} \boldsymbol{b}$$

Among which, $\boldsymbol{\tau}$ is the penalty coefficient, $\boldsymbol{I}$ is the identity matrix.
The optimization framework is:

$$\min\left(||\boldsymbol{Xb}-\boldsymbol{y}||_2^2+\tau||\boldsymbol{b}||_2^2+||\boldsymbol{Lb}-\boldsymbol{y}_L||_2^2\right)$$

The solution of this equation is:

$$\widehat{\boldsymbol{b}}=(\boldsymbol{X}^{\mathrm{t}}\boldsymbol{X}+\tau^2\boldsymbol{I}+\lambda^2\boldsymbol{L}^{\mathrm{t}}\boldsymbol{L})^{-1}(\boldsymbol{X}^{\mathrm{t}}\boldsymbol{y}+\lambda^2\boldsymbol{L}^{\mathrm{t}}\boldsymbol{y}_{\mathrm{L}})$$

Regarding to the model update and transfer strategy of Tikhonov's regularization, Kunz et al. (2010, 2010) discussed the influence of standard design selection. Shahbazikhah and Kalivas, (2013) proposed a consensus modeling strategy to optimize the selection of regularization parameters. While, Tencate et al., (2016) brought forward a method for selecting model update parameters based on the fusion strategy. Farrell et al., (2012) used Tikhonov's regularization strategy to update the drug NIRS model under different conditions, and obtained satisfactory results. Similar study was done for sharing the model on different NIRS instruments based on Tikhonov's regularization strategy (Hu Y. et al., 2020).

Except for sample amplification, model update can also take samples and features to amplify at the same time, which can be expressed as (Andries et al., 2018):

$$\begin{pmatrix}\boldsymbol{y}\\ \boldsymbol{0}\\ \lambda\boldsymbol{y}_{\mathrm{L}}\end{pmatrix}=\begin{pmatrix}\boldsymbol{X}\,\boldsymbol{0}\\ \boldsymbol{0}\,\boldsymbol{X}\\ \lambda\boldsymbol{L}\,\lambda\boldsymbol{L}\end{pmatrix}\begin{pmatrix}\boldsymbol{b}_m\\ \boldsymbol{b}_s\end{pmatrix}$$

Rudnitskaya et al., (2017) compared the effects of DS algorithm, Tikhonov regularization and Joint-Y PLS on model transfer and update for the potential sensor array instrument, proving that Tikhonov regularization and Joint-Y PLS can get better results. On the basis of the least absolute shrinkage and selection operator (LASSO), Kunz and She, (2013) proposed a robust fused LASSO algorithm (RFL) for the maintenance and transfer of the calibration model. Genetic algorithm was performed by Guo et al., (2017) to correct the wavelength variables from multiple Raman spectra, and the updated model by Tikhonov regularization obtained satisfactory results. Based on the ridge regression updating, Zhang F. et al., (2019) combined the prediction optimization and the 2-norm constraint of model coefficients, realized the update of model coefficients, and solved the problem of deterioration of model prediction ability and reliability caused by instrument drift or sample changes.

Based on the Lambert Beer's law, Sulub and Small, (2007) proposed a spectral simulation calculation method. Spectra of the mixture can be generated by the absorption signal of the pure substance and the background signal of the instrument. In this way, whether for host or slave instrument, the calibration set can be quickly established by measuring the spectra of the pure substance in the mixture. Haaland, (2000) synthesized spectra by adding the temperature-affected background signal to the calibration spectra as to solve the problem of the model's adaptability to temperature. The temperature-affected background signal was obtained by the least square expansion of the temperature spectra. By PDS, Sulub et al., (2008) realized the spectral transfer of multiple NIRS instruments (including raster scan and FT instruments) to determine the content of active ingredients in medicines, and updated the slave model by preparing a placebo. This method can analyze 30 medicines in 12 minutes, instead of 5 hours by HPLC method. Saiz-Abajo et al., (2005) established a robust calibration model by adding different types of noise and interference to the calibration spectra, as well as the ensemble modeling strategy. Pierna et al., (2011) added the spectral changes of different instruments and samples with different states (water content and particle size, etc.) to the spectra of 10 calibration samples, generating hundreds of virtual modeling spectra, and obtaining a robust prediction model.

Cooper et al., (2011) took the spectra of several calibration samples from the host as the target, and obtained the same number of virtual spectra through mathematical mixing calculation to form a virtual spectral matrix. The host model predicted the virtual spectral matrix of host and slave, and established the SBC calibration curve based on the predicted value of host and slave. In the further study, they also used the spectra of 13 pure compounds to convert the NIRS of jet fuel by the above method, which can accurately predict the properties and composition of aromatics, hydrogen content, density, viscosity, etc (Cooper, Larkin and Abdelkader, 2011). Rauscher et al., (2018) also refered this method to transfer the spectra of the non-dispersive infrared spectrometer to monitor the quality of oil. Similarly, Abdelkader et al., (2012) used the spectra of 15 pure compounds and calculated the virtual spectra in segments, improving the transfer effect of the method. Silva et al., (2017) established the spectral conversion coefficient on the virtual spectral matrix using DS algorithm, and realized the transfer of oil from the desktop NIR spectrometer to the handheld instrument.

Ni, Brown and Man, (2011) combined the superimposed PLS method with model update to establish a robust calibration model for the sharing of multiple NIRS instruments. Honorato et al., (2005) selected wavelength variables by the continuous projection algorithm (SPA) to establish a robust MLR model shared by multiple instruments, and the results were slightly better than PDS-PLS. With establishment of a local calibration model, Igne and Hurburgh, (2010) realized the model sharing of four NIRS instruments from two brands. Liu J. et al., (2018) first constructed the 3D spectral cube collected from multiple instruments, and employed Tchebichef discrete orthogonal moments to extract chemical features. Finally, a stepwise regression method was conducted to establish a universal model on multiple instruments.

Moving window correction (MSC) was employed by Kramer et al., (2008) for the preprocessing of jet fuel spectra on two NIRS instruments, and a better result was obtained by optimizing the size of the moving window. Also, Liu, Yu and MacGregor, (2007) mainly used the MSC algorithm to standardize the line-scan NIR imaging system. Sahni, Isaksson and Næs, (2005) compared the effects of MSC, OSC, FIR, PDS and global models on the correction of optical path changes of optical fiber probes, and results revealed that PDS and global models proved to be the better solutions. Extended multiplicative signal correction (EMSC) was used for the preprocessing of bacterial Raman spectra on multiple instruments. The PLS-DA algorithm can successfully establish a discriminant model for different bacterial species (Guo et al., 2018). Preys, Roger and Boulet, (2008) combined orthogonal signal correction (OSC) with external parameter orthogonalisation (EPO) to establish a robust calibration model, which solved the problem that OSC can not consider the influence of external interference and the predictive performance degradation of EPO when the influence of external factors on the target value is too high.

Wijewardane, Ge and Morgan, (2016, 2016) modelled the air-dried ground soil to predict the organic and inorganic carbon content at different moisture levelsu by EPO, DS and a global model. Ackerson, Ge and Morgan, (2017) and Roudier et al., (2017) also solved similar problems through EPO and DS algorithms. Amat-Tosello, Dupuy and Kister, (2009) used the EPO algorithm to simultaneously realize the sharing of gasoline short-wave NIRS on four instruments. EPO was employed by Hans and Allison, (2019) to reduce the influence of temperature and humidity on the NIRS to predict the calorific value of biomass, and to establish a model that was not sensitive to temperature and humidity. Similarly, the temperature mixing model strategy can eliminate the influence of temperature on NIRS prediction of reducing sugar and moisture content in longan honey. Thygesen and Lundqvist, (2000) and Thamasopinkul et al., (2017) successfully built a temperature mixing model for predicting moisture content in wood using NIRS, which was more effective than single PDS algorithm.

Luoma et al., (2018) attempted to use additive partial least squares modeling strategy for model maintenance. According to different measurement conditions, the residual PLS model was established to solve the problem of model inapplicability by addition. Elizalde, Asua and Leiza, (2005) also adopted a similar strategy to solve the problem of model inapplicability caused by changes in the online Raman spectrometer. Nouri et al., (2017) established a hybrid model of soil NIRS from lab and airborne hyperspectral images, and updated the model with the difference spectra and zero concentration of the standard spectra. This model can be used for prediction of airborne hyperspectral imaging.

Local model that is based on the local kernel function is often established in the SVR method, which may not work when the new spectrum from the new instrument is added to the calibration set. Based on the transfer learning idea, Yu and Ji, (2015) used regularized multi-task learning (RMTL) to estimate the relationship between the SVR model in the new condition and the original model, and successfully improved SVM regression. Highlight of this method is that the most important support vectors of SVR model can be selected as the sample for model update.

Progress of Applications

SBC Method

Brito et al., (2016) used the SBC algorithm to correct the model prediction values of the total suspended solids (TSS) and chemical oxygen demand (COD) of waste water by UV spectroscopy, and the effect was better than the SSC algorithm and the SLRDS algorithm. Brouckaert et al., (2018) also used the SBC algorithm to correct the predicted values of the Raman spectroscopy established in the laboratory for the determination of the content of the two components of the liquid detergent, and used it for industrial-level online analysis, and obtained satisfactory results. Dambergs et al., (2012) used the SBC algorithm to quickly analyze the predicted value of the tannin content in red wine by ultraviolet spectroscopy and corrected it on multiple instruments, and obtained relatively consistent prediction results.

According to the PCA classification model of different types of coffee beans in the NIR spectroscopy, Myles et al., (2006) used the SBC algorithm to convert the PCA scores on the two instruments, and obtained a better transmission effect. Wang J. et al., (2018) conducted a similar study on the hyperspectral prediction models of lamb protein from different origins, and the result is that the SBC algorithm has a better effect. Li T. et al., (2018) used SBC algorithm to correct the acid value and peroxide value of edible oil predicted on two NIRS instruments.

SSC Method

Pierna et al., (2010) used the spectral difference correction algorithm (SSC) to transfer the spectrum of more than 9000 feed samples on the dispersive desktop NIR spectrometer to the MEMS handheld spectrometer using 25 transfer standard samples to establish the determination of feed fat, fiber, and protein, and starch PLS model. Zamora-Rojas et al., (2012) used the SSC algorithm to transfer the pork spectrum collection on the desktop NIR spectrometer to the handheld instrument to realize the routine quality analysis beside the pork processing line. Daikos et al., (2019) used a method similar to the SSC algorithm to subtract the background of the substrate, and transferred the near-infrared imaging spectrum of the coating on one substrate material to another substrate material, realizing the sharing of PLS models. Smith et al., (2002, 2004) used the SSC algorithm to realize the spectral transmission of different near-infrared spectroscopy instruments to predict the content of active ingredients in the whole paracetamol tablets. This article also tried to use six standard materials to establish a spectral response correction model through the SLRDS algorithm, which was slightly inferior to the SSC algorithm.

Hayes, Walsh and Greensill, (2016) calibrated the wavelength of the array detector type short-wave near-infrared spectrometer, and then used the SSC algorithm to update the model. In predicting the soluble solid content of apples, the results are equivalent to PDS, but the implementation process is simple and easy. Xu and Li, (2017) aimed at the lack of universality in the multi-channel grading detection model for fruit quality, and adopted the DS algorithm (MSSC-DS) for correction of the average spectral difference to transfer the online detection spectrum of the crown pear sugar content between the two spectrometers, the prediction accuracy of the model can meet the actual requirements of production (less than 0.5 °Brix). Roggo et al., (2002) used the SSC algorithm to transfer the near-infrared reflectance spectrum of sugar beet, and good results can be obtained no matter whether the actual sample or the general sample is used as the transfer standard sample. Saranwong and Kawano, (2004) used the SSC algorithm for the NIRS transmission of the online fruit screening machine, which greatly improved the MLR and PLS correction models, and the compensation of the average spectral difference of the standard sample set was better than the result of linear fitting or polynomial fitting.

Soldado et al., (2013) combined the SSC algorithm with the transfer by orthogonal projection (TOP) method to transfer the silage spectrum from the desktop NIR spectrometer to a portable instrument, the built model can accurately predict the content of dry matter, neutral detergent fiber and crude protein. Sun H. et al., (2019) first screened the characteristic variables through regression coefficients, and then compensated the absorbance through the SSC algorithm, and established a universal NIR analysis model for the quality of fresh jujubes between different instruments.

Shenk's Method

Qin and Gong, (2016) compared the transfer effects of PDS, DWPDS and Shenk's algorithms on the NIRS of tobacco leaf and smoke powder. Shenk's algorithm is superior to other methods. It is also a feasible way to establish a mixed model of tobacco leaf and smoke powder NIRS. Based on Shenk's algorithm, Garcia-Olmo et al., (2001) investigated the transfer effects of four standard sample components on fatty acid liquid near-infrared spectroscopy, and the results showed that the closer the composition of the sample to the test sample, the better the transfer effect. Roza-Delgado et al., (2017) used Shenk's algorithm to transfer the milk spectrum on desktop and portable near-infrared spectrometers, and established a model that can quickly detect milk composition on site. Masahiro and Yukihiro, (2005) used a similar Shenk's algorithm to calibrate the wavelength and absorbance of the NIRS respectively, to solve the problem of spectrum changes before and after the online instrument maintenance, and to achieve the correction and maintenance of the vinyl content prediction model in molten polypropylene. Perezmarin, Garridovaro and Guerreroginel, (2006) used Shenk's algorithm to successfully transmit the NIRS of feed on different scanning instruments.

DS Method

Milanez et al., (2016) used the DS algorithm to transfer the NIRS of the adulterated ethanol gasoline on two instruments, and a 100% recognition success rate can be obtained for PLS-DA discriminant analysis. Silva et al., (2017) used the DS algorithm to transfer the Fourier transform NIRS of the drug to multiple handheld spectrometers, and established an analysis model for predicting the content of different crystal forms. Brito et al., (2017) used the DS algorithm to transfer the NIRS of the flour from the desktop instrument to the handheld instrument, and then re-established the PLS model on the handheld instrument, which can obtain better results. Ji and Han, (2014) used the UVE-SPA wavelength screening method combined with the DS algorithm to realize the transmission of the NIRS of the apple on the same model and different models of fruit portable instruments. Hu and Xia, (2012) realized model transfer of navel orange total sugar prediction by near-infrared spectroscopy using DS algorithm.

Chen Y. et al., (2015) used the DS algorithm to transfer the NIR hyperspectral spectra of soil under different humidity, and realized that the model based on the air-dried soil spectrum could be used for soil samples with different moisture content. Wang S. et al., (2019) used the DS algorithm to convert the NIRS of soil under different humidity to eliminate the interference of moisture on the prediction of soil organic matter content. Ji et al., (2016) used the DS algorithm to transfer the NIRS of the water-bearing untreated soil, and made predictions through a quantitative calibration model established by dried and crushed samples. Silva et al., (2015) used the DS algorithm to convert the gasoline spectra measured by three mid-infrared spectrometers, and based on the global modeling strategy, the PLS-DA or SIMCA method can be used to distinguish gasoline from different origins. Liu C. et al., (2017) used the DS algorithm to transfer two edible oil spectra above the NIRS, and established an analysis model for predicting acid value and peroxide value. Lopez-Moreno, Palanco and Laserna, (2005) used the DS algorithm to transfer the iso-LIBS spectrum at room temperature to the iso-spectrum at high temperature, and established a model for predicting the metal content in a high-temperature environment.

Khaydukova et al., (2017) used the DS algorithm to convert the volt-ampere signal and the potential signal in the electronic tongue sensor, so that the regression model based on the potential data can be used for the prediction of the volt-ampere signal. Weng H. et al., (2018) used the DS algorithm to transfer citrus spectra between different models of hyperspectral imagers, and the extreme learning machine discriminant model of citrus canker has an accuracy of 86.2% on the slave spectra. Fonollosa et al., (2015, 2016) used DS, PDS and other algorithms to transfer the signal of the metal oxide gas sensor matrix on multiple sensors and obtained satisfactory results. Panchuk et al., (2017) used the DS algorithm to convert different types of spectra, such as the conversion between energy-dispersive XFS and vis-UV, and the conversion between different NIRS wavelength ranges. Vaughan et al., (2012) used the DS combined with PLS algorithm to convert the spectra of two LC-MS, and initially solved the problem of metabolomics data fusion. Morais and Lima, (2015) realized the conversion of digital imaging on two devices using DS algorithm for predictive analysis of serum creatinine content. Khoshkam et al., (2012, 2013) embedded the DS algorithm in the multivariate resolution analysis of the extended matrix for the study of reaction kinetics and obtained good results. Surkova et al., (2020) successfully transferred the spectrum from the UV-Vis spectrometer to the optical multi-sensor system composed of four LEDS by using DS algorithm, and realized the transmission between optical multi-sensor systems.

PDS Method

Barreiro et al., (2008) used the PDS algorithm to transfer the spectrum from the desktop NIR spectrometer to the portable spectrometer, and established an analytical model that can detect the olive breeding process in the field. Alamar et al., (2007) used the PDS algorithm to transfer the FT-NIRS of 447 Jonagold apples to the matrix spectrometer, and before the spectrum transfer, the wavelength of the Fourier transform spectrum was normalized by the segmented cubic Hermite interpolation, and then the analytical model of the soluble solid content was established. Sulub et al., (2008) used the PDS algorithm and the wavelet hybrid direct correction algorithm (WHDS) to achieve spectral conversion on five NIR spectrometers for rapid analysis of the uniformity of the effective ingredient content of the tablet. Luo et al., (2017, 2017) combined wavelength selection and PDS algorithm for the conversion of bovine blood near-infrared spectroscopy, which can achieve rapid diagnosis of bovine anemia on multiple instruments. Cen H. et al., (2016) used PDS to transmit the hyperspectral spectrum of citrus canker, and then established a least squares support vector machine discriminant model. The recognition rate of the prediction set of the model increased from 26% before transmission to 97%.

Pereira et al., (2016) used the dual-window PDS algorithm (DWPDS) to transfer the NIRS of the drug powder to the spectrum of the tablet, which provides a feasible way to quickly obtain a sample of the drug spectrum correction set. Sohn, Barton and Himmelsbach, (2007) also used the DWPDS algorithm to transfer the NIRS for analyzing the cellulose content of flax between different instruments, and achieved good results. Galvan et al., (2020) used the DWPDS algorithm to transfer the low-field nuclear magnetic resonance spectrum of gasoline between different instruments, among which the resolution of low-field nuclear magnetic instruments is different.

Yang Y. et al., (2018) used DS and PDS algorithms to effectively transfer the surface-enhanced Raman prediction model of potassium sorbate in sweet-scented osmanthus wine to bayberry wine to realize the transfer of the same analyte prediction model among different species. Similar examples of transferring models between different sample types include the application of quality analysis of chilled pork and eggs (Liu J. et al., 2014, 2015; Dong et al., 2019). Boiret et al., (2011) used the PDS algorithm to transfer the transmission NIRS of the coated tablets between two Fourier transform spectrometers of the same model, and the predicted standard deviations of the effective ingredient content before and after the transfer were 4.0% and 2.4%, respectively. Sales, Callao and Rius, (2000) used the PDS algorithm to transfer the signal of the potential sensor under the two test conditions, and selected the transfer standard sample based on the K-S algorithm, and obtained satisfactory results. Marchesini et al., (2017) used the PDS algorithm to transfer the NIRS from the desktop instrument of the undried whole corn stover to two portable instruments.

Ge Y. et al., (2011) used the PDS algorithm to convert the diffuse reflectance spectra of the soil measured on multiple NIRS instruments of different models, and established a model to predict the organic carbon content in the soil. Rodrigues et al., (2017) used the PDS algorithm to transfer the MIRS of crude oil on two instruments, and then used the orthogonal projections to latent structures (OPLS) to establish a crude oil density prediction model, and got better results. Li X. et al., (2011) used the PDS algorithm to convert the spectrum of propylene glycol-water solution measured on multiple hand held NIR spectrometers, and realized the rapid identification of ethylene glycol adulteration. Gryniewicz-Ruzicka et al., (2011) used the PDS algorithm to transmit the spectra of multiple Raman spectrometers measuring the content of propylene glycol in glycerol. Thygesen and Van Den Berg, (2011) used algorithms such as DS and PDS to transfer excitation-emission three-dimensional fluorescence spectra, and then took advantage of PARAFAC to obtain satisfactory results. Sanllorente et al., (2019) also used a similar method to transfer the three-dimensional fluorescence spectrum between the portable fluorescence spectrometer of the LED light source and the fluorescence spectrometer of the xenon light source. Sun X. et al., (2019) used the PDS algorithm to transfer the three-dimensional fluorescence spectra on the two instruments, and then used the self-weighted alternating normalized residual fitting algorithm (SWANRF) to establish a three-linear decomposition multidimensional quantitative model. The results show that PDS can maintain the "second-order advantage" of the second-order tensor correction method.

Wang M. et al., (2015) used PDS to realize the transfer of NIRS of leaves picked from different tree species and different periods, and solved the problem of standard spectra based on chlorophyll content through linear interpolation. Watari and Ozaki, (2004) also used a similar method to convert the NIRS of random polypropylene and block polypropylene in the molten state, so that the model of ethylene content in one type of polypropylene can predict another type of polypropylene. Li Y. et al., (2020) discussed the influence of the number of PLS latent variables on the transfer performance of the model when the PLS model is established by the NIRS after PDS transmission. The results show that too high number of latent variables can easily cause over-fitting, affect the robustness of the model, and make the error of model transmission larger.

Sun Z. et al., (2018) used the PDS algorithm to convert the NIRS of the plasma alcohol precipitation process between two different types of instruments, and established an analytical model for predicting the content of total protein, albumin and globulin. Xiao H. et al., (2017) combined linear interpolation with the PDS algorithm to convert the Fourier transform NIRS of a single grape into a portable grating spectroscopic spectrum, and established a model for predicting the content of soluble solids. Fernandez et al., (2016) compared DS, PDS, OSC, GLSW and other methods to transmit the signal of the gas sensor matrix at different temperatures, and the results of the PDS algorithm are better. Hoffmann et al., (2016) used the PDS algorithm to transfer the FT-NIRS to the linear gradient filter handheld instrument, and the quantitative and qualitative models obtained good results. Di Anibal et al., (2012) used the PDS algorithm to transmit the UV spectrum used to determine whether illegal substances are added to the fragrance, and combined with the PLS-DA method to obtain satisfactory results. Zheng Y. et al., (2020) used the PDS method to transfer the NIRS of fish meal on the grating-type instrument to the Fourier-type instrument, and there was no significant difference in predicting the content of crude protein, crude fat, methionine, lysine and other components.

Pu et al., (2017) used the PDS algorithm to successfully transfer the banana spectrum from the handheld NIR spectrometer to the hyperspectral imaging instrument, and established a model for predicting the soluble solid content. Xi C. et al., (2014) used the PDS algorithm to transfer the NIRS of amoxicillin capsules and the NIRS of its contents, so that the amoxicillin capsule quantitative model can accurately predict and analyze the powder spectrum of the content, and proposed an index to judge whether the spectrum is successfully transmitted. Gislason, Chan and Sardashti, (2001) used the PDS algorithm to realize the transfer of process nuclear magnetic resonance spectrum on different instruments, and compared the effect of DS combined with SSC algorithm. Monakhova and Diehl, (2016) used the PDS algorithm to transfer the spectrum of the sunflower lecithin and soybean lecithin mixture obtained by three high-resolution NMR instruments, and the result was better than the DS algorithm and the establishment of a mixed model.

Chen, Brown and Lo, (1997) used the PDS algorithm to transfer the UV-visible spectrum of the cuvette with a 10mm optical path to a fiber optic probe with a 2mm optical path. Before using the PDS algorithm for conversion, the spectrum was subjected to Fourier transform processing. Lin, Lo and Brown, (1997) used the PDS algorithm to transfer the spectrum from the scanning NIRS instrument (cuvette) to the Fourier transform (fiber probe) instrument. Shi G. et al., (2013) used the PDS algorithm to better solve the transfer of the NIRS of the mixture of fish meal and soybean meal on two different spectroscopic types of instruments. Tortajada-Genaro, Campíns-Falcó and Bosch-Reig, (2001) used the PDS algorithm for the conversion of chemiluminescence signals on two instruments, and established a model for the rapid determination of Cr(III), Cr(VI) and total Cr content in water through the PLS method. Griffiths et al., (2006) used the PDS algorithm combined with the variable selection method to solve the problem of the failure of the multivariate calibration model caused by the drift of the ICP-AES instrument over time.

Wang, Huck and Yang, (2019) used the PDS to transfer the NIRS of Poria cocos samples measured on two different brands of NIR spectrometers, and established a model to predict alkali-soluble polysaccharides in Poria cocos. Morais et al., (2019) used the DS and the PDS to standardize the MIR spectrochemical database of complex biological tissues and established a complete spectrum standardization process. Grelet et al., (2015, 2017) used the PDS algorithm to standardize the instruments in the European dairy MIR spectroscopy network, which can convert spectra on spectrometers of different brands into spectra on the host computer to realize the sharing of quantitative calibration models. Ji W. et al., (2015) used the PDS algorithm to eliminate the influence of moisture and environment on the NIRS of the field soil, and the transmitted spectrum can be accurately predicted by the model established in the laboratory. Pierna et al., (2013) designed and produced a standard sample pool for the transfer of NIRS microscopic imaging instruments. Different parts of the standard sample pool are equipped with meat and bone meal of different animals, and the spectrum transfer of multiple imaging instruments is realized through the PDS algorithm.

CCA Method

Zheng K. et al., (2013) used the CCA algorithm to convert the NIRS between different times and different brands of milk, and the content of dimethyl fumarate in milk can be predicted by the sample enrichment-NIRS measurement method. Liu Y. et al., (2019, 2019) used the DS algorithm and the CCA algorithm to realize the transfer of the near-infrared spectroscopy analysis model of wood lignin content between different models of portable spectrometers. Yang J. et al., (2019) compared the transfer effects of algorithms such as CCA, SST, CTWM, ICA, and PDS on the transmission of tobacco NIRS on desktop, portable and handheld instruments, and the results showed that the CCA algorithm is superior to other methods. Luo J. et al., (2016) used the CCA algorithm to transfer the NIRS that predicts the content of polyester in textiles, and the results are better than the PDS method. Fan P. et al., (2018) used CCA or PDS algorithm combined with linear interpolation method for the transfer of soil near-infrared spectroscopy. A soil nutrient content model can be used to accurately predict soil nutrient content in different regions.

Establishment of Global Model

Eliaerts et al., (2020) used the S/B algorithm, the PDS algorithm and the method of establishing a hybrid global model to transfer the cocaine classification and quantitative SVM models on desktop and portable infrared spectrometers. The results showed that the method of establishing the hybrid model has better results. Yang Z. et al., (2019) transferred the model established in the laboratory to feed production enterprises for online application by using the method of hybrid modeling of near-infrared spectroscopy and online spectroscopy, and the predicted values of moisture content and crude protein content were in good agreement with the actual measured values, which can meet the requirements of online analysis. Chen and Wang, (2019) used TrAdaBoost algorithm based on principal component analysis and weighted extreme learning machine to establish a global model to realize the sharing of models on multiple instruments. Ni L. et al., (2019) used the establishment of a global mixing model of multiple instruments to realize the sharing of tobacco models on different near-infrared spectroscopy instruments.

Pereira et al., (2008) compared the transmission effects of DS, PDS, OSC and model update methods based on the NIRS of gasoline on different instruments. The results show that DS combined with model update strategy can get better results. Fernandez-Ahumada et al., (2008) used Shenk's algorithm and PDS algorithm to transfer the feed spectrum measured on the laboratory raster-scan near-infrared spectrometer to the online matrix instrument, and then through the model update, the transfer of the model can be better realized. Debus et al., (2019) used the method of establishing a hybrid model to solve the problem of sharing the multi-element calibration model of mid-infrared spectroscopy evaluation environment carbon-containing particulate matter among multiple instruments of the same model.

Krapf et al., (2013) used the PDS algorithm to transfer the laboratory NIRS of the samples during the anaerobic digestion process of energy crops to the online analysis instrument, and the problem of online analysis model establishment was better solved through model update. Li X. et al., (2017) used the combination of Shenk's, PDS and CCA algorithms and hybrid modeling technology to establish a hybrid model based on a homogeneous tobacco powder model, which was successfully applied to the prediction of nicotine content in non-uniform cut tobacco samples and tobacco slice samples. Clavaud et al., (2017) constructed a global calibration set of more than 3000 spectra of various types of freeze-dried drugs and their moisture content on two NIRS of the same model, and established a global model by PLS, SVR, Bayesian Ridge regression, KNN and other methods. The results show that the predictive ability of SVR was better. Ozdemir and Williams, (1999) used a hybrid global model combined with genetic regression to establish a model of four vis-UV spectrometers, one of which is a matrix spectrometer, and the other three are dual-beam scanning instruments.

Kupyna et al., (2010) used a global model to solve the problem of the application of acoustic multivariate quantitative correction models under different test conditions (temperature, flow rate, etc.). Igne and Hurburgh, (2008) compared the effects of multiple transmission methods on the same model and different models of NIRS instruments, and the results proved that establishing a stable and sound local model is a better strategy. On the basis of Shenk's algorithm, Fontaine, Hörr and Schirmer, (2004) can accurately predict and analyze the model of amino acid content in feed ingredients on dozens of instruments in the NIR network through the model update strategy. Steinbach et al., (2017) established a hybrid calibration model for the drug transmission Raman spectra measured on the two instruments, and the model built can be accurately applied to the spectra obtained on the two instruments.

Other Applications

Xu Z. et al., (2019) realized the transfer between NIR spectra of rice single grains and rice flours of different varieties by spectral space transformation. Online monitoring of tobacco nicotine and total sugar was realized through the SST from offline model (Wu J. et al., 2017). In addition, model transfer by SST algorithm has also been applied to edible oil acid value and peroxide value (Liu C. et al., 2020), rice (Xu Z., 2020), hyperspectral image on plant phenotype (Rehman et al., 2020) and Terahertz spectral instruments (Zhou, Zhu and Wei, 2020).

Yang H. et al., (2018) successfully converted the apple spectra on two portable NIR spectrometers using simple linear regression direct standardization (SLRDS) algorithm. Salguero-Chaparro et al., (2013) used Transfer by Orthogonal Projection (TOP) algorithm to transfer the olive spectra from the raster-scan NIR spectrometer to the array portable spectrometer, established an analytical model to predict fat, free acid and moisture content. Liu, Han and Yang, (2011) compared the transfer effects of SBC, OSC, DS, PDS, and local centralization to the silage NIR spectra on the same and different types of spectrometers. They further studied the influence of different temperatures and measuring accessories on the NIRS of rice straw and found local centralization method can eliminate the influences on the spectra to a certain extent (Liu, Huang and Han, 2015). Similar study on pharmerceutical samples was done by Bergman et al. and result revealed local centralization can be realized with little samples (Bergman et al., 2006).

Li X. et al., (2018) combined the wavelet multiscale piecewise direct standardization (WMPDS) with SBC algorithm to realize the transfer of NIR spectra of different types of soils and the correction of prediction for total nitrogen and total carbon content. Walczak, Bouveresse and Massart, (1997); Greensill and Walsh, (2002) compared the effects of DS, PDS, DWPDS, OSC, FIR, and WT on the transfer of citrus spectra between micro-array detector NIR spectrometers. Research showed WT model update can get better results.

Martins, Galvão and Pimentel, (2010) combined SPA wavelength variable algorithm with the multi-model consensus strategy, and established calibration model on MLR, which was proved to be more effective than PDS-PLS model. Yoon, Chung and Han, (2001) first performed first-order derivative and OSC processing on NIRS of gasoline, and then converted the spectra on the two instruments through PDS algorithm, eventually established the model for predicting the benzene content in gasoline. Yahaya et al., (2015) established an analytical method for rapid prediction of mango acidity by the variable screening coupled with MLR, realizing the application to multiple instruments by optimization of variable selection.

References

Abdelkader, M. F., Cooper, J. B. and Larkin, C. M. (2012). Calibration Transfer of Partial Least Squares Jet Fuel Property Models Using a Segmented Virtual Standards Slope-Bias Correction Method. *Chemometrics and Intelligent Laboratory Systems*, 110(1): 64-73.

Ackerson, J. P., Morgan, C. L. S. and Ge, Y. (2017). Penetrometer-Mounted VisNIR Spectroscopy: Application of EPO-PLS to in Situ VisNIR Spectra. *Geoderma*, 286(1): 131-138.

Adhihetty, Indira S., Joseph A. McGuire, Boonsri. Wangmaneerat, Thomas M. Niemczyk and David M. Haaland. (1991). Achieving Transferable Multivariate Spectral Calibration Models: Demonstration with Infrared Spectra of Thin-Film Dielectrics on Silicon. *Analytical Chemistry*, 63: 2329-2338.

Alamar, Carmen, Els Bobelyn, Jeroen Lammertyn, Bart Nicolaï and Enrique Molto. (2007). Calibration Transfer between NIR Diode Array and FT-NIR Spectrophotometers for Measuring the Soluble Solids Contents of Apple. *Postharvest Biology and Technology*, 45(1): 38-45.

Aldridge, P. K., C. L. Evans, H. W. Ward, S. T. Colgan, N. Boyer and P. J. Gemperline. (1996). Near-IR Detection of Polymorphism and Process-Related Substances. *Analytical Chemistry*, 68: 997-1002.

Amat-Tosello, S., Dupuy, N. and Kister, J. (2009). Contribution of External Parameter Orthogonalisation for Calibration Transfer in Short Waves-Near Infrared

Spectroscopy Application to Gasoline Quality. *Analytica Chimica Acta*, 642(1-2): 6-11.

Anderson, C. E. and Kalivas, J. H. (1999). Fundamentals of Calibration Transfer through Procrustes Analysis. *Applied Spectroscopy*, 53(10): 1268-1276.

Andrew, A. and Fearn, T. (2004). Transfer by Orthogonal Projection: Making Near-Infrared Calibrations Robust to between-Instrument Variation. *Chemometrics and Intelligent Laboratory Systems*, 72(1): 51-56.

Andrews, D. T. and Wentzell, P. D. (1997). Applications of Maximum Likelihood Principal Component Analysis: Incomplete Data Sets and Calibration Transfer. *Analytica Chimica Acta*, 350(3): 341-352.

Andries, E. (2017). Penalized Eigendecompositions: Motivations from Domain Adaptation for Calibration Transfer. *Journal of Chemometrics*, 31(4): e2818-e2831.

Andries, E., Kalivas, J. H., Gurung, A., (2018). Sample and Feature Augmentation Strategies for Calibration Updating. *Journal of Chemometrics*, 33(1): e3038.

Anibal, C. V., Ruisánchez, I., Fernández, M., Forteza, R., Cerdà, V., and Pilar Callao, M. (2012). Standardization of UV-Visible Data in a Food Adulteration Classification Problem. *Food Chemistry*, 134(4): 2326-2331.

Bakeev, K. A. and Kurtyka, B. (2005). Sources of Measurement Variability and Their Effect on the Transfer of Near Infrared Spectral Libraries. *Journal of Near Infrared Spectroscopy*, 13(6): 339-348.

Barnes, Susan E., Thomas Thurston, Jo Ann Coleman, Ann Diederich, Darryl Ertl, James Rydzak, Patrick Ng, Katherine Bakeev and Dharmesh Bhanushali. (2010). NIR Diffuse Reflectance for On-scale Monitoring of the Polymorphic Form Transformation of Pazopanib Hydrochloride (GW786034); Model Development and Method Transfer. *Analytical Methods*, 2: 1890-1899.

Barreiro, P., Herrero, D., Hernández, N., Gracia, A. and León, L. (2008). Calibration Transfer between Portable and Laboratory NIR Spectrophotometers. *Acta Horticulturae*, 802: 373-378.

Barring, H. K., Boelens, H. F. M., De Noord, O. E. and Smilde, A. K. (2001). Optimizing Meta-Parameters in Continuous Piecewise Direct Standardization. *Applied Spectroscopy*, 55(4):458-466.

Bergman, E. L., Brage, H., Josefson, M., Svensson, O. and Sparén, A. (2006). Transfer of NIR Calibrations for Pharmaceutical Formulations between Different Instruments. *Journal of Pharmaceutical and Biomedical Analysis*, 41(1): 89-98.

Bergman, E. L., Brage, H., Leion, H., Svensson, O. and Sparén, A. (2003). Transfer of NIR Calibrations between Sites and Different Instruments. *NIR News*, 14(4): 6-7.

Bin, J., Li, X., Fan, W., Zhou, J. H. and Wang, C. W. (2017). Calibration Transfer of Near-Infrared Spectroscopy by Canonical Correlation Analysis Coupled with Wavelet Transform. *Analyst*, 142(12): 2229-2238.

Blanco, M., Coello. J., Iturriaga, H., Maspoch, S. and Rovira, E. (1995). Wavelength calibration transfer between diode array UV-visible spectrophotometers. *Applied Spectroscopy*, 49(5): 593-597.

Blanco, M., Cueva, M. R. and Peguero, A. (2011). NIR Analysis of Pharmaceutical Samples without Reference Data: Improving the Calibration. *Talanta*, 85(4): 2218-2225.

Blanco, M. and Peguero, A. (2010). Analysis of Pharmaceuticals by NIR Spectroscopy without a Reference Method. *Tr AC: Trends in Analytical Chemistry*, 29(10): 1127-1136.

Boiret, M., Meunier, L. and Ginot, Y. M. (2011). Tablet Potency of Tianeptine in Coated Tablets by Near Infrared Spectroscopy: Model Optimisation, Calibration Transfer and Confidence Intervals. *Journal of Pharmaceutical and Biomedical Analysis*, 54(1): 510-516.

Boucher, T., Dyar, M. D. and Mahadevan, S. (2017). Proximal Methods for Calibration Transfer. *Journal of Chemometrics*, 31(4): e2877-e2885.

Bouveresse, E., Massart, D. and Dardenne, P. (1994). Calibration Transfer across Near-Infrared Spectrometric Instruments Using Shenk's Algorithm: Effects of Different Standardisation Samples. *Analytica Chimica Acta*, 297(3): 405-416.

Bouveresse, E. and Massart, D. L. (1996). Standardisation of Near-Infrared Spectrometric Instruments: A Review. *Vibrational Spectroscopy*, 11(1): 3-15.

Bouveresse, E., Massart, D. L. and Dardenne, P. (1995). Modified Algorithm for Standardization of Near-Infrared Spectrometric Instruments. *Analytical Chemistry*, 67: 1381-1389.

Bouveresse, E., Hartmann, C., Massart, D. L., Last, I. R. and Prebble, K. A. (1996). Standardization of Near-Infrared Spectrometric Instruments. *Analytical Chemistry*, 68: 982-990.

Brito, A. L. B., Santos, A. V. P., Milanez, K. T. M., Pontes, M. J. C. and Pontes, L. F. B. L. (2017). Calibration Transfer of Flour NIR Spectra between Benchtop and Portable Instruments. *Analytical Methods*, 9(21): 3184-3190.

Brito, R. S., Pinheiro, H. M., Ferreira, F., Matos, J. S., Pinheiro, A. and Lourenço, N. D. (2016). Calibration Transfer Between a Bench Scanning and a Submersible Diode Array Spectrophotometer for in Situ Wastewater Quality Monitoring in Sewer Systems. *Applied Spectroscopy*, 70(3): 443-454.

Brouckaert, D., Uyttersprot, J. S., Broeckx, W. and De Beer, T. (2018). Calibration Transfer of a Raman Spectroscopic Quantification Method for the Assessment of Liquid Detergent Compositions from At-Line Laboratory to In-Line Industrial Scale. *Talanta*, 179: 386-392.

Busch, K. W., Soyemi, O., Rabbe, D., Humphrey, K., Dundee, B. and Busch, M. A. (2000). Wavelength calibration of a dispersive near-infrared spectrometer using trichloromethane as a calibration standard. *Applied Spectroscopy*, 54(9):1321-1326.

Cao, Y. T., Yuan, H. F. and Zhao, Z. (2018). A New Spectra Transfer Method for Multivariate Calibration Model of Molecular Spectroscopy Analysis. *Spectroscopy and Spectral Analysis*, 38(3): 973-981.

Capron, X., Walczak, B., De Noord, O. E. and Massart, D. L. (2005). Selection and Weighting of Samples in Multivariate Regression Model Updating. *Chemometrics and Intelligent Laboratory Systems*, 76(2): 205-214.

Cen, H. Y., Weng, H. Y. and He, Y. A method for delivering hyperspectral models of citrus canker: ZL 201610260903.2 [P]. 2016-09-21.

Chen, C. S., Brown, C. W. and Lo, S. C. (1997). Calibration Transfer from Sample Cell to Fiber-Optic Probe. *Applied Spectroscopy*, 51: 744-748.

Chen, D., Lu, F. and Li, Q. F. (2017). Development of Multi-Scale Modeling Methods for Calibration Transfer in Near Infrared Spectroscopy. *Nanotechnology and Precision Engineering*, 15(2): 121-126.

Chen, H., Zhang, Z. M., Miao, L., Zhan, D. J., Zheng, Y. B., Liu, Y., Lu, F. and Liang, Y. Z. (2015). Automatic Standardization Method for Raman Spectrometers with Applications to Pharmaceuticals. *Journal of Raman Spectroscopy*, 46(1): 147-154.

Chen, W. R., Bin, J., Lu, H. M., Zhang, Z. M. and Liang, Y. Z. (2016). Calibration Transfer via an Extreme Learning Machine Auto-Encoder. *Analyst*, 141(6): 1973-1980.

Chen, X. S., Jiao, Y. P. and Su, M. A solution to strange peaks in near-infrared spectroscopy model transfer: ZL 201811189178.X [P]. 2019-03-08.

Chen, Y. Y., Qi, K., Liu, Y. L., He, J. H. and Jiang, Q. H. (2015). Transferability of Hyperspectral Model for Estimating Soil Organic Matter Concerned with Soil Moisture. *Spectroscopy and Spectral Analysis*, 35(6): 17051708.

Chen, Y. Y. and Wang, Z. B. (2019). Cross Components Calibration Transfer of NIR Spectroscopy Model through PCA and Weighted ELM-Based Tradaboost Algorithm. *Chemometrics and Intelligent Laboratory Systems*, 192: 103824.

Chen, Z. P., Li, L. M., Yu, R. Q., Littlejohn, D., Nordon, A., Morris, J., Dann, A. S., Jeffkins, P. A. and Stimpson, S. L. (2011). Systematic Prediction Error Correction: A Novel Strategy for Maintaining the Predictive Abilities of Multivariate Calibration Models, *Analyst*, 136(1): 98-106.

Chen, Z. P. and Morris, J. (2008). Improving the Linearity of Spectroscopic Data Subjected to Fluctuations in External Variables by the Extended Loading Space Standardization. *Analyst*, 133(7): 914-922.

Chen, Z. P., Morris, J. and Martin, E. (2005). Correction of Temperature-Induced Spectral Variations by Loading Space Standardization. *Analytical Chemistry*, 77(5): 1376-1384.

Choquette, S. J., Etz, E. S., Hurst, W. S., Blackburn, D. H. and Leigh, S. D. (2007). Relative Intensity Correction of Raman Spectrometers: NIST SRMS 2241 through 2243 for 785nm, 532nm, and 488nm/514.5nm Excitation. *Applied Spectroscopy*, 61(2), 117-129.

Chu, X. L., Yuan, H. F. and Lu, W. Z. (2001). Model Transfer in Multivariate Calibration. *Spectroscopy and Spectral Analysis*, 21(6): 881-885.

Chu, X. L., Yuan, H. F. and Lu, W. Z. (2002). Calibration Transfer of Spectra from Near Infrared Spectrometers. *Chinese Journal of Analytieal Chemistry*, 30(1): 114-119.

Cinier, R. and Guilment, J. (1998). High Precision Measurements: From the Laboratory to the Plant. *Journal of Near Infrared Spectroscopy*, 6(1): 291-297.

Clark, R. D. (1997). Optisim: An Extended Dissimilarity Selection Method for Finding Diverse Representative Subsets. *Journal of Chemical Information and Computer Science*, 37(6)1181-1188.

Clavaud, M., Roggo, Y., Degardin, K., Sacré, P. Y., Hubert, P. and Ziemons, E. (2017). Global Regression Model for Moisture Content Determination Using Near-Infrared Spectroscopy. *European Journal of Pharmaceutics and Biopharmaceutics*, 119(1): 343-352.

Cogdill, R. P., Anderson, C. A .and Drennen, J. K. (2002). Process Analytical Technology Case Study, Part III: Calibration Monitoring and Transfer. *AAPS Pharm Sci Tech*, 6(2): 284-297.

Coleman, M. D., Brewer, P. J., Smith, I. M., Harris, P. M. Clift, M. G. and Milton, M. (2007). Calibration Transfer Strategy to Compensate for Instrumental Drift in Portable Quadrupole Mass Spectrometers. *Analytica Chimica Acta*, 601(2): 189-195.

Cooper, J. B., Larkin, C. M. and Abdelkader, M. F. (2011). Calibration Transfer of Near-IR Partial Least Squares Property Models of Fuels Using Virtual Standards. *Journal of Chemometrics*, 25(9): 496-505.

Cooper, J. B., Larkin, C. M. and Abdelkader, M. F. (2011) Virtual Standard Slope and Bias Calibration Transfer of Partial Least Squares Jet Fuel Property Models to Multiple Near Infrared Spectroscopy Instruments. *Journal of Near Infrared Spectroscopy*, 19 (2): 139-150.

Dabros, M., Amrhein, M., Gujral, P. and Stockar, U. V. (2007). On-line Recalibration of Spectral Measurements Using Metabolite Injections and Dynamic Orthogonal Projection. *Applied Spectroscopy*, 61(5): 507-513.

Daikos, O., Heymann, K. and Scherzer, T. (2019). Development of a PLS Approach for the Determination of the Conversion in UV-Cured White-Pigmented Coatings by NIR Chemical Imaging and Its Transfer to Other Substrates. *Progress in Organic Coatings*, (132): 116-124.

Dambergs, R. G., Mercurio, M. D., Kassara, S., Cozzolino, D. and Smith, P. A. (2012). Rapid Measurement of Methyl Cellulose Precipitable Tannins Using Ultraviolet Spectroscopy with Chemometrics: Application to Red Wine and Inter-Laboratory Calibration Transfer. *Applied Spectroscopy*, 66(6): 656-664.

Dardenne, P. (2002). Calibration Transfer in near Infrared Spectroscopy. *NIR News*, 13(4): 3-7.

Dardenne, P and Welle, R. (1998). New Approach for Calibration Transfer from a Local Database to a Global Database. *Journal of Near Infrared Spectroscopy*, 6(1): 55-60.

Debus, B., Takahama, S., Weakley, A. T., Seibert, K. and Dillner, A. M. (2019). Long-Term Strategy for Assessing Carbonaceous Particulate Matter Concentrations from Multiple Fourier Transform Infrared (FT-IR) Instruments: Influence of Spectral Dissimilarities on Multivariate Calibration Performance. *Applied Spectroscopy*, 73(3): 271-283.

Deshmukh, S., Kamde, K., Jana, A., Korde, S., Bandyopadhyay, R., Sankar, R., Bhattacharyya, N. and Pandey, R. A. (2014). Calibration Transfer between Electronic Nose Systems for Rapid in Situ Measurement of Pulp and Paper Industry Emissions. *Analytica Chimica Acta*, 841(1): 58-67.

Despagne, F., Walczak, B. and Massart, D. L. (1998). Transfer of Calibrations of Near-Infrared Spectra Using Neural Networks. *Applied Spectroscopy*, 52(5): 732-745.

Dong, X. G., Dong, J., Li, Y. L., Xu, H., and Tang, X. (2019). Maintaining the Predictive Abilities of Egg Freshness Models on New Variety Based on VIS-NIR Spectroscopy Technique. *Computers and Electronics in Agriculture*, 156: 669-676.

Dreassi, E., Ceramelli, G., Perruccio, P. L. and Corti, P. (1998). Transfer of Calibration in Near-Infrared Reflectance Spectrometry. *Analyst*, 123: 1259-1264.

Drennen, J. (2003). Calibration Transfer: A Critical Component of Analytical Method Validation. *NIR News*, 14(5): 14–15.

Du, W., Chen, Z. P., Zhong, L. J., Wang, S. X., Yu, R. Q., Nordon, A., Littlejohn, D. and Holden, M. (2011). Maintaining the Predictive Abilities of Multivariate Calibration Models by Spectral Space Transformation. *Analytica Chimica Acta*, 690(1): 64-70.

Duponche, L., Ruckebusch, C., Huvenne, J. P. and Legrand, P. (1999). Standardization of Near-IR Spectrometers Using Artificial Neural Networks. *Journal of Molecular Structure*, 480-481: 551-556.

Eliaerts, J., Meert, N., Dardenne, P., Durme, F. V. and Wael, K. D. (2020). Evaluation of a Calibration Transfer between a Benchtop and Portable Mid-Infrared Spectrometer for Cocaine Classification and Quantification. *Talanta*, 209: 120481.

Elizalde, O., Asua, J. M. and Leiza, J. R. (2005). Monitoring of Emulsion Polymerization Reactors by Raman Spectroscopy: Calibration Model Maintenance. *Applied Spectroscopy*, 59(10): 1280-1285.

Eskildsen, C., Hansen, P., Skov, T., Marini, F. and NøRgaard, L. (2016). Evaluation of Multivariate Calibration Models Transferred between Spectroscopic Instruments: Applied to Near Infrared Measurements of Flour Samples. *Journal of Near Infrared Spectroscopy*, 24(2): 151-156.

Fan, P. P., Li, X. Y., Lv, M. R., Wu, L. N. and Liu, Y. (2018). Vis-NIR Model Transfer of Total Nitrogen Between Different Soils. *Spectroscopy and Spectral Analysis*, 38(10): 3210-3214.

Fan, S. X., Li, J. B., Xia, Y., Tian, X., Guo, Z. M. and Huang, W. Q. (2019). Long-Term Evaluation of Soluble Solids Content of Apples with Biological Variability by Using Near-Infrared Spectroscopy and Calibration Transfer Method. *Postharvest Biology and Technology*, 151(1): 79-87.

Fan, X. Q., Lu, H. M. and Zhang, Z. M. (2018). Direct Calibration Transfer to Principal Components via Canonical Correlation Analysis. *Chemometrics and Intelligent Laboratory Systems*, 181(1): 21-28.

Farrell, J. A., Higgins, K. and Kalivas, J. H. (2012). Updating a Near-Infrared Multivariate Calibration Model Formed with Lab-Prepared Pharmaceutical Tablet Types to New Tablet Types in Full Production. *Journal of Pharmaceutical and Biomedical Analysis*, 61(1): 114-121.

Fearn, T. (1997). Calibration Transfer without Standards. *NIR News*, 8(5): 7-8.

Fearn, T. (2001). Standardization and Calibration Transfer for Near Infrared Instruments: A Review. *Journal of Near Infrared Spectroscopy*, 9(4): 229-244.

Fearn, T., Eddison, C., Withey, R. and Cowe, I. A. (1996). A Method for Wavelength Standardisation in Filter Instruments. *Journal of Near Infrared Spectroscopy*, 4(1): 111-118.

Fernandez-Ahumada, E., Garrido-Varo, A., Guerrero, J. E., Perez-Marin, D. and Fearn, T. (2008). Taking NIR Calibrations of Feed Compounds from the Laboratory to the Process: Calibration Transfer between Predispersive and Postdispersive Instruments. *Journal of Agricultural and Food Chemistry*, 56(21): 10135-10141.

Fernandez, L., Guney, S., Gutierrez-Galvez, A., and Marco, S. (2016). Calibration Transfer in Temperature Modulated Gas Sensor Arrays. *Sensors and Actuators B: Chemical*, 231(1): 276-284.

Feudale, R. N., Woody, N. A., Tan, H., Myles, A. J., Brown, S. D. and Ferré, J (2002). Transfer of Multivariate Calibration Models: A Review. *Chemometrics and Intelligent Laboratory Systems*, 64(2): 181-192.

Folch-Fortuny, A., Vitale, R., De Noord, O. E., and Ferrer, A. (2017). Calibration Transfer between NIR Spectrometers: New Proposals and a Comparative Study. *Journal of Chemometrics*, 31(3): e2874-e2884.

Fonollosa, J., Fernández, L., Gutiérrez-Gálvez, A., Huerta, R. and Marco, S. (2016). Calibration Transfer and Drift Counteraction in Chemical Sensor Arrays Using Direct Standardization. *Sensor and Actuator B*, 236(1): 1044-1053.

Fonollosa, J., Neftci, E., Huerta, R., and Marco, S. (2015). Evaluation of Calibration Transfer Strategies between Metal Oxide Gas Sensor Arrays. *Procedia Engineering*, 120(1): 261-264.

Fontaine, J., Hörr, J. and Schirmer, B. (2003). Amino acid contents in raw materials can be precisely analyzed in a global network of near-infrared spectrometers: collaborative trials prove the positive effects of instrument standardization and repeatability files. *Journal of Agricultural and Food Chemistry*, 52(4): 701-708.

Forina, M., Drava, G., Armanino, C., Boggia, R., Lanteri, S., Leardi, R., Corti, P., Giangiacomo, R., Galliena, C., Bigoni, R., Quartari, I., Serra, D., Leoni, O. and Lazzeri, L. (1995). Transfer of Calibration Function in Near-Infrared Spectroscopy. *Chemometrics and Intelligent Laboratory Systems*, 27: 189-203.

Hoffmann, U. and Zanier-Szydlowski, N. (1999). Portability of Near Infrared Spectroscopic Calibrations for Petrochemical Parameters. *Journal of Near Infrared Spectroscopy*, 7(1): 33-45.

Hopkins, D. W. (2003). Shoot-out 2002: Transfer of Calibration for Content of Active in a Pharmaceutical Tablet. *NIR News*, 14(5): 10-13.

Hutsebaut, D., Vandenabeele, P. and Moens, L. (2005). Evaluation of An Accurate Calibration and Spectral Standardization Procedure for Raman Spectroscopy. *Analyst*, 130(8): 1204-1214.

Galvan, D., Bona, E., Borsato, D., Danieli, E. and Killner, M. H. M. (2020). Calibration Transfer of Partial Least Squares Regression Models between Desktop Nuclear Magnetic Resonance Spectrometers. *Analytical chemistry*, 92: 12809-12816.

Galvao, R. K. H., Soares, S. F. C., Martins, M. N., Pimentel, M. F. and Araújo, M. (2015). Calibration Transfer Employing Univariate Correction and Robust Regression. *Analytica Chimica Acta*, 864(1): 1-8.

Garcia-Olmo, J., Garrido-Varo, A. and De Pedro, E. (2001). The Transfer of Fatty Acid Calibration Equations Using Four Sets of Unsealed Liquid Standardization Samples. *Journal of Near Infrared Spectroscopy*, 9(1): 49-62.

Ge, Y. F., Morgan, C. L. S., Grunwald, S., Brown, D. J. and Sarkhot, D. V. (2011). Comparison of Soil Reflectance Spectra and Calibration Models Obtained Using Multiple Spectrometers. *Geoderma*, 161(3-4): 202-211.

Gemperline, P. J., Cho, J. H., Aldridge, P. K. and Sekulic, S. S. (1996). Appearance of Discontinuities in Spectra Transformed by the Piecewise Direct Instrument Standardization Procedure. *Analytical Chemistry*, 68: 2913-2915.

Gislason, J., Chan, H. and Sardashti, M. (2001). Calibration Transfer of Chemometric Models Based on Process Nuclear Magnetic Resonance Spectroscopy. *Applied Spectroscopy*, 55(11): 1553-1560.

Greensill, C. V. and Walsh, K. B. (2002). Calibration Transfer between Miniature Photodiode Array-Based Spectrometers in the near Infrared Assessment of Mandarin Soluble Solids Content. *Journal of Near Infrared Spectroscopy*, 10(1): 27-35.

Greensill, C. V., Wolfs, P. J., Spiegelman, C. H. and Walsh, K. B. (2001). Calibration Transfer between PDA-Based NIR Spectrometers in the NIR Assessment of Melon Soluble Solids Content. *Applied Spectroscopy*, 55(5): 647-653.

Grelet, C., Fernández Pierna, J. A., Dardenne, P., Baeten, V. and Dehareng, F. (2015). Standardization of Milk Mid-Infrared Spectra from a European Dairy Network. *Journal of Dairy Science*, 98(4): 2150-2160.

Grelet, C., Fernández Pierna, J. A., Dardenne, P., Soyeurt, H., Vanlierde, A., Colinet, F., Bastin, C., Gengler, N., Baeten, V. and Dehareng, F. (2017). Standardization of Milk Mid-Infrared Spectrometers for the Transfer and Use of Multiple Models. *Journal of Dairy Science*, 100(10): 7910-7921.

Griffiths, M. L., Svozil, D., Worsfold, P. and Evans, E. H. (2006). The Application of Piecewise Direct Standardisation with Variable Selection to the Correction of Drift in Inductively Coupled Atomic Emission Spectrometry. *Journal of Analytical Atomic Spectrometry*, 21(10): 1045-1052.

Gryniewicz-Ruzicka, C. M., Arzhantsev, S., Pelster, L. N., Westenberger, B. J., Buhse, L. F. and Kauffman, J. F. (2011). Multivariate Calibration and Instrument Standardization for the Rapid Detection of Diethylene Glycol in Glycerin by Raman Spectroscopy. *Applied Spectroscopy*, 65(3)334-341.

Guo, S. X., Heinke, R., Stöckel, S., Rösch, P., Bocklitz, T., and Popp, J. (2017). Towards an Improvement of Model Transferability for Raman Spectroscopy in Biological Applications. *Vibrational Spectroscopy*, 91(1): 111-118.

Guo, S. X., Koher, A., Zimmermann, B., Heinke, R., Stöckel, S., Rösch, P., Popp, J. and. Bocklitz, T. (2018). Extended Multiplicative Signal Correction Based Model Transfer for Raman Spectroscopy in Biological Applications. *Analytical Chemistry*, 90(16): 9787-9795.

Haaland, D. M. (2000). Synthetic Multivariate Models to Accommodate Unmodeled Interfering Spectral Components during Quantitative Spectral Analyses. *Applied Spectroscopy*, 54(2): 246-254.

Hacisalihoglu, G., Gustin, J. L., Louisma, J., Armstrong, P., Peter, G. F., and Walker, A. R., Settles, A. M. Enhanced Single Seed Trait Predictions in Soybean (Glycine max) and Robust Calibration Model Transfer with Near-Infrared Reflectance Spectroscopy. *Journal of Agricultural and Food Chemistry*, 2016, 64: 1079-1086.

Hans, G. and Allison, B. (2019). Temperature and Moisture Insensitive Prediction of Biomass Calorific Value from Near Infrared Spectra Using External Parameter Orthogonalization. *Journal of Near Infrared Spectroscopy*, 27(4): 1-11.

Hayes, C. J., Walsh, K. B. and Greensill, C. V. (2016). Improving Calibration Transfer between Shortwave near Infrared Silicon Photodiode Array Instruments. *Journal of Near Infrared Spectroscopy*, 24(1): 59-68.

Hoffmann, U., Pfeifer, F., Hsuing, C. and Siesler, H. W. (2016). Spectra Transfer between a Fourier Transform Near-Infrared Laboratory and a Miniaturized Handheld Near-Infrared Spectrometer. *Applied Spectroscopy*, 70(5): 852-860.

Hong, S. J., Huang, W., Zhang, L. G., Ge, J., Luan, S. R. and Ni, L. J. (2020). A Near Infrared Spectroscopy Calibration Model Transfer Method Based on Scale Invariant Feature Transform to Select Stable Characteristic Wavelengths. *Journal of Instrumental Analysis*, 38(10): 1260-1266.

Hong, T. L., Tsai, S. J. and Tsou, S. C. S. (1994). Development of a Sample Set for Soya Bean Calibration of near Infrared Reflectance Spectroscopy. *Journal of Near Infrared Spectroscopy*, 2(4): 223-227.

Hong, Y. S., Chen, Y., Zhang, Y., Liu, Y. F., Liu, Y. L., Yu, L. and Cheng, H. (2018). Transferability of Vis-NIR Models for Soil Organic Carbon Estimation between two Study Areas by Using Spiking. *Soil Science Society of America Journal*, 82(5): 1231-1242.

Honorato, F. A., Galvao, R. K. H., Pimentel, M. F., De Barros Neto, B., Araújo, M. C. U. and De Carvalho, F. R. (2005). Robust Modeling for Multivariate Calibration Transfer by the Successive Projections Algorithm. *Chemometrics and Intelligent Laboratory Systems*, 76(1): 65-72.

Hu, M. and Li, Q. L. (2019). An Efficient Model Transfer Approach to Suppress Biological Variation in Elastic Modulus and Firmness Regression Models Using Hyperspectral Data. *Infrared Physics and Technology*, 99(1): 140-151.

Hu, R. W. and Xia, J. F. (2012). Transfer of NIRS Calibration Model for Determining Total Sugar Content in Navel Orange. *Food Science*, 33(3): 28-32.

Hu, Y., Li, B. Y., Zhang, J., Peng, Q. R. (2020). A New NIR Calibration Transfer Method Based on Parameter Correction. *Spectroscopy and Spectral Analysis*, 40(6): 1804-1808.

Hu, Y., Peng, S. L., Bi, Y. M., Tang, L. (2012). Calibration Transfer Based on Maximum Margin Criterion for Qualitative Analysis Using Fourier Transform Infrared Spectroscopy. *Analyst*, 137(24): 5913-5918.

Huang, G. G., Chen, X. J., Li, L. M., Chen, X., Yuan, L. M. and Shi, W. (2020). Domain Adaptive Partial Least Squares Regression. *Chemometrics and Intelligent Laboratory Systems*, 201: 103986.

Isabelle, M., Dorney, J., Lewis, A., Lloyd, G. R., Old, O. Shepherd, N., Rodriguez-Justo, M., Barr, H., Lau, K., Bell, I., Ohrel, S., Thomas, G., Stone, O. and Kendall, C. (2016). Multi-Centre Raman Spectral Mapping of Oesophageal Cancer Tissues: A Study to Assess System Transferability. *Faraday Discussions*, 187: 87-103.

Isaksson, T., Yang, H., Kemeny, G. J., Jackson, R. S., Wang, Q., Alam, M. K., Alam, M. K. and Griffiths, P. R. (2003). Accurate Wavelenuth Measurements of a Putative Standard for Near-Infrared Diffuse Reflection Spectrometry. *Applied Spectroscopy*, 57(2): 176-185.

Igne, B. and Hurburgh, C. R. (2008). Standardisation of Near Infrared Spectrometers: Evaluation of Some Common Techniques for Intra- and Inter-Brand Calibration Transfer. *Journal of Near Infrared Spectroscopy*, 16(6): 539-550.

Igne, B. and Hurburgh, C. R. (2010). Local Chemometrics for Samples and Variables: Optimizing Calibration and Standardization Processes. *Journal of Chemometrics*, 24(2): 75-86.

Igne, B. and Hurburgh, C. R. (2010). Using the Frequency Components of near Infrared Spectra: Optimising Calibration and Standardisation Processes. *Journal of Near Infrared Spectroscopy*, 18(1): 39-47.

Igne, B., Roger, J. M., Roussel, S., Bellon-Maurel, V. and Hurburgh, C. R. (2009). Improving the Transfer of Near Infrared Prediction Models by Orthogonal Methods. *Chemometrics and Intelligent Laboratory Systems*, 99(1): 57-65.

Jaworski, A., Wikiel, H. and Wikiel, K. (2017). Temperature Compensation by Calibration Transfer for an AC Voltammetric Analyzer of Electroplating Baths. *Electroanalysis*, 29(1): 67-76.

Jaworski, A., Wikiel, H. and Wikiel, K. (2018). Temperature Compensation by Embedded Temperature Variation Method for an AC Voltammeric Analyzer of Electroplating Baths. *Electroanalysis*, 30(1): 1-12.

Ji, N. Y. and Han, D. H. (2014). Study on near-infrared prediction model transfer for apples. *Journal of Food Safety and Quality*, 5(3): 712-717.

Ji, W., Viscarra Rossel, R. A. and Shi, Z. (2015). Improved Estimates of Organic Carbon Using Proximally Sensed Vis-NIR Spectra Corrected by Piecewise Direct Standardization. *European Journal of Soil Science*, 66(4): 670-678.

Ji, W. J., Li, S., Chen, S. C., Shi, Z., Rossel, R. A. V. and Mouazen, A. M. (2016). Prediction of Soil Attributes Using the Chinese Soil Spectral Library and Standardized Spectra Recorded at Field Conditions. *Soil and Tillage Research*, 155: 492-500.

Kalivas, J. H., Siano, G. S., Andries, E. and Goicoechea, H. C. (2009). Calibration Maintenance and Transfer Using Tikhonov Regularization Approaches. *Applied Spectroscopy*, 63(7): 800-809.

Kauppinen, A., Toiviainen, M., Lehtonen, M., Järvinen, K., Paaso, J. and Juuti, M., Ketolainen, J. (2014). Validation of a Multipoint Near-Infrared Spectroscopy Method for In-Line Moisture Content Analysis during Freeze-Drying. *Journal of Pharmaceutical and Biomedical Analysis*, 95(1): 229-237.

Khaydukova, M., Medina-Plaza, C., Rodriguez-Mendez, M. L., Panchuk, V., Kirsanov, D. and Legin, A. (2017). Multivariate Calibration Transfer between two Different Types of Multisensor Systems. *Sensors and Actuators B: Chemical*, 246(1): 994-1000.

Khaydukova, M., Panchuk, V., Kirsanov, D. and Legin, A. (2017). Multivariate Calibration Transfer between Two Potentiometric Multisensor Systems. *Electroanalysis*, 29(9): 2161-2166.

Khoshkam, M. and Kompany-Zareh, M. (2013). Calibration Transfer in Model Based Analysis of Second Order Consecutive Reactions. *Chemometrics and Intelligent Laboratory Systems*, 120(1): 15-24.

Khoshkam, M., Van Den Berg, F. and Kompany-Zareh, M. (2012). Achieving Bilinearity in Non-Bilinear Augmented First Order Kinetic Data Applying Calibration Transfer. *Chemometrics and Intelligent Laboratory Systems*, 115(1): 1-8.

Koehler, F. W., Small, G. W., Combs, R. J., Knapp, R. B. and Kroutil, R. T. (2000). Calibration Transfer Algorithm for Automated Qualitative Analysis by Passive Fourier Transform Infrared Spectrometry. *Analytical Chemistry*, 72(7): 1690-1698.

Koehler, F. W., Small, G. W., Combs, R. J., Knapp, R. B. and Kroutil, R. T. (2000). Calibration Transfer in the Automated Detection of Acetone by Passive Fourier Transform Infrared Spectrometry. *Applied Spectroscopy*, 54(5): 706-714.

Kompany-Zareh, M. and Berg, F. V. D. (2010). Multi-Way Based Calibration Transfer between Two Raman Spectrometers. *Analyst*, 135(6): 1382-1388.

Kramer, K. E., Morris, R. E. and Rose-Pehrsson, S. L. (2008). Comparison of two Multiplicative Signal Correction Strategies for Calibration Transfer without Standards. *Chemometrics and Intelligent Laboratory Systems*, 92(1): 33-43.

Kramer, K. E. and Small, G. W. (2007). Blank Augmentation Protocol for Improving the Robustness of Multivariate Calibrations. *Applied Spectroscopy*, 61(5): 497-506.

Krapf, L. C., Nast, D., Gronauer, A., Schmidhalter, U. and Heuwinkel, H. (2013). Transfer of a Near Infrared Spectroscopy Laboratory Application to an Online Process Analyser for in Situ Monitoring of Anaerobic Digestion. *Bioresource Technology*, 129(1): 39-50.

Kunz, M. R., Kalivas, J. H. and Andries, E. (2010). Model Updating for Spectral Calibration Maintenance and Transfer Using 1-Norm Variants of Tikhonov Regularization. *Analytical Chemistry*, 82(9): 3642-3649.

Kunz, M. R., Ottaway, J., Kalivas, J. H. and Andries, E. (2010). Impact of Standardization Sample Design on Tikhonov Regularization Variants for Spectroscopic Calibration Maintenance and Transfer. *Journal of Chemometrics*, 24(3-4): 218-229.

Kunz, M. R. and She, Y. Y. (2013). Multivariate Calibration Maintenance and Transfer through Robust Fused LASSO. *Journal of Chemometrics*, 27(9): 233-242.

Kupyna, A., Rukke, E. O., Schüller, R. B., and Isaksson, T. (2010). The Effect of Flow Rate in Acoustic Chemometrics on Liquid Flow: Transfer of Calibration Models. *Chemometrics and Intelligent Laboratory Systems*, 100(2): 110-117.

Laref, R., Losson, E., Sava, A. and Siadat, M. (2018). Support vector machine regression for calibration transfer between electronic noses dedicated to air pollution monitoring. *Sensors*, 18(11): 3716.

Leion, H., Folestad, S., Josefson, M., and Sparén, A. (2005). Evaluation of Basic Algorithms for Transferring Quantitative Multivariate Calibrations Between Scanning Grating and FT NIR Spectrometers. *Journal of Pharmaceutical and Biomedical Analysis*, 37(1): 47-55.

Li, H., Wang, J. X., Xing, Z. N., Shen, G. (2011). Influence of Improved Kennard/Stone Algorithm on the Calibration Transfer in Near-Infrared Spectroscopy. *Spectroscopy and Spectral Analysis*, 31(2): 362-365.

Li, J., Yu, X. N., Ge, W. Z., Ge, W. Z., and An, D. (2019). Qualitative Analysis of Maize Haploid Kernels Based on Calibration Transfer by Near-Infrared Spectroscopy. *Analytical Letters*, 52(2): 249-267.

Li, Q. F., Sun, X. Q., Ma, X. Y., Li, B., Wang, H. J., Lv, H. Y., Wang, Q., Xu, K. and Chen, Da. (2019). A Calibration Transfer Methodology for Standardization of Raman Instruments with Different Spectral Resolutions Using Double Digital Projection Slit. *Chemometrics and Intelligent Laboratory Systems*, 191(1): 143-147.

Li, T. R., Liu, C. L., Wei, L., Wu, J., Sun, X. (2018). Near-Infrared Spectral Model Transfer of Acid Value and Peroxide Value of Edible Oil by Slope Intercept Correction Algorithm. *Journal of the Chinese Cereals and Oils Association*, 33(1): 118-124.

Li, X., Arzhantsev, S., Kauffman, J. F., and Spencer, J. A. (2011). Detection of Diethylene Glycol Adulteration in Propylene Glycol-Method Validation through a Multi-Instrument Collaborative Study. *Journal of Pharmaceutical and Biomedical Analysis*, 54(5): 1001-1006.

Li, X., Bin, J., Fan, W., Zhou, J. H. and Chen, W. R. (2017). Near Infrared Spectral Hybrid Model Quantitative Analysis on Samples with Different Physical States. *Chinese Journal of Analytical Chemistry*, 45(7): 958-964.

Li, X. Y., Cai, W. S. and Shao, X. G. (2011). Correcting Multivariate Calibration Model for Near Infrared Spectral Analysis Without Using Standard Samples. *Journal of Near Infrared Spectroscopy*, 23(5): 285-291.

Li, X. Y., Liu, Y., Lv, M. R., Zou, Y. and Fan, P. P. (2018). Calibration Transfer of Soil Total Carbon and Total Nitrogen between Two Different Types of Soils Based on Visible-Near-Infrared Reflectance Spectroscopy. *Journal of Spectroscopy*, 1-10.

Li, X. Y., Zhang, H. G., Lu, J. G., Shi, Y. Z. and Chen, J. S. (2018). A new method of model transfer in near infrared spectral quantitative analysis. *Computers and Applied Chemistry*, 35(1): 27-36.

Li, Y. Q., Hong, S. J., Huang, W., Ge, J. Luan, S. R. and Ni, L. J. (2020). Effect of Number of Latent Variables for Partial Least Square Model Based on Near Infrared Spectroscopy on Models Transfer Performance. *Journal of Instrumental Analysis*, 38(10): 1231-1238.

Liang, C., Yuan, H. F., Zhao, Z., Song, C. F. and Wang, J. J. (2016). A New Multivariate Calibration Model Transfer Method of Near-Infrared Spectral Analysis. *Chemometrics and Intelligent Laboratory Systems*, 153(1): 51-57.

Lima, F. S. G. amd Borge, L. E. P. (2002). Evaluation of Standardization Methods of Near Infrared Calibration Models. *Journal of Near Infrared Spectroscopy*, 10(4): 269-278.

Lin, J. (1998). Near-IR Calibration Transfer Between Different Temperatures. *Applied Spectroscopy*, 52: 1591-1596.

Lin, J., Lo, S. C. and Brown, C. W. (1997). Calibration Transfer from a Scanning Near-IR Spectrophotometer to a FT-Near-IR Spectrophotometer. *Analytica Chimica Acta*, 349(1-3): 263-269.

Lin, Z. Z., Xu, B., Li, Y., Shi, X. Y. and Qiao, Y. J. (2013). Application of Orthogonal Space Regression to Calibration Transfer without Standards. *Journal of Chemometrics*, 27(11): 406-413.

Liu, C. L., Li, T. R., Wei, L. N., Xu, Y. Y. and Wu, J. Z. (2017). Research on Application of Direct Standardization Algorithm in Near-Infrared Spectrum Calibration Transfer of Acid Value and Peroxide Value of Edible Oil. *Spectroscopy and Spectral Analysis*, 37(10): 3042-3050.

Liu, C. L., Liu, H. Y., Sun, X. R., Wu, J. Z. and Yang, L. F. (2020). Transfer of Near-infrared Spectroscopy Model of Edible Oil Acid Value and Peroxidation Value. *Transactions of the Chinese Society for Agricultural Machinery*, 51(9): 344-349.

Liu, C. L., Zhou, Z. Y., Li, T. R., Xu, Y. Y., Sun, X. R. and Wu, J. Z. (2019). Application of Migration Learning in Transfer of Oil Spectral Model. *Journal of Food Science and Technology*, 37(4): 95-102.

Liu, J. J., Li, B. Q., Zhai, H. L., Lu, S. H. and Li, S. S. (2018). The Common Quantitative Model for the Determination of Multiple Near Infrared Spectrometers. *Chemometrics and Intelligent Laboratory Systems*, 182(1): 117-123.

Liu, J., Li, X. Y., Guo, X. X., Jin, R., Xu, S. M. and Ku, J. (2014). *Transactions of the Chinese Society of Agricultural Engineering*, 30(17): 276-284.

Liu, J., Li, X. Y., Jin, R., Xu, S. M. and Ku, J. (2015). Extending Hyperspectral Detecting Model of pH in Fresh Pork to New Breeds. *Spectroscopy and Spectral Analysis*, 35(7): 1973-1979.

Liu, X., Han, L. J. and Yang, Z. L. (2011). Transfer of Near Infrared Spectrometric Models for Silage Crude Protein Detection between Different Instruments. *Journal of Dairy Science*, 94(11): 5599-5610.

Liu, X., Huang, C. J. and Han, L. J. (2015). Calibration Transfer of Near-Infrared Spectrometric Model for Calorific Value Prediction of Straw Using Different Scanning Temperatures and Accessories. *Energy and Fuels*, 29(10): 6450-6455.

Liu, Y., Cai, W. S. and Shao, X. G. (2014). Standardization of Near Infrared Spectra Measured on Multi-Instrument. *Analytica Chimica Acta*, 836(1): 18-23.

Liu, Y., Cai, W. S. and Shao, X. G. (2016). Linear Model Correction: A Method for Transferring a Near-Infrared Multivariate Calibration Model without Standard Samples. *Spectrochimica Acta Part A: Molecular and Biomolecular Spectroscopy*, 169(1): 197-201.

Liu, Y., Xu, H., Xia, Z. Z. and Guo, Z. Y. (2018). Multi-Spectrometer Calibration Transfer Based on Independent Component Analysis. *Analyst*, 143(5): 1274-1280.

Liu, Y. Y., Xiong, Z. X., Wang, Y., Liang, L. and Fang, G. (2019). Study on the transform of near-infrared calibration models for lignin determination between different types of portable near-infrared spectrometers. *Journal of Forestry Engineering*, 4(4): 93-98.

Liu, Y. Y., Yang, H., Xiong, Z. X., Liang, L. and Gui, F. G. (2019). Study on Near-infrared Calibration Model Transfer for Lignin Content in Pulpwood. *Transactions of China Pulp and Paper*, 34(3): 43-49.

Liu, Z., Yu, H. L. and MacGregor, J. F. (2007). Standardization of Line-Scan NIR Imaging Systems. *Journal of Chemometrics*, 21(3-4): 88-95.

Liu, Z. W., Xu, L. J. and Chen, X. J. (2020). Near Infrared Spectroscopy Transfer Based on Deep Autoencoder. *Spectroscopy and Spectral Analysis*, 40(7): 2313-2318.

Lopez-Moreno, C., Palanco, S. and Laserna, J. J. (2005). Calibration Transfer Method for the Quantitative Analysis of High-temperature Materials with Stand-off Laser-induced Breakdown Spectroscopy. *Journal of Analytical Atomic Spectrometry*, 20: 1275-1279.

Lu, H. X., Wu, P. F., Yang, H. H. and Liu, Z. B. (2019). A NIR Model Transfer Method Based on Least Angle Regression Combined with Simple Linear Regression Direct Standardization. *Journal of Instrumental Analysis*, 38(1): 39-45.

Luo, J., Nie, M. F., Wu, S. H., Fan, W. and Xu, M. (2016). Research on the Transfer and Sharing of Fast and Nondestructive Calibration Model for Textiles. *China Fiber Inspection*, 10(1): 79-81.

Luo, X., Ikehata, A., Sashida, K., Piao, S. J., Okura, T. and Terada, Y. (2017). Calibration Transfer across Near Infrared Spectrometers for Measuring Hematocrit in the Blood of Grazing Cattle. *Journal of Near Infrared Spectroscopy*, 25(1): 15-25.

Luo, X., Ikehata, A., Sashida, K., Piao, S. J., Okura, T. and Terada, Y. (2017). Transfer of Calibration Model between Near-Infrared Spectrometers for Hematocrit Measurement of Grazing Cattle. *NIR news*, 28(7): 16-21.

Luoma, P., Natschläger, T., Malli, B., Pawliczek, M. and Brandstetter, M. (2018). Additive Partial Least Squares for Efficient Modelling of Independent Variance Sources Demonstrated on Practical Case Studies. *Analytica Chimica Acta*, 1007: 10-15.

Malli, B. and Birlutiu, A. (2017). Natschlager T. Standard-Free Calibration Transfer-An Evaluation of Different Techniques. *Chemometrics and Intelligent Laboratory Systems*, 161(1): 49-60.

Mann, C. K. and Vickers, T. J. (1999). Instrument-to-instrument transfer of Raman spectra. *Applied Spectroscopy*, 53(7): 856-861.

Marchesini, G., Serva, L., Garbin, E., Mirisola, M. and Andrighetto, I. (2017). Near-Infrared Calibration Transfer for Undried Whole Maize Plant between Laboratory and On-Site Spectrometers. *Italian Journal of Animal Science*, 17(1): 66-72.

Martens, H., Hoy, M., Wise, B. M., Bro, R., and Brockhoff, P. B. (2003). Pre-Whitening of Data by Covariance-Weighted Pre-Processing. *Journal of Chemometrics*, 17(3): 153-165.

Martins, M. N., Galvão, R. K. H. and Pimentel, M. H. (2010). Multivariate Calibration Transfer Employing Variable Selection and Subagging. *Journal of the Brazilian Chemical Society*, 21(1): 127-134.

Martinsen, P., Jordan, B., McGlone, A., Gaastra, P. and Laurie, T. (2008). Accurate and Precise Wavelength Calibration for Wide Bandwidth Array Spectrometers. *Applied Spectroscopy*, 62(9): 1008-1012.

Martinsen, P., McGlone, V. A., Jordan, R. B. and Gaastra, P. (2010). Temporal Sensitivity of the Wavelength Calibration of a Photodiode Array Spectrometer. *Applied Spectroscopy*, 64(12): 1325-1329.

Masahiro, W. and Yukihiro, O. (2005). Practical Calibration Correction Method for the Maintenance of an On-Line Near-Infrared Monitoring System for Molten Polymers. *Applied Spectroscopy*, 59(4): 487-495.

Milanez, K. D. T. M., Nobrega, T. C. A., Nascimento, D. S., Galvão, R. K. H. and Pontes, M. J. C. (2017). Selection of Robust Variables for Transfer of Classification Models Employing the Successive Projections Algorithm. *Analytica Chimica Acta*, 984(1): 76-85.

Milanez, K. D. T. M., Silva, A. C., Paz, J. E. M., Galvão, R. K. H. and Pontes, M. J. C. (2016). Standardization of NIR Data to Identify Adulteration in Ethanol Fuel. *Microchemical Journal*, 124(1): 121-126.

Mishra, P. and Nikzad-Langerodi, R. (2020). Partial Least Square Regression versus Domain Invariant Partial Least Square Regression with Application to Near-infrared Spectroscopy of Fresh Fruit. *Infrared Physics and Technology*, 111: 103547.

Monakhova, Y. B. and Diehl, B. W. K. (2016). Transfer of Multivariate Regression Models between High-Resolution NMR Instruments: Application to Authenticity Control of Sunflower Lecithin. *Magnetic Resonance in Chemistry*, 54(9): 712-717.

Morais, C. D. L. M. and De Lima, K. M. G. (2015). Determination and Analytical Validation of Creatinine Content in Serum Using Image Analysis by Multivariate Transfer Calibration Procedures. *Analytical Methods*, 7: 6904-6910.

Morais, C. L. M., Paraskevaidi, M., Cui, L., Fullwood, N. J., Isabelle, M., Lima, K. M., Martin-Hirsch, P. L., Sreedhar, H., Trevisan, J.,Walsh, M. J., Zhang, D. Y., Zhu, Y. G. and Martin, F. L. (2019). Standardization of Complex Biologically Derived Spectrochemical Datasets. *Nature protocols*, 14(5): 1546-1577.

Mou, Y., Zhou, L., Yu, S., Chen, W. Z. Zhao, X. and You, X. (2016). Robust Calibration Model Transfer. *Chemometrics and Intelligent Laboratory Systems*, 156(1): 62-71.

Munoz, S. G., Macgregor, J. F. and Kourti, T. (2005). Product Transfer between Sites Using Joint-Y PLS. *Chemometrics and Intelligent Laboratory Systems*, 79(1-2): 101-114.

Myles, A. J., Zimmerman, T. A. and Brown, S. D. (2006). Transfer of Multivariate Classification Models between Laboratory and Process Near-Infrared Spectrometers for the Discrimination of Green Arabica and Robusta Coffee Beans. *Applied Spectroscopy*, 60 (10): 1198-1203.

Ni, L. J., Dong, X. X., and Zhang, L. G. Standard sample preparation method for near-infrared model transfer of natural plants and its application: ZL CN201711264567.X [P]. 2018-07-17.

Ni, L. J., Han, M. Y., Luan, S. and Zhang, L. (2019). Screening Wavelengths with Consistent and Stable Signals to Realize Calibration Model Transfer of Near Infrared Spectra. *Spectrochimica Acta Part A: Molecular and Biomolecular Spectroscopy*, 206(1): 350-358.

Ni, L. J., Xiao, L. X., Yao, H. M., Ge, J., Zhang, L. G. and Luan, S. R. (2019). Construction of Global and Robust Near-Infrared Calibration Models Based on Hybrid Calibration Sets Using Partial Least Squares (PLS) Regression. *Analytical Letters*, 52(7): 1177-1194.

Ni, L. J., Xiao, L. X. and Zhang, L. G. (2018). Calibration Transfer of Near Infrared Spectral Models without Standards based on Spectrum Ratio Analysis. *Journal of Instrumental Analysis*, 37(5): 539-546.

Ni, W. D., Brown, S. D. and Man, R. L. (2010). Data Fusion in Multivariate Calibration Transfer. *Analytica Chimica Acta*, 661(2): 133-142.

Ni, W. D., Brown, S. D. and Man, R. L. (2011). Stacked PLS for Calibration Transfer without Standards. *Journal of Chemometrics*, 25(3): 130-137.

Nikzad-Langerodi, R., Zellinger, W., Lughofer, E. and Saminger-Platz, S. (2018). Domain-Invariant Partial-Least-Squares Regression. *Analytical Chemistry*, 90(11): 6693-6701.

Noord, O. E. (1994). Multivariate Calibration Standardization. *Chemometrics and Intelligent Laboratory Systems*, 25(2): 85-97.

Norgaard, L. (1995). Direct Standardisation in Multi Wavelength Fluorescence Spectroscopy. *Chemometrics and Intelligent Laboratory Systems*, 29(2): 283-293.

Nouri, M., Gomez, C., Gorretta, N. and Roger, J. M. (2017). Clay Content Mapping from Airborne Hyperspectral VIS-NIR Data by Transferring a Laboratory Regression Model. *Geoderma*, 298(1): 54-66.

Oliveri, P., Casolino, M. C., Casale, M., Medini, L., Mare, F. and Lanteri, S. (2013). A Spectral Transfer Procedure for Application of a Single Class-Model to Spectra Recorded by Different Near-Infrared Spectrometers for Authentication of Olives in Brine. *Analytica Chimica Acta*, 761(1): 46-52.

Ottaway, J., Kalivas, J. H. (2015). Feasibility Study for Transforming Spectral and Instrumental Artifacts for Multivariate Calibration Maintenance. *Applied Spectroscopy*, 69(3): 407-416.

Ozdemir, D., Mosley, M. and Williams, R. (1998). Hybrid Calibration Models: An Alternative to Calibration Transfer. *Applied Spectroscopy*, 52(4): 599-603.

Ozdemir, D. and Williams, R. (1999). Multi-instrument Calibration with Genetic Regression in UV-Visible Spectroscopy. *Applied Spectroscopy*, 53(2): 210-217.

Panchuk, V., Kirsanov, D., Oleneva, E., Semenov, V. and Legin, A. (2017). Calibration Transfer between Different Analytical Methods. *Talanta*, 170(8): 457-463.

Pavón, J. L. P., Sánchez, M. D. N., Pinto, C. G., Laespada, M. E. F. and Cordero, B. M. (2003). Calibration Transfer for Solving the Signal Instability in Quantitative Headspace-Mass Spectrometry. *Analytical Chemistry*, 75(22): 6361-6367.

Peng, J. G., Peng, S. L., Jiang, A., Jiang, A. and Tan, J. (2011). Near-Infrared Calibration Transfer Based on Spectral Regression. *Spectrochimica Acta Part A: Molecular and Biomolecular Spectroscopy*, 78(4): 1315-1320.

Pereira, C. F., Pimentel, M. F., Galvao, R. K. H., Honorato, F. A., Stragevitch, L. and Martins, M. N. (2008). A Comparative Study of Calibration Transfer Methods for Determination of Gasoline Quality Parameters in three Different Near Infrared Spectrometers. *Analytica Chimica Acta*, 611(1): 41-47.

Pereira, L. S. A., Carneiro, M. F., Botelho, B. G. and Sena, M. M. (2016). Calibration Transfer from Powder Mixtures to Intact Tablets: A New Use in Pharmaceutical Analysis for a Known Tool. *Talanta*, 147(1): 351-357.

Pérezmarín, D., Garridovaro, A. and Guerreroginel, J. (2006). Remote Near Infrared Instrument Cloning and Transfer of Calibrations to Predict Ingredient Percentages in Intact Compound Feedstuffs. *Journal of Near Infrared Spectroscopy*, (3): 81-91.

Pierna, J. A. F., Chauchard, F., Preys, S., Roger, J. M., Galtier, O., Baeten, V. and Dardenne, P. (2011). How to Build a Robust Model against Perturbation Factors with Only a Few Reference Values: A Chemometric Challenge at 'ChimiomÉTrie 2007'. *Chemometrics and Intelligent Laboratory Systems*, 106(2): 152-159.

Pierna, J. A. F., Sanfeliu, A. B., Slowikowski, B., von Holst, C., Maute, O., Han, L., Amato, G., de la Roza-Delgado, B., Perez-Marin, D. C., Lilley, G., Dardenne, P. and Baeten, V. (2013). Standardization of NIR Microscopy Spectra Obtained from Inter-Laboratory Studies by Using a Standardization Cell. *Biotechnology Agronomy Society and Environment*, 17(4): 547-555.

Pierna, J. A. F., Vermeulen, P., Lecler, B., Baeten, V. and Dardenne, P. (2010). Calibration Transfer from Dispersive Instruments to Handheld Spectrometers. *Applied Spectroscopy*, 64(6): 644-648.

Pissarda, A., Marques, E. J. N., Dardenne, P., Lateur, M., Pasquini, C., Pimentel, M. F., Pierna, J. A. F. and Baeten, V (2021). Evaluation of a Handheld Ultra-compact NIR Spectrometer for Rapid and Non-destructive Determination of Apple Fruit Quality. *Postharvest Biology and Technology*, 172: 111375.

Poerio, D. V. and Brown, S. D. (2017). Dual-Domain Calibration Transfer Using Orthogonal Projection. *Applied Spectroscopy*, 72(3): 378-391.

Preys, S., Roger, J. M. and Boulet, J. C. (2008). Robust Calibration Using Orthogonal Projection and Experimental Design. Application to the Correction of the Light

Scattering Effect on Turbid NIR Spectra. *Chemometrics and Intelligent Laboratory Systems*, 91(1): 28-33.

Pu, Y. Y., Sun, D. W., Riccioli, C., Buccheri, M., Grassi, M., Cattaneo, T. M. and Gowen, A. (2017). Calibration Transfer from Micro NIR Spectrometer to Hyperspectral Imaging: A Case Study on Predicting Soluble Solids Content of Bananito Fruit (Musa Acuminata). *Food Analytical Methods*, 11(4): 1021-1033.

Qin, Y. H. and Gong, H. L. (2016). NIR Models for Predicting Total Sugar in Tobacco for Samples with Different Physical States. *Infrared Physics and Technology*, 77: 239-243.

Rauscher, M. S., Krump, M., Schardt, M., Koehler, M. H. and Koch, A. W. (2018). Multivariate Calibration Methods for a Non-Dispersive Infrared Sensor for Engine Oil Condition Monitoring. *Technisches Messen*, 85(6): 395-409.

Ray, K. G. and McCreery, R. L. (1997). Simplified Calibration of Instrument Response Function for Raman Spectrometers Based on Luminescent Intensity Standards. *Applied Spectroscopy*, 51(1): 108-116.

Rehman, T. U., Zhang, L. B., Ma, D. D., Wang, L. J. and Jin, J. (2020). Calibration Transfer across Multiple Hyperspectral Imaging-based Plant Phenotyping Systems: I – Spectral Space Adjustment. *Computers and Electronics in Agriculture*, 176: 105685.

Ridder, T. D., Steeg, B. J. V. and Price, G. L. (2014). Robust Calibration Transfer in Noninvasive Ethanol Measurements, Part I: Mathematical Basis for Spectral Distortions in Fourier Transform Near-Infrared Spectroscopy (FT-NIR). *Applied Spectroscopy*, 68(8): 852-864.

Ridder, T. D., Ver, Steeg, B. J., Laaksonen, B. D., and Radigan, W. T. (2014). Robust Calibration Transfer in Noninvasive Ethanol Measurements, Part II: Modification of Instrument Measurements by Incorporation of Expert Knowledge (Mimik). *Applied Spectroscopy*, 68(8): 865-878.

Rodgers, J. E., Ghosh, S. and Cardwell, W. D. (2001). Measuring Nylon Carpet Yarn Heat History by Remote NIR Spectroscopy. Part II: Applying Remote Fiber Optic NIR Techniques to the Manufacturing Environment. *Textile Research Journal*, 71(2): 135-144.

Rodrigues, R. R. T., Rocha, J. T. C., Oliveira, L. M. S. L., Dias, J. C. M., Müller, E. I., Castro, E. V. and Filgueiras, P. R. (2017). Evaluation of Calibration Transfer Methods Using the ATR-FTIR Technique to Predict Density of Crude Oil. *Chemometrics and Intelligent Laboratory Systems*, 166(1): 7-13.

Rodriguez, J. D., Westenberger, B. J., Buhse, L. F. and Kauffman, J. F. (2011). Standardization of Raman Spectra for Transfer of Spectral Libraries across Different Instruments. *Analyst*, 136(20): 4232-4240.

Roggo, Y., Duponchel, L., Noe, B. and Huvenne, J. P. (2002). Sucrose Content Determination of Sugar Beets by Near Infrared Reflectance Spectroscopy. Comparison of Calibration Methods and Calibration Transfer. *Journal of Near Infrared Spectroscopy*, 10(1): 137-150.

Roudier, P., Hedley, C. B., Lobsey, C. R., Rossel, R. V. and Leroux, C. (2017). Evaluation of two Methods to Eliminate the Effect of Water from Soil Vis-NIR Spectra for Predictions of Organic Carbon. *Geoderma*, 296(1): 98-107.

Roza-Delgado, B., Garrido-Varo, A., Soldado, A., Arrojo, A. G., Valdés, M. C., Maroto, F. and Pérez-Marín, D. (2017). Matching Portable NIRS Instruments for in Situ Monitoring Indicators of Milk Composition. *Food Control*, 76(1): 74-81.

Rudnitskaya, A., Costa, A. M. S. and Delgadillo, I. (2017). Calibration Update Strategies for an Array of Potentiometric Chemical Sensors. *Sensors and Actuators B Chemical*, 238(1): 1181-1189.

Rukundo, I. R., Danao, M. G. C., Weller, C. L., Wehling, R. L. and Eskridge, K. M. (2020). Use of a Handheld Near Infrared Spectrometer and Partial Least Squares Regression to Quantify Metanil Yellow Adulteration in Turmeric Powder. *Journal of Near Infrared Spectroscopy*, 28(2): 81–92.

Sahni, N. S., Isaksson, T. and Næs, T. (2005). Comparison of Methods for Transfer of Calibration Models in Near-Infared Spectroscopy: A Case Study Based on Correcting Path Length Differences Using Fiber-Optic Transmittance Probes in In-Line Near-Infrared Spectroscopy. *Applied Spectroscopy*, 59(4): 487-495.

Saiz-Abajo, M. J., Mevik, B. H., Segtnan, V. H., and Næs, T. (2005). Ensemble Methods and Data Augmentation by Noise Addition Applied to the Analysis of Spectroscopic Data. *Analytica Chimica Acta*, 533(2): 147-159.

Sales, F., Callao, M. P. and Rius, F. X. (1999). Multivariate Standardization Techniques on Ion-Selective Sensor Arrays. *Analyst*, 124: 1045-1051.

Sales, F., Callao, M. P. and Rius, F. X. (2000). Multivariate Standardization for Correcting the Ionic Strength Variation on Potentiometric Sensor Arrays. *Analyst*, 125(5): 883-888.

Salguero-Chaparro, L., Palagos, B., Pena-Rodríguez, F. and Roger, J. M. (2013). Calibration Transfer of Intact Olive NIR Spectra Between a Pre-Dispersive Instrument and a Portable Spectrometer. *Computers and Electronics in Agriculture*, 96: 202-208.

Sanllorente, S., Rubio, L., Ortiz, M. C. and Sarabia, L. A. (2019). Signal Transfer with Excitation-Emission Matrices between a Portable Fluorimeter Based on Light-Emitting Diodes and a Master Fluorimeter. *Sensors and Actuators B: Chemical*, 285: 240-247.

Saranwong, S. and Kawano, S. (2004). A Simple Method of Instrument Standardisation for a Near Infrared Sorting Machine: The Utilisation of Average Spectra as Input Vectors. *Journal of Near Infrared Spectroscopy*, 12(1): 359-365.

Seichter, F., Vogt, J., Radermacher, P. and Mizaikoff, B. (2017). Nonlinear Calibration Transfer Based on Hierarchical Bayesian Models and Lagrange Multipliers: Error Bounds of Estimates via Monte Carlo E Markov Chain Sampling. *Analytica Chimica Acta*, 951(1): 32-45.

Seichter, F., Vogt, J., Radermacher, P. and Mizaikoff, B. (2017). Response-Surface Fits and Calibration Transfer for the Correction of the Oxygen Effect in the Quantification of Carbon Dioxide via FTIR Spectroscopy. *Analytica Chimica Acta*, 972(1): 16-27.

Shahbazikhah, P. and Kalivas, J. H. (2013). A Consensus Modeling Approach to Update a Spectroscopic Calibration. *Chemometrics and Intelligent Laboratory Systems*, 120(1): 142-153.

Shan, P., Zhao, Y. H., Wang, Q. Y., Ying, Y. and Peng, S. L. (2020). Principal Component Analysis or Kernel Principal Component Analysis Based Joint Spectral Subspace

Method for Calibration Transfer. *Spectrochimica Acta Part A: Molecular and Biomolecular Spectroscopy*, 227: 117653.

Shenk, J. S. and Westerhaus, M. O. (1991). New Standardization and Calibration Procedures for NIRS Analytical Systems. *Crop Science*, 31: 1694-1696.

Shenk, J. S., Westerhaus, M. O. and Templeton, W. C. (1985). Calibration transfer between near infrared reflectance spectrophotometers. *Crop Science*, 25: 159-161.

Shi, G. T., Han, L. J., Yan, Z. L., Chen, L. G. and Liu, X. (2013). Near Infrared Calibration Transfer for Quantitative Analysis of Fish Meal Mixed with Soybean Meal. *Journal of Near Infrared Spectroscopy*, 18(3): 509-522.

Shi, G. Y., Cao, J. and Zhang, Y. Z. (2020). Near infrared wood defects detection based on transfer learning. *Electric Machines and Control*, 24(10): 159-166.

Shi, X. Z., Wang, Z. G., Du, W., Yi, H., Zhang, J., Chen, Z. P. and Yu, R. Q. (2014). On-line Quantitative Monitoring and Control of Tobacco Flavors by Near Infrared Spectroscopy Combined with Advanced Calibration Transfer Method. *Chinese Journal of Analytical Chemistry*, 42(11): 1673-1678.

Shi, Y. Y., Li, J. Y. and Chu, X. L. (2019). Progress and Applications of Multivariate Calibration Model Transfer Methods. *Chinese Journal of Analytical Chemistry*, 47(4): 479-487.

Siano, G. G. and Goicoechea, H. C. (2007). Representative Subset Selection and Standardization Techniques. A Comparative Study Using NIR and a Simulated Fermentative Process UV Data. *Chemometrics and Intelligent Laboratory Systems*, 88(2)204-212.

Silva, N. C., Cavalcanti, C. J., Honorato, F. A., Amigo, J. M. and Pimentel, M. F. (2017). Standardization from a Benchtop to a Handheld NIR Spectrometer Using Mathematically Mixed NIR Spectra to Determine Fuel Quality Parameters. *Analytica Chimica Acta*, 954(1): 32-42.

Silva, N. C., Pimentel, M. F., Honorato, R. S., Talhavini, M., Maldaner, A. O. and Honorato, F. A. (2015). Classification of Brazilian and Foreign Gasolines Adulterated with Alcohol Using Infrared Spectroscopy. *Forensic Science International*, 253(1): 33-42.

Silva, V. H., Da Silva, J. J. and Pereira, C. F. (2017). Portable Near-Infrared Instruments: Application for Quality Control of Polymorphs in Pharmaceutical Raw Materials and Calibration Transfer. *Journal of Pharmaceutical and Biomedical Analysis*, 134(1): 287-294.

Siska, J. J. and Hurburgh, C. R. (2001). The Standardisation of Near Infrared Instruments Using Master Selection and Wiener Filter Methods. *Journal of Near Infrared Spectroscopy*, 9: 107-116.

Skotare, T., Nilsson, D., Xiong, S., Geladi, P. and Trygg, J. (2019). Joint and Unique Multiblock Analysis for Integration and Calibration Transfer of NIR Instruments. *Analytical Chemistry*, 91(5): 3516-3524.

Small, G. W., Harms, A. C., Kroutil, R. T., Ditillo, J. T. and Loerop, W. R. (1990). Design of Optimized Finite Impulse Response Digital Filters for Use with Passive Fourier Transform Infrared Interferograms. *Analytical Chemistry*, 62(17): 1768-1777.

Smith, M. R. (2004). Calibration Transfer in Pharmaceutical near Infrared Spectroscopy. *NIR News*, 15(6): 13-15.

Smith, M. R., Jee, R. D. and Moffat, A. C. (2002). The Transfer Between Instruments of a Reflectance Near-Infrared Assay for Paracetamol in Intact Tablets. *Analyst*, 127(12): 1682-1692.

Smith, M. R., Jee, R. D., Moffat, A. C., Rees, D. R. and Broad, N. W. (2004) A Procedure for Calibration Transfer Between Near-Infrared Instruments- A Worked Example Using a Transmittance Single Tablet Assay for Piroxicam in Intact Tablets. *Analyst*, 129(9): 806-816.

Sohn, M., Barton, F. E. and Himmelsbach, D. S. (2007). Transfer of Near-Infrared Calibration Model for Determining Fiber Content in Flax: Effects of Transfer Samples and Standardization Procedure. *Applied Spectroscopy*, 61(4): 414-418.

Sohn, M., Himmelsbach, D. S., Barton, F. E. and De Haseth, J. A. (2009). Transfer of Calibrations for Barley Quality from Dispersive Instrument to Fourier Transform Near-Infrared Instrument. *Applied Spectroscopy*, 63(10): 1190-1196.

Soldado, A., Fearn, T., Martinez-Fernandez, A. and De la Roza-Delgado, B. (2013). The Transfer of NIR Calibrations for Undried GrassSilage from the Laboratory to On-Site Instruments: Comparison of two Approaches. *Talanta*, 105(1): 8-14.

Song, H. Y. and Qin, G. (2015). Study on the Calibration Transfer of Near Infrared Spectroscopy Model for Soil Organic Matter Content Prediction by Using FIR. *Spectroscopy and Spectral Analysis*, 35(12): 3360-3363.

Soyemi, O., Rabbe, D., Busch, M. A., Igne, B., Bondi, R. W and Drennen, J. K. (2001). Design of a Modular, Dispersive Spectrometer for Fundamental Studies in Near-Infrared Spectroscopy. *Spectroscopy*, 16(4): 24-33.

Steinbach, D., Anderson, C. A., McGeorge, G., Igne, B., Bondi, R. W. and Drennen, J. K. (2017). Calibration Transfer of a Quantitative Transmission Raman PLS Model: Direct Transfer vs. Global Modeling. *Journal of Pharmaceutical Innovation*, 12(4): 347-356.

Stork, C. L. and Kowalski, B. R. (1999) Weighting Schemes for Updating Regression Models-a Theoretical Approach. *Chemometrics and Intelligent Laboratory Systems*, 48(2): 151-166.

Stout, F. and Kalivas, J. H. (2006). Tikhonov Regularization in Standardized and General Form for Multivariate Calibration with Application towards Removing Unwanted Spectral Artifacts. *Journal of Chemometrics*, 20(1-2): 22-33.

Sulub, Y., Lobrutto, R., Vivilecchia, R. and Wabuyele, B. W. (2008). Content Uniformity Determination of Pharmaceutical Tablets Using five Near-Infrared Reflectance Spectrometers: A Process Analytical Technology (PAT) Approach Using Robust Multivariate Calibration Transfer Algorithms. *Analytica Chimica Acta*, 611(2): 143-150.

Sulub, Y., LoBrutto, R., Vivilecchia, R. and Wabuyele, B. W. (2008). Near-Infrared Multivariate Calibration Updating Using Placebo: A Content Uniformity Determination of Pharmaceutical Tablets. *Vibrational Spectroscopy*, 46(2): 128-134.

Sulub, Y. and Small, G. W. (2007). Spectral Simulation Methodology for Calibration Transfer of Near-Infrared Spectra. *Applied Spectroscopy*, 61(4): 406-413.

Sum, S. T. and Brown, S. D. (1998). Standardization of Fiber-Optic Probes for Near-Infrared Multivariate Calibrations. *Applied Spectroscopy*, 52(6): 869-877.

Sun, H. X., Zhang, S. J., Xue, J. X., Zhao, X. T., Xing, S. H., Chen, C. H., and LI, C. J. (2019). Model Transfer Method of Fresh Jujube Soluble Solids Detection Using Variables Optimization and Correction Algorithms. *Spectroscopy and Spectral Analysis*, 39(4): 1041-1046.

Sun, L., Hsiung, C., Pederson, C. G., Zou, P., Smith, V., von Gunten, M. and O'Brien, N. A. (2016). Pharmaceutical Raw Material Identification Using Miniature Near-Infrared (MicroNIR) Spectroscopy and Supervised Pattern Recognition Using Support Vector Machine. *Applied Spectroscopy*, 70(5): 816-825.

Sun, L., Hsiung, C. and Smith, V. (2019). Investigation of Direct Model Transferability Using Miniature Near-Infrared Spectrometers. *Molecules*, 24(10): 1997.

Sun, X. D., Wu, H. L., Chen, Y., Chen, J. C. and Yu, R. Q. (2019). Chemometrics-Assisted Calibration Transfer Strategy for Determination of three Agrochemicals in Environmental Samples: Solving Signal Variation and Maintaining Second-Order Advantage. *Chemometrics and Intelligent Laboratory Systems*, 194: 103869.

Sun, Z. Y., Wang, J. Y., Nie, L., Li, L., Cao, D. W., Fan, J. J. Wang, H. Y., Liu, R. C., Zhang, Y. R. and Zang, H. C. (2018). Calibration Transfer of Near Infrared Spectrometers for the Assessment of Plasma Ethanol Precipitation Process. *Chemometrics and Intelligent Laboratory Systems*, 181(1): 64-71.

Surkova, A., Bogomolov, A., Legin, A. and Kirsanov, D. (2020). Calibration Transfer for LED-Based Optical Multisensor Systems. *ACS Sensors*, 5: 2587-2595.

Swierenga, H., Haanstra, W. G., Weijer, A. P. D. and Buydens, L. M. C. (1998). Comparison of Two Different Approaches toward Model Transferability in NIR Spectroscopy. *Applied Spectroscopy*, 52(1): 7-16.

Tan, C. and Li, M. L. (2007). Calibration Transfer Between Two Near-Infrared Spectrometers Based on a Wavelet Packet Transform. *Analytical Sciences*, 23(2): 201-206.

Tan, H., Sum, S. T., and Brown, S. D. (2002). Improvement of a Standard-Free Method for Near-Infrared Calibration Transfer. *Applied Spectroscopy*, 56(8): 1098-1106.

Tan, H. W. and Brown, S. D. (2001). Wavelet Hybrid Direct Standardization of Near-Infrared Multivariate Calibrations. *Journal of Chemometrics*, 15(8): 647-663.

Tao, C., Wang, Y., Cui, W., Zou, B., Zou, Z., R. and Tu, Y. L. (2019). A Transferable Spectroscopic Diagnosis Model for Predicting Arsenic Contamination in Soil. *Science of the Total Environment*, 669: 964-972.

Tencate, A. J., Kalivas, J. H. and White, A. J. (2016). Fusion Strategies for Selecting Multiple Tuning Parameters for Multivariate Calibration and Other Penalty Based Processes: A Model Updating Application for Pharmaceutical Analysis. *Analytica Chimica Acta*, 921(1): 28-37.

Terrell, M. (2015). Two Case Studies of the Transfer of Near Infrared Methods for the Analysis of Pharmaceutical Solid Dosage Forms. *NIR News*, 26(5): 8-9.

Thamasopinkul, C., Ritthiruangdej, P., Kasemsumran, S., Suwonsichon, T., Haruthaithanasan, V. and Ozaki, Y. (2017). Temperature Compensation for Determination of Moisture and Reducing Sugar of Longan Honey by Near Infrared Spectroscopy. *Journal of Near Infrared Spectroscopy*, 25(1): 36-44.

Thygesen, L. and Lundqvist, S. P. (2000). NIR Measurement of Moisture Content in Wood under Unstable Temperature Conditions. Part 2. Handling Temperature Fluctuations. *Journal of Near Infrared Spectroscopy*, 8(1): 191-199.

Thygesen, J. and Van Den Berg, F. W. J. (2011). Calibration Transfer for Excitation-Emission Fluorescence Measurements. *Analytica Chimica Acta*, 705(1-2): 81-87.

Tillmann, P., Reinhardt, T. C. and Paul, C. (2000). Networking of near Infrared Spectroscopy Instruments for Rapeseed Analysis: A Comparison of Different Procedures. *Journal of Near Infrared Spectroscopy*, 8(2): 101-107.

Tortajada-Genaro, L. A., Campíns-FalcóP and Bosch-Reig, F. (2001). Calibration Transfer in Chemiluminescence Analysis: Application to Chromium Determination by Luminol-Hydrogen Peroxide Reaction. *Analytica Chimica Acta*, 446(1): 383-390.

Vaughan, A. A., Dunn, W. B., Allwood, J. W., Wedge, D. C., Blackhall, F. H., Whetton, A. D., Dive, C. and Goodacre, R. (2012). Liquid Chromatography-Mass Spectrometry Calibration Transfer and Metabolomics Data Fusion. *Analytical Chemistry*, 84(22): 9848-9857.

Via, B. K., So, C. L., Shupe, T. F., Eckhardt, L. G., Stine, M. and Groom, L. H. (2005). Prediction of Wood Mechanical and Chemical Properties in the Presence and Absence of Blue Stain Using Two near Infrared Instruments. *Journal of Near Infrared Spectroscopy*, 13(4):201-212.

Vogt, F. and Booksh, K. (2004). Influence of Wavelength-Shifted Calibration Spectra on Multivariate Calibration Models. *Applied Spectroscopy*, 58(5): 624-635.

Walczak, B., Bouveresse, E. and Massart, D. L. (1997). Standardization of Near-Infrared Spectra in the Wavelet Domain. *Chemometrics and Intelligent Laboratory Systems*, 36(1): 41-51.

Wang, A. D., Wu, Z. S, Jia, Y. F., Zhang, Y. Y. and Zhan, X. Y. (2018). Model Transfer of On-Line Pilot-Scale Near Infrared Quantitative Model Based on Orthogonal Signal Regression. *Spectroscopy and Spectral Analysis*, 38(4): 1082-1088.

Wang, A. D., Yang, P., Chen, J., Wu, Z. S., Jia, Y. F., Ma, C. H. and Zhan, X. Y. (2019). A New Calibration Model Transferring Strategy Maintaining the Predictive Abilities of NIR Multivariate Calibration Model Applied in Different Batches Process of Extraction. *Infrared Physics and Technology*, 103: 103046.

Wang, H., Lin, Z. X., Wu, B. L., Xu, S. H., Yuan, L. F. and Luo, C. Spectral Analysis Model Transfer Technology Based on Radial Basis Neural Network: ZL 201610396494.9 [P]. 2016-11-02.

Wang, J., Guo, Z. H., He, F. J. and Liang, X. Y. (2018). Maintenance Methods of Protein Detection Model for Mutton of Different Areas of Ningxia Based on Hyperspectral. *The Food Industry*, 39(6): 118-121.

Wang, J. J., Li, Z. F., Wang, Y., Liu, Y., Cai, W. S. and Shao, X. G. (2016). A Dual Model Strategy to Transfer Multivariate Calibration Models for Near-Infrared Spectral Analysis. *Spectroscopy Letters*, 49(5): 348-354.

Wang, J. J., Zhe, W., Liu, Y., Cai, W. S. and Shao, X. G. (2014). A calibration transfer method for NIR model based on extended spectrum. *Acta Tabacaria Sinica*, 20(6): 1-6.

Wang, M., Zheng, K. Y., Yang, G. J., Ma, W. J., Wang, J. H., Wang, T. Y. and Du, Y. P., (2015). A Robust Near-Infrared Calibration Model for the Determination of

Chlorophyll Concentration in tree Leaves with a Calibration Transfer Method. *Analytical Letters*, 48(11): 1707-1719.

Wang, Q., DeJesus, S., Conzen, J. P., Schmidt, A. and Weiler, H. (1998). Calibration Transfer in near Infrared Analysis of Liquids and Solids. *Journal of Near Infrared Spectroscopy*, 6(A): A201-A205.

Wang, Q. B., Yang, H. H., Pan, X. P. and Li, L. Q. (2019). A Near Infrared Spectroscopy Model Transfer Method Based on Wavelet Transform Combined with Dynamic Time Warping. *Journal of Instrumental Analysis*, 38(12): 1423-1429.

Wang, Q. B., Yang, H. H., Pan, X. P., Liang, G. and Cheng, X. (2020). A model transfer method based on Random Forest-Direct Orthogonal Signal Correction. *Laser and Infrared*, 50(9): 1081-1087.

Wang, S. F., Han, P., Song, H. Y., Liang, G. and Cheng, X. (2019). Application of Slope/Bias and Direct Standardization Algorithms to Correct the Effect of Soil Moisture for the Prediction of Soil Organic Matter Content Based on the Near Infrared Spectroscopy. *Spectroscopy and Spectral Analysis*, 39(6): 1986-1992.

Wang, W. H., Huck, C. W. and Yang, B. (2019). NIR Model Transfer of Alkali-Soluble Polysaccharides in Poria Cocos with Piecewise Direct Standardization. *NIR News*, 30(5-6): 6-14.

Wang, Y. B., Yuan, H. F. and Lu, W. Z. (2005). A New Calibration Transfer Method Based on Target Factor Analysis. *Spectroscopy and Spectral Analysis*, 25(3): 398-401.

Wang, Y. D. and Kowalski, B. R. (1992). Calibration Transfer and Measurement Stability of Near-Infrared Spectrometers. *Applied Spectroscopy*, 46(5): 764-771.

Wang, Y. D. and Kowalski, B. R. (1993). Temperature-compensating Calibration Transfer for Near-infrared Filter Instruments. *Analytical Biochemistry*, 65: 1301-1303

Wang, Y. D., Lysaght, M. J. and Kowalski, B. R. (1992). Improvement of Multivariate Calibration through Instrument Standardization. *Analytical Chemistry*, 64(5): 562-565.

Wang, Y. D., Veltkamp, D. J. and Kowalski, B. R. (1991). Multivariate Instrument Standardization. *Analytical Chemistry*, 63(23): 2750-2756.

Wang, Y. H., Hu, W. Y., Song, P. F., Shu, R. X., Kai, Y., Wang, L. P., Zhao, L. L. and Li, J. H. (2019). Model Transfer Between Different Fourier Instruments and the Analysis of Error. *Spectroscopy and Spectral Analysis*, 39(3): 308-312.

Wang, Z. Y., Dean, T., and Kowalski, B. R. (1995). Additive Background Correction in Multivariate Instrument Standardization. *Analytical Chemistry*, 67: 2379-2385.

Watari, M. and Ozaki, Y. (2004). Prediction of Ethylene Content in Melt-State Random and Block Polypropylene by Near-Infrared Spectroscopy and Chemometrics: Comparison of a New Calibration Transfer Method with a Slope/Bias Correction Method. *Applied Spectroscopy*, 58(10): 1210-1218.

Wei, F., Liang, Y. Z., Yuan, D. L. and Wang, J. J. (2008). Calibration Model Transfer for Near-Infrared Spectra Based on Canonical Correlation Analysis. *Analytica Chimica Acta*, 623(1): 22-29.

Weng, H. Y., Cen, H. Y. and He, Y. (2018). Hyperspectral Model Transfer for Citrus Canker Detection Based on Direct Standardization Algorithm. *Spectroscopy and Spectral Analysis*, 38(1): 235-239.

Wijewardane, N. K., Ge, Y. and Morgan, C. L. S. (2016). Moisture Insensitive Prediction of Soil Properties from VNIR Reflectance Spectra Based on External Parameter Orthogonalization. *Geoderma*, 267(1): 92-101.

Wijewardane, N. K., Ge, Y. and Morgan, C. L. S. (2016). Prediction of Soil Organic and Inorganic Carbon at Different Moisture Contents with Air-Dry Ground VNIR: A Comparative Study of Different Approaches. *European Journal of Soil Science*, 67(5): 605-615.

Workman, J. J. (2016). First Principles of Instrument Calibration. *NIR News*, 27(3)12-15.

Workman, J. J. (2018). A Review of Calibration Transfer Practices and Instrument Differences in Spectroscopy. *Applied Spectroscopy*, 72(3): 340-365.

Wu, J. Z., Li, J., Du, W., Peng, S. M. and Wang, Z. G. (2017). Establishment and application of NIR models for online cut tobacco quality monitoring in primary processing. *Tobacco Science and Technology*, 50(10): 69-73.

Wulfert, F., Kok, W. T, Noord, O. E. D. and Smilde, A. K. (2000). Correction of Temperature-Induced Spectral Variation by Continuous Piecewise Direct Standardization. *Analytical Chemistry*, 72(7): 1639-1644.

Xi, C. C., Feng, Y. C. and Hu, C. Q. (2014). Evaluation of Piecewise Direct Standardization Algorithm for Near Infrared Quantitative Model Updating. *Chinese Journal of Analytical Chemistry*, 42(9): 1307-1313.

Xiao, H., Sun, K., Sun, Y., Wei, K. L., Tu, K. and Pan, L. Q. (2017). Comparison of Benchtop Fourier-Transform (FT) and Portable Grating Scanning Spectrometers for Determination of Total Soluble Solid Contents in Single Grape Berry (Vitis vinifera L.) and Calibration Transfer. *Sensors*, 17(11): 2693.

Xin, X. W., Gong, H. L., Ding, X. Q., Zeng, J. X. and Liu, Q. Y. (2017). Study on Calibration Model Transfer for the Near Infrared Spectrum Based on Improved S/B Algorithm. *Spectroscopy and Spectral Analysis*, 37(12): 3709-3713.

Xu, H. R. and Li, Q. Q. (2017). Calibration Model Transfer between Visible/NIR Spectrometers in Sugar Content On-line Detection of Crown Pears. *Transactions of the Chinese Society for Agricultural Machinery*, (9): 317-322.

Xu, J. L., Dorrepaal, R. M., Martinez-Gonzalez, J., Tsenkova, R. and Gowen, A. A. (2020). Near-infrared Multivariate Model Transfer for Quantification of Different Hydrogen Bonding Species in Aqueous Systems. *Journal of Chemometrics*, 34: e3274.

Xu, Z. P. (2020). Study on calibratrion transfer methods of rapid and nondestructive near infrared detection for individual crop kernels. *Beijing: University of Science and Technology of China.*

Xu, Z. P., Fan, S., Cheng, W. M., Liu, J., Zhang, P. F., Yang, Y., Xu, C., Liu, B. M., Liu, J., Wang, Q. and Wu, Y. J. (2020). A Correlation-Analysis-Based Wavelength Selection Method for Calibration Transfer. *Spectrochimica Acta Part A: Molecular and Biomolecular Spectroscopy*, 230: 118053.

Xu, Z. P., Fan, S., Liu, J., Liu, B. M., Tao, L. Z, Wu, J., Hu, S. P., Zhao, L. P., Wang, Q. and Wu, Y. J. (2019). A Calibration Transfer Optimized Single Kernel Near-Infrared Spectroscopic Method. *Spectrochimica Acta Part A: Molecular and Biomolecular Spectroscopy*, 220: 117098.

Yahaya, O. K. M., MatJafri, M. Z., Aziz, A. A. and Omar, A. F. (2015). Visible Spectroscopy Calibration Transfer Model in Determining pH of Sala Mangoes. *Journal of Instrumentation*, 10(5): T05002.

Yan, K. and Zhang, D. (2015). Improving the Transfer Ability of Prediction Models for Electronic Noses. *Sensors and Actuators B: Chemical*, 220(1): 115-124.

Yan, K. and Zhang, D. (2016). Calibration Transfer and Drift Compensation of E-Noses via Coupled Task Learning. *Sensors and Actuators B Chemical*, 225(1): 288-297.

Yang, H., Isaksson, T., Jackson. R. S. and Griffiths, P. R. (2003). Effect of Resolution on the Wavenumber Determination of a Putative Standard to Be Used for Near Infrared Diffuse Reflection Spectra Measured on Fourier Transform Near Infrared Spectrometers. *Journal of Near Infrared Spectroscopy*, 11(4): 241-255.

Yang, H., Xiong, Z. X. and Chen, T. (2018). Study on near-infrared calibration model transfer for soluble solid content in apple. *Chinese Journal of Analysis Laboratory*, 37(2): 163-167.

Yang, H. H., Zhang, X. F., Fan, Y. X., Xie, P. M. and Chu, X. L. (2014). Near Infrared Spectroscopic Model Transfer Based on Simple Linear Regression. *Chinese Journal of Analytical Chemistry*, 42(9): 1229-1234.

Yang, J. X., Lou, X. P., Yang, H. Q., Yang, H. B., Liu, C. Y., Wu, J. J. and Bin, J. (2019). Improved Calibration Transfer between Near-Infrared (NIR) Spectrometers Using Canonical Correlation Analysis. *Analytical Letters*, 52(14): 2188-2202.

Yang, P., Chen, J., Wu, C. Y., Zhan, X. Y. and Zang, H. C. (2019). Achievement of Moisture Transfer of Near Infrared Quantitative Model from Small-test Preparation Process to Pilot-test by Directed Direct Orthogonal Signal Correction Combined with Slope/Bias Correction. *Journal of Instrumental Analysis*, 38(9): 1044-1050.

Yang, Y., Peng, Y. K., Li, Y. Y., Fang, X. Q., Zhen, C., Wang, W. X. and Zheng, X. C. (2018). Calibration Transfer of Surface-Enhanced Raman Spectroscopy Quantitative Prediction Model of Potassium Sorbate in Osmanthus Wine to Other Wine. *Spectroscopy and Spectral Analysis*, 38(3): 824-829.

Yang, Z. L., Yang, Q. K., Shen, G. H., Mei, J. Q., Huang, Y. P. and Han. L. J. (2019). Online Application of Soybean Meal NIRS Quantitative Analysis Model from Laboratory to Factory. *Transactions of the Chinese Society for Agricultural Machinery*, 50(8): 358-365.

Yoon, J., Chung, H. and Han, C. (2001). Calibration Transfer Algorithm for NIR Spectroscopy as an Online Analyzer. *IFAC Proceedings Volumes*, 34(27): 303-308.

Yoon, J., Lee, B. and Han, C. (2002). Calibration Transfer of Near-Infrared Spectra Based on Compression of Wavelet Coefficients. *Chemometrics and Intelligent Laboratory Systems*, 64(1): 1-14.

Yu, B. F. and Ji, H. B. (2015). Near-Infrared Calibration Transfer Via Support Vector Machine and Transfer Learning. *Analytical Methods*, 7(6): 2714-2725.

Yu, B. F., Ji. H, B. and Kang, Y. (2016). Standardization of Near Infrared Spectra Based on Multi-Task Learning. *Spectroscopy Letters*, 49(1): 23-29.

Zamora-Rojas, E., Perez-Marin, D., Pedro-Sanz, E. D., Guerrero-Ginel, J. E., and Garrido-Varo, A. (2012). Handheld NIRS analysis for Routine Meat Quality Control: Database Transfer from At-line Instruments. *Chemometrics and Intelligent Laboratory Systems*, 114(1): 30-35.

Zeaiter, M., Roger, J. M. and Bellon-Maurel, V. (2005). Robustness of Models Developed by Multivariate Calibration: Part II: The Influence of Pre-Processing Methods. *Trends in Analytical Chemistry*, 24(5): 437-445.

Zeaiter, M., Roger, J. M. and Bellon-Maurel, V. (2006). Dynamic Orthogonal Projection. A New Method to Maintain the On-Line Robustness of Multivariate Calibrations. Application to NIR-based Monitoring of Wine Fermentations. *Chemometrics and Intellignent Laboratory Systems*, 80(2): 225-236.

Zeaiter, M., Roger, J. M., Bellon-Maurel, V., and Rutledge, D. N. (2004). Robustness of Models Developed by Multivariate Calibration. Part I: The Assessment of Robustness. Trends in Analytical Chemistry, 23(2): 157-170.

Zhang, F. Y., Chen, W., Zhang, R. Q., Ding, B. Y., Yao, H. M., Ge, J., Ju, L., Yang, W. Y. and Du, Y. P. (2017). Sampling Error Profile Analysis for Calibration Transfer in Multivariate Calibration. *Chemometrics and Intelligent Laboratory Systems*, 171(1): 234-240.

Zhang, F. Y., Zhang, R. Q., Ge, J., Chen, W. C., Yang, W. Y. and Du, Y. P. (2018). Calibration Transfer Based on the Weight Matrix (CTWM) of PLS for Near Infrared (NIR) Spectral Analysis. *Analytical Methods*, 10(18): 2169-2179.

Zhang, F. Y., Zhang, R. Q., Wang, W. M., Yang, W. Y., Li, L., Xiong, Y. R., Kang, Q. D. and Du, Y. P. (2019). Ridge Regression Combined with Model Complexity Analysis for Near Infrared (NIR) Spectroscopic Model Updating. *Chemometrics and Intelligent Laboratory Systems*, 195: 103896.

Zhang, J., Cai, W. S. and Shao, X. G. (2017). New Algorithms for Calibration Transfer in Near Infrared Spectroscopy. *Progress in Chemistry*, 29(8): 101-109.

Zhang, J., Cui, X. Y., Cai, W. S. and Shao, X. G. (2018). Modified Linear Model Correction: A Calibration Transfer Method without Standard Samples. *NIR news*, 29(8): 24-27.

Zhang, J., Guo, C., Cui, X. Y., Cai, W. S. and Shao, X. G. (2019). A Two-Level Strategy for Standardization of Near Infrared Spectra by Multi-Level Simultaneous Component Analysis. *Analytica Chimica Acta*, 1050(1): 25-31.

Zhang, L., Small, G. W. and Arnold, M. A. (2002). Calibration Standardization Algorithm for Partial Least-Squares Regression: Application to the Determination of Physiological Levels of Glucose by Near-Infrared Spectroscopy. *Analytical Chemistry*, 74(16): 4097-4108.

Zhang, L., Small, G. W. and Arnold, M. A. (2003). Multivariate Calibration Standardization across Instruments for the Determination of Glucose by Fourier Transform Near-Infrared Spectrometry. *Analytical Chemistry*, 75(21): 5905-5915.

Zhang, L., Tian, F., Kadri, C., Xiao, B., Li, H. G., Pan, L. N. and Zhou, H. W. (2011). On-Line Sensor Calibration Transfer among Electronic Nose Instruments for Monitoring Volatile Organic Chemicals in Indoor Air Quality. *Sensor and Actuators B: Chemistry*, 160(1): 899-909.

Zhang, L. G., Li, Y. Q., Huang, W., Ni, L. J. and Ge, J. (2020). The Method of Calibration Model Transfer by Optimizing Wavelength Combinations Based on Consistent and Stable Spectral Signals. *Spectrochimica Acta Part A: Molecular and Biomolecular Spectroscopy*, 227(1): 117647.

Zhang, X. B., Feng, Y. C. and Hu, C. Q. (2009). Progress in calibration transfer of near-infrared multivariate model. *Chinese Journal of Pharmaceutical Analysis*, 29(8): 1390-1399.

Zhao, Y. H., Yu, J. L., Shan, P., Zhao, Z. H., Jiang, X. Y. and Gao, S. L. (2019). PLS Subspace-Based Calibration Transfer for Near-Infrared Spectroscopy Quantitative Analysis. *Molecules,* 249(7): 1289-1306.

Zhao, Y. H., Zhao, Z. H., Shan, P., Peng, S. L., Yu, J. L. and Gao, S. L. (2019). Calibration Transfer Based on Affine Invariance for NIR without Transfer Standards. *Molecules*, 24(9): 1802.

Zheng, K. Y., Feng, T., Zhang, W., Huang, X. W., Li, Z. H., Zhang, D., Shi, J. Y., Marunaka, Y. and Zou, X. B. (2020). Refining Transfer Set in Calibration Transfer of Near Infrared Spectra by Backward Refinement of Samples. *Analytical Methods*, (12): 1495-1503.

Zheng, K. Y., Feng, T., Zhang, W., Huang, X. W., Li, Z. H., Zhang, D., Yao. Y. and Zou, X. B. (2019). Variable Selection by Double Competitive Adaptive Reweighted Sampling for Calibration Transfer of Near Infrared Spectra. *Chemometrics and Intelligent Laboratory Systems*, 191: 109-117.

Zheng, K. Y., Xiang, C. L., Cao, P., Geng, J. P. and Du, Y. P. (2013). Correcting NIR Spectra of Dimethyl Fumarate in Milk Measured for Different Brands and in Different Dates. *European Food Research and Technology*, 237(5): 787-794.

Zheng, K. Y., Zhang, X., Iqbal, J., Fan, W., Wu, T., Du, Y. P. and Liang, Y. Z. (2014). Calibration Transfer of Near-Infrared Spectra for Extraction of Informative Components from Spectra with Canonical Correlation Analysis. *Journal of Chemometrics*, 28(10): 773-784.

Zheng, W. R., Li, S. W., Han, Y. L., Shi, S. Q., Zhu, X. Z. and Jin, X. (2020). Study on Transfer Learning Prediction Methods for Soil Available Phosphorus NIR. *Journal of Instrumental Analysis*, 38(10): 1274-1281.

Zheng, Y. H., Song, T., Zhang, S., Guo, L. J. and Zhang, F. P. (2020). Spectral Transfer of Near-infrared Spectrometric Model for Fish Meal. *Journal of Instrumental Analysis*, 38(11): 1378-1384.

Zhou, S. L., Zhu, S. P. and Wei, X. (2020). Improving the Transfer Ability of Calibration Model for Terahertz Spectroscopy. *Spectroscopy Letters*, 53(6): 448-457.

Zhou, Z. K., Li, C. X., Wang, Z., Liu. R., Chen, W. L., and Xu, K. X. (2020). Study on Cefradine Granules Component Analysis and Calibration Transfer Method Based on Near-Infrared Spectroscopy. *Spectroscopy and Spectral Analysis*, 40(11): 3562-3566.

Zhu, Y., Fearn, T., Samuel, D., Dhar, A., Hameed, O., Bown, S. G., and Lovat, L. B. (2008). Error Removal by Orthogonal Subtraction (EROS): A Customised Pre-Treatment for Spectroscopic Data. *Journal of Chemometrics*, 22(1): 130-134.

Zou, C. M., Zhu, H. M., Shen, J. R., He, Y., Su, J. E., Fan, X. Q., Lu, H. M., Zhang, Z. M. and Chen, Y. (2019). Scalable Calibration Transfer without Standards via Dynamic Time Warping for Near-Infrared Spectroscopy. *Analytical Methods*, 35(11): 4481-4493.

Biographical Sketch

Yue Huang

Affiliation: China Agricultural University

Education: Ph.D. Food Safety (September, 2007 — July, 2012), College of Science, China Agricultural University, Beijing, P. R. China, B.S. Applied Chemistry (September, 2003— July, 2007), Department of Chemistry, China Agricultural University, Beijing, P. R. China

Business Address: Qinghua East Road 17#, Beijing, China

Research and Professional Experience: Nondestructive detection on food quality and safety by spectroscopies combined with chemometrics. Intelligent detection on foods with imaging and sensory approaches.

Professional Appointments: Associate Professor

Honors: Top 100 Young Scholars of Industrial Research in China (2021)

Publications from the Last 3 Years:

1) Exploring the use of Near-infrared spectroscopy as a tool to predict quality attributes in prickly pear (Rosa roxburghii Tratt) with chemometrics variable strategy. *Journal of Food Composition and Analysis*, 105(2022): 104225.
2) A review on the application of spectroscopy to the condiments detection: from safety to authenticity. *Critical Reviews in Food Science and Nutrition*, (2021): doi.org/10.1080/10408398.2021.
1901257.
3) Predicting heavy metals in dark sun-cured tobacco by near-infrared spectroscopy modeling based on the optimized variable selections. *Industrial Crops and Products*, 172 (2021): 114003.
4) Rapid spectral analysis of agro-products using an optimal strategy: dynamic backward interval PLS-competitive adaptive reweighted sampling. *Analytical and Bioanalytical Chemistry*, 412 (2020): 2795-2804.
5) Near infrared spectral variable optimization by final complexity adapted models combined with uninformative variables elimination-a validation study. *Optics*, 203 (2020): 164019.
6) Optimal modeling pattern of variables selection on analog complex using UVE-PLS regression. *IOP SciNotes*, 1 (2020): 014201.

7) Spectral interval optimization on rapid determination of prohibited addition in pesticide by ATR-FTIR. *Pest Management Science*, 75 (2019): 1743-1749.
8) Moving window smoothing on the ensemble of competitive adaptive reweighted sampling algorithm. *Spectrochimica Acta A*, 214 (2019): 129-138.
9) Rapid quantification of analog complex using partial least squares regression on mass spectrum. *Chemical Papers*, 73 (2019): 1003-1012.
10) Spectral interval combination optimization (ICO) on rapid quality assessment of Solanaceae plant: a validation study. *Journal of Food Science and Technology*, 56 (2019): 2158-2166.

Chapter 2

¹H-NMR Fingerprinting and Pattern Recognition Stepwise Strategy for Quality and Authenticity Control of Olive Oil

Gabriela E. Viacava[1]**, Blanca Gallo**[2]**,**
Luis A. Berrueta[2] **and Rosa M. Alonso-Salces**[3,*]
[1]Consejo Nacional de Investigaciones Científicas y Técnicas (CONICET), Grupo de Investigación en Ingeniería en Alimentos (GIIA), INCyTAA, Department of Chemical and Food Engineering, Faculty of Engineering, Universidad Nacional de Mar del Plata (UNMdP), Mar del Plata, Argentina
[2]Department of Analytical Chemistry, Faculty of Science and Technology, Universidad del País Vasco/Euskal Herriko Unibertsitatea (UPV/EHU), Bilbao, Spain
[3]Consejo Nacional de Investigaciones Científicas y Técnicas (CONICET), CIAS-IIPROSAM, Faculty of Exact and Natural Sciences, Universidad Nacional de Mar del Plata (UNMdP), Mar del Plata, Argentina

Abstract

The high price of olive oil, its distinctive sensory profile and its reputation as a healthy source of dietary fats make olive oil a target for fraud. The most common types of olive oil fraud are illegal blending with other vegetable oils (VOs) or low-quality olive oils, deliberate mislabeling of less expensive classes of olive oils, other VOs or their blends with olive oils, and mislabeling of geographical origin or Protected Designation of Origin (PDO) declaration. Olive oil adulteration, being one of the biggest financial frauds in the agricultural sector, evidenced the need to update and harmonize analytical methods for quality and authenticity control of olive oil. A novel stepwise strategy

* Corresponding Author's Email: rosamaria.alonsosalces@gmail.com.

In: Chemometrics
Editor: Larry D. Crenshaw
ISBN: 979-8-88697-274-0
© 2022 Nova Science Publishers, Inc.

based on the ^{1}H-NMR fingerprint of edible oils and multivariate data analysis has been developed in order to assure the authenticity and traceability of olive oils and their declared blends with VOs. This approach provides an analytical tool to detect fraud when olive oil is illegally blended with VOs or a 'legal' blend is falsely labelled respect to the botanical nature of the oils mixed and/or the percentage of each oil in the declared mixture. ^{1}H-NMR spectral data of olive and virgin olive oils and their mixtures with the VOs most commonly used to make blends, i.e., sunflower, high oleic sunflower, corn, virgin and refined avocado, virgin and refined hazelnut, soybean oil, refined palm olein and desterolised high oleic sunflower oils, was analysed by pattern recognition techniques to develop multivariate classification and regression models, which are organised in a decision tree to afford a stepwise strategy for the aimed purposes. Partial least squares discriminant analysis (PLS-DA) provided satisfactory, stable and robust binary classification models with recognition and prediction abilities of 90–100% of correct hits for most of the models to distinguish the type of olive oil and identify the VO in the blend. Partial least squares regression (PLS-R) afforded regression models with excellent precisions and acceptable accuracies to determine the percentage of VO in the mixture. The effectiveness of the proposed strategy was tested with blind samples, the results of which were satisfactory and confirmed its potential to support regulations and control bodies.

Keywords: olive oil, nuclear magnetic resonance, multivariate data analysis, decision tree, adulteration, authentication

Introduction

The high price of olive oil, the distinctive sensory profile, and its reputation as a healthy source of dietary fats make olive oil a target for fraud. The most common types of olive oil fraud are illegal blending with other vegetable oils (VOs) or low quality olive oils, deliberate mislabelling of less expensive classes of olive oils, other VOs or their blends with olive oils, and mislabelling of geographical origin or Protected Designation of Origin (PDO) declaration. As a result, olive oil adulteration for the purpose of financial gain has become one of the biggest sources of agricultural fraud in the European Union, as pointed out by the European Parliament (EC, 2020; European Parliament, 2014). Both the EU being the world's largest olive oil producer and consumer, accounting for 70% of global production (IOC, 2019), as well as the enlarged competitiveness, highlight the need to update and harmonize analytical methods for assuring quality and authenticity of olive oil. In the long-term, a lack of trust in the quality and authenticity of olive oil has the potential to damage the reputation and competitiveness of the European olive oil sector. In this context, the so called OLEUM Project was supported by the European Commission with the overall objective of improving existing analytical methods and developing new strategies of analysis for assuring the quality and authenticity of olive oil (OLEUM Project, 2016).

According to the Reg. (EU) 29/2012, olive oils, of any edible category, can be mixed with VOs and marketed highlighting the presence of olive oil on the labelling outside of the list of ingredients, by words, images or graphic representations, only in the case that it accounts for at least 50% of the blend (EC, 2012). Amendments to this regulation also established that Member States have the possibility to allow or prohibit the production in their territory of blends of olive oil and other VOs for internal consumption, the marketing in their territory of such blends coming from other countries, and/or the production in their territory of such blends for marketing in another Member State or for exportation. In this context, the terms of 'legal' and 'illegal' blends arise alluding to admixtures that comply or not with the EU regulation, as well as those adopted by each Member State. Moreover, it is noteworthy that the regulation and its amendments do not refer to any analytical parameter or method to control the percentage of olive oil in the admixture or the botanical origin of oil. Indeed, the need of analytical methods in order to confirm the presence of olive oil in the blend, to distinguish pure and adulterated olive oils, to identify the adulterant oils in the blend, as well as to determine the percentage of olive oil and/or the adulterants in the blend, is evidenced and is an issue of major concern in order to implement the established regulations (Conte, Bendini, Valli, Lucci, Moret, Maquet, et al., 2019). The adulteration of olive oil with less expensive oils in order to take an economic advantage is the most commonly performed with VOs such as sunflower, soybean, corn, rapeseed, hazelnut and peanut oils. Virgin olive oil (VOO) is also mixed with less expensive commercial categories such as refined olive oil and olive pomace oil, and even defective olive oils submitted to fraudulent practices such as deodorization (Gómez-Coca, Pérez-Camino, Bendini, Gallina Toschi, & Moreda, 2020; Meenu, Cai, & Xu, 2019; Osorio, Haughey, Elliott, & Koidis, 2014).

A recent review discussed the analytical techniques used for the detection and quantification of adulteration of extra virgin olive oil (EVOO) with low-grade olive oils and other VOs (Meenu et al., 2019). Two types of approaches are the most widely studied, those focused on the determination of specific chemical markers by chromatographic techniques (H. S. Green, Li, De Pra, Lovejoy, Steiner, Acworth, et al., 2020; Luo, Gao, Zhang, & Yu, 2020; Yang, Ferro, Cavaco, & Liang, 2013); and those based on spectroscopic techniques, which are usually coupled to multivariate data analysis to develop classification and/or regression models. Vibrational spectroscopies, mainly FTIR (Maggio, Cerretani, Chiavaro, Kaufman, & Bendini, 2010) and Raman (Jiménez-Carvelo, Osorio, Koidis, González-Casado, & Cuadros-Rodríguez, 2017; Zhang, Zou, Qi, Liu, Zhang, & Yin, 2011), UV-visible spectrophotometry (Jabeur, Zribi, & Bouaziz, 2016), mass spectrometry (Ozcan-Sinir, 2020), fluorescence spectroscopy (Tan, Li, Jiang, Shi, Xiao, Jia, et al., 2018), and NMR (Guyader, Thomas, Portaluri, Jamin, Akoka, Silvestre, et al., 2018; Vigli, Philippidis, Spyros, & Dais, 2003), as well as chromatographic fingerprinting (De La Mata-Espinosa, Bosque-Sendra, Bro, & Cuadros-Rodríguez, 2011; Jiménez-Carvelo, González-Casado, & Cuadros-Rodríguez, 2017; Priego Capote, Ruiz Jiménez, & Luque De Castro, 2007), achieved promising results discriminating pure oils of different botanical origins, blends of EVOO with VOs at different proportions, as well as quantifying the percentage of these oils in a blend, usually focused at low percentage levels. Indeed, in literature, very few works deals with the verification of the percentage of olive oil in fraudulent blends with VOs with regard to the labelling compliance of Reg. (EU) 29/2012, i.e., blends with percentages of olive oil close to the 50%. Among them, Gomez-Coca et al. (2020) successfully proposed the combination of four purity parameters, some described in legislation (Commission regulation (EEC) 2568/91), organized in decisional trees to discern olive oil concentrations, using sunflower oil as a model of VO to blend with olive oil (Gómez-Coca, Pérez-Camino, Martínez-Rivas, Bendini, Gallina Toschi, & Moreda, 2020). The potential of the fatty acid composition of the oil determined using the official method and multivariate data analysis was also proved for blends with sunflower oil (Monfreda, Gobbi, & Grippa, 2012) and other VOs (Monfreda, Gobbi, & Grippa, 2014). Spectroscopic techniques as TD-NMR and FTIR combined with chemometrics were investigated for the same goal (De la Mata, Dominguez-Vidal, Bosque-Sendra, Ruiz-Medina, Cuadros-Rodríguez, & Ayora-Cañada, 2012; Santos, Kock, Santos, Lobo, Carvalho, & Colnago, 2017).

Most of the studies handled the adulteration of EVOO with VOs, and several ones, with refined olive oil (Azizian, Mossoba, Fardin-Kia, Karunathilaka, & Kramer, 2016; Caponio, Summo, Bilancia, Paradiso, Sikorska, & Gomes, 2011; Carranco, Farrés-Cebrián, Saurina, & Núñez, 2018;

Izquierdo, Lastra-Mejías, González-Flores, Cancilla, Aroca-Santos, & Torrecilla, 2020; Jergović, Peršurić, Saftić, & Kraljević Pavelić, 2017; Jiang, Li, Chen, & Weng, 2018; Mignani, Ciaccheri, Ottevaere, Thienpont, Conte, Marega, et al., 2011; Torrecilla, Cancilla, Matute, Díaz-Rodríguez, & Flores, 2013; Torrecilla, García, García, & Rodríguez, 2011). However, only few research works studied the mixtures of olive oil (OO), i.e., blends of virgin and refined olive oil, with other VOs (De La Mata-Espinosa et al., 2011; De la Mata et al., 2012; Jiménez-Carvelo, González-Casado, et al., 2017; Jiménez-Carvelo, Osorio, et al., 2017; Monfreda et al., 2012, 2014), or the challenged adulteration of refined olive oil with refined hazelnut oil (Agiomyrgianaki, Petrakis, & Dais, 2010; Grošelj, Vračko, Pierna, Baeten, & Novič, 2008; Mannina, D'Imperio, Capitani, Rezzi, Guillou, Mavromoustakos, et al., 2009).

Chemical methods traditionally used for quality and authenticity control of olive oil are laborious, time-consuming, involves expensive and toxic chemicals with high environmental impact, require sample preparation and skilled operators. New and complementary analytical techniques will overcome some of these drawbacks and/or even support currently used methods to accomplish the complex task of the detection and quantification of olive oil mixtures with other oils. In this sense, metabolomic approaches, which allow rapid determination of several classes of chemical components using efficient advanced instrumental techniques coupled to chemometrics, are given great attention. MS-based methodologies and NMR spectroscopy are widely used in nowadays research in food analysis for quality control and traceability (Lioupi, Nenadis, & Theodoridis, 2020). Different NMR techniques, i.e., ^{1}H-NMR, ^{13}C-NMR, ^{31}P-NMR and/or ^{19}F-NMR have been used to characterise olive oils with authenticity and traceability purposes (R. M. Alonso-Salces, Segebarth, Garmón-Lobato, Holland, Moreno-Rojas, Fernández-Pierna, et al., 2015; Dais & Spyros, 2007; Guillén & Ruiz, 2001; Jiang et al., 2018; Mannina, Sobolev, & Segre, 2003; Sacchi, Addeo, & Paolillo, 1997; Vigli et al., 2003). Most of the NMR approaches developed for olive oil authentication, detection of olive oil adulteration and/or determination of olive oil blends with VOs were based on measuring signals of the NMR spectrum that give quantitative information of certain compounds or are used to calculate some parameters and ratios (i.e., profiling) (Agiomyrgianaki et al., 2010; García-González, Mannina, D'Imperio, Segre, & Aparicio, 2004; Jiang et al., 2018; Mannina et al., 2009; Popescu, Costinel, Dinca, Marinescu, Stefanescu, & Ionete, 2015; Vigli et al., 2003; Vlahov, 1997, 2009; Zamora, Alba, & Hidalgo, 2001). Instead, NMR fingerprinting approaches were only reported in few studies using low-field NMR spectroscopy (Parker, Limer, Watson, Defernez, Williamson, & Kemsley, 2014; Santos et al., 2017; Wang, Wang, Hou, & Nie, 2020). To the authors' knowledge, high field ^{1}H-NMR fingerprinting combined with chemometrics has only been used by Fang et al. (2013) to discriminate among different pure edible oils and fats, and to detect the adulteration of canola oils with animal fats (Fang, Goh, Tay, Lau, & Li, 2013). The present study was the first one that used high-field NMR fingerprinting to study mixtures of olive oil with other VOs. Taking into account that ^{1}H-NMR technique is characterised by its operational ease, simple sample preparation, the use of small volumes of organic solvents, rapid analysis times, being non-destructive and high-throughput, and providing high amount of chemical information (Guillén et al., 2001; Lioupi et al., 2020; Meenu et al., 2019), the development of a ^{1}H-NMR fingerprinting approach to detect olive oil adulteration and determine the composition of oil blends would contribute significantly to the field. The present chapter gathers the results of the proposed analytical strategy based on ^{1}H-NMR fingerprinting together with multivariate classification and regression models organised in a decision tree scheme to determine the composition of an oil blend from both points of view, the botanical nature of the oils as well as the percentage of each oil in the blend (Rosa María Alonso-Salces, Berrueta, Quintanilla-Casas, Vichi, Tres, Collado, et al., 2022). Moreover, the research work here presented addressed some issues that had not been considered in previous studies: (*i*) the discrimination between blends made with VOO and those with olive oil; (*ii*) the distinction of pure VOOs and their blends with VOs; (*iii*) the study of a large sample set with pure oils and blends of the most common VOs used for olive oil adulteration; and (*iv*) a wide range of percentages of the VOs in the blend, including those for the verification of the labelling compliance of Reg. (EU) 29/2012.

^{1}H-NMR Spectra of Vegetable Oils

The ^{1}H-NMR spectra of the vegetable oils studied, both pure and blended (binary mixtures of VO with VOO or OO) oils, were similar, presenting apparently only differences in the signal intensities. However, a careful observation of the extended spectra revealed characteristic patterns, interesting for the determination of the botanical origin of oils, as well as the composition of blended oils. The chemical shifts of ^{1}H-NMR signals of the major and some minor compounds in the oil samples, as well as their assignments to protons of the different functional groups are gathered in Table 1. The ¹H signals of triacylglycerides, the major components of edible oils, are well-known (Mannina & Segre, 2002; Sacco, Brescia, Liuzzi, Reniero, Guillou, Ghelli, et al., 2000). Triacylglycerides present different substitution patterns depending on the length, degree and kind of unsaturation of the acyl groups (Aparicio & Harwood, 2013; Guillén et al., 2001). The ¹H signals of minor components are only observed in the ¹H-NMR spectrum when they are not overlapped with those of the main components, and their intensities are high enough to be detected. Thus, sterols such as cycloartenol, β-sitosterol and stigmasterol; hydrocarbons as squalene; diglycerides as *sn*-1,2 diacylglycerides and *sn*-1,3 diacylglycerides; terpenes; phenolics as secoiridoid derivatives, saturated and unsaturated aldehydes as (*E*)-2-hexenal; and other volatile compounds are detected using high field NMR instruments (R. M. Alonso-Salces, Héberger, Holland, Moreno-Rojas, Mariani, Bellan, et al., 2010; Rosa M. Alonso-Salces, Holland, & Guillou, 2011; Rosa M. Alonso-Salces, Moreno-Rojas, Holland, Reniero, Guillou, & Heberger, 2010; R. M. Alonso-Salces et al., 2015; Christophoridou & Dais, 2009; D'Imperio, Mannina, Capitani, Bidet, Rossi, Bucarelli, et al., 2007; Guillén et al., 2001; Guillén & Ruiz, 2003; Mannina et al., 2002; Mannina et al., 2003; Owen, Mier, Giacosa, Hull, Spiegelhalder, & Bartsch, 2000; Papadia, Del Coco, Muzzalupo, Rizzi, Perri, Cesari, et al., 2011; Pérez-Trujillo, Gómez-Caravaca, Segura-Carretero, Fernández-Gutiérrez, & Parella, 2010; Rotondo, Salvo, Giuffrida, Dugo, & Rotondo, 2011; Sacchi, Patumi, Fontanazza, Barone, Fiordiponti, Mannina, et al., 1996; Shi, Zhu, Zhou, Huo, Long, Zeng, et al., 2019; Ün & Ok, 2018). The changes in the ¹H-NMR fingerprints of authentic olive oil samples observed when they are blended with VOs, are mainly due to their different compositions in triglycerides, diglycerides and some minor constituents of the unsaponifiable fraction. The intensity and the area of the ¹H-NMR signals are proportional to the number of protons in each functional group and the number of functional groups present in the sample, therefore quantitative data of the oil sample composition is provided. Indeed, ¹H-NMR profiling with medium (400 MHz) and high (500 MHz and 600 MHz) field NMR instruments have been used to calculate several quantitative parameters in order to distinguish different VOs from olive oils, as well as their admixtures. Among these parameters, the iodine value (IV), the proportions of oleic, linoleic, linolenic or saturated acyl groups (Agiomyrgianaki et al., 2010; García-González et al., 2004; Guillén et al., 2003; Popescu et al., 2015; Vigli et al., 2003), different ¹H signal ratios (Fauhl, Reniero, & Guillou, 2000; Mannina et al., 2009), and the proportions of *sn*-1,2 diacylglycerides, *sn*-1,3 diacylglycerides and 1-monoacylglycerides by direct integration of the ^{1}H signal (San Martín, Avenoza, Peregrina, & Busto, 2020) have led to the following conclusions. A percentage of linolenic acyl groups higher than 0.9% is indicative of the presence of vegetable oil in olive oil (Mannina, Patumi, Fiordiponti, Emanuele, & Segre, 1999). Olive oil presents high content of monounsaturated fatty acids, mainly oleic acid; sunflower oil, high concentrations of linoleic acid; and VOO respect to other VOs, the lowest amounts of linoleic acid (Popescu et al., 2015; San Martín et al., 2020). The IV indicates the unsaturation level of the oil, which is closely related to its fatty acids composition. Thus, high content of saturated or monosaturated fatty acids (oleic acid) results in low IVs; whereas high content of linolenic acid (with three double bonds), in high IVs. Indeed, olive oil presents the lowest IV in comparison with other VOs (Agiomyrgianaki et al., 2010; Popescu et al., 2015). VOO contains about 2–3% of diglycerides, mainly *sn*-1,2 diacylglycerides. The presence of significant amounts of *sn*-1,3 diacylglycerides indicates the loss of freshness of the VOO and/or the addition of refined olive oil. β-sitosterol content is lower than 0.2% in virgin and refined olive oils, whereas in vegetable oils are higher (Guillén et al., 2001). The fatty acid profiles of refined olive and refined hazelnut oils present obvious differences, especially regarding linolenic acid and saturated fatty acids (Agiomyrgianaki et al., 2010). Indeed, hazelnut oils contain remarkably lower amounts of linolenic acid, saturated fatty acids, squalene and β-sitosterol than olive oils, while oleic acid and linoleic acid are slightly more abundant in hazelnut oils than in olive oil (Mannina et al., 2009). Sunflower oil is characterized by high levels of linoleic acid, and may contain up to 2% of linolenic acid (Fauhl et al., 2000). The content of polyunsaturated fatty acids (linoleic and linolenic) is high in soybean, corn and sunflower, and low in olive, hazelnut and palm oil. High amounts of linolenic acid are also observed in soybean and rapeseed oils (Vigli et al., 2003). The combination of several of these parameters allowed to detect mixtures of olive oil with proportions of sunflower oil higher than 10%; and hazelnut oil higher than 25% (Fauhl et al., 2000), at 10% in conjunction with pattern recognition techniques (Mannina et al., 2009), and higher than 8% with ^{13}C-NMR and artificial neural networks (García-González et al., 2004). ^{1}H and ^{31}P-NMR profiling together with multivariate data analysis permitted the correct discrimination of refined olive oils, refined hazelnut oils, and adulterated refined olive oils with 1% to 95% (w/w) of refined hazelnut oils (Agiomyrgianaki et al., 2010); as well as to detect the adulteration of fresh virgin olive oil with different vegetable oils at percentages as low as 5% (w/w) (Vigli et al., 2003). Low field NMR instruments have been used for oil profiling or fingerprinting, achieving about 80% of correct classification to discriminate pure and blended oils (Santos et al., 2017; Wang et al., 2020), and detect the adulteration of olive oil with hazelnut oil at concentrations higher than 13% (w/w) (Gouilleux, Marchand, Charrier, Remaud, & Giraudeau, 2018; Parker et al., 2014). Hence, previous research works reported in literature support that NMR spectroscopy, and in particular ¹H-NMR, provide useful information for the aim of current work.

Table 1. Chemical shift assignments of the ¹H-NMR signals of the main components in olive oil

#	Chemical shift (ppm)	Multiplicity[a]	Functional group	Attribution
1	0.318	d	-CH_2- (cyclopropanic ring)	cycloartenol
2	0.527	s	-CH_2-	alcohol, sterol
3	0.543	d	-CH_2- (cyclopropanic ring)	cycloartenol
4	0.669	s	-CH_3 (C18-steroid group)	β-sitosterol
5	0.687	s	-CH_3 (C18-steroid group)	stigmasterol
6	0.740	t	-CH_3 (^{13}C satellite of signal at 0.87 ppm, acyl group)	
7	0.80-1.04	t	-CH_3 (acyl group)	
7a	0.83	t	-CH_3 (acyl group)	saturated
7b	0.866	t	-CH_3 (acyl group)	oleic (or ω-9)
7c	0.89	t	-CH_3 (acyl group)	linoleic (or ω-6)
7d	0.960	t	-CH_3 (acyl group)	linolenic (or ω-3)
8	0.987	t	-CH_3 (^{13}C satellite of signal at 0.87 ppm, acyl group)	
9	1.19-1.44		-$(CH_2)_n$- (acyl group)	
9a	1.243		-$(CH_2)_n$- (acyl group)	saturated
9b	1.256		-$(CH_2)_n$- (acyl group)	oleic (or ω-9)
9c	1.288		-$(CH_2)_n$- (acyl group)	linoleic (or ω-6) and linolenic (or ω-3)
10	1.51-1.65		-OCO-CH_2-CH_2- (acyl group)	
10a	1.57		-OCO-CH_2-CH_2- (acyl group)	saturated
10b	1.58		-OCO-CH_2-CH_2- (acyl group)	oleic (or ω-9)
10c	1.59		-OCO-CH_2-CH_2- (acyl group)	linoleic (or ω-6) and linolenic (or ω-3)
11	1.662	s	-CH_3	squalene
12	1.96-2.07		-CH_2-CH=CH- (acyl group)	
12a	1.97		-CH_2-CH=CH- (acyl group)	oleic (or ω-9)
12b	2.01-2.03		-CH_2-CH=CH- (acyl group)	linoleic (or ω-6) and linolenic (or ω-3)
12c	2.05-2.07		-CH2-CH=CH- (acyl group)	linolenic (or ω-3)
13	2.22-2.32	m	-OCO-CH2- (acyl group)	

Table 1. (Continued)

#	Chemical shift (ppm)	Multiplicity[a]	Functional group	Attribution
13a	2.24	m	-OCO-CH_2- (acyl group)	saturated
13b	2.25	m	-OCO-CH_2- (acyl group)	oleic (or ω-9)
13c	2.27	m	-OCO-CH_2- (acyl group)	linoleic (or ω-6)
13d	2.31	m	-OCO-CH_2- (acyl group)	linolenic (or ω-3)
14	2.40-2.45	m	-OCO-CH_2- (^{13}C satellite of signal at 2.26-2.32 ppm, acyl group)	
15	2.72-2.82		=CH-CH_2-CH= (acyl group)	
15a	2.754	t	=CH-CH_2-CH= (acyl group)	linoleic (or ω-6)
15b	2.789	t	=CH-CH_2-CH= (acyl group)	linolenic (or ω-3)
16	3.69-3.73	d	-CH_2OH (glyceryl group)	*sn*-1,2-diacylglycerides
17	4.05-4.09	q	>C*H*-OH (glyceryl group)	*sn*-1,3-diacylglycerides
18	4.09-4.32		-CH_2OCOR (glyceryl group)	triacylglycerides
19	4.571	d		terpene
20	4.648	s		terpene
21	4.699	s		terpene
22	5.05-5.15	m	>C*H*OCOR (glyceryl group)	*sn*-1,2-diacylglycerides
23	5.22-5.28	m	>C*H*OCOR (glyceryl group)	triacylglycerides
24	5.28-5.38	m	-C*H*=C*H*- (acyl group)	
25	5.52-5.43	m	-C*H*=C*H*- (^{13}C satellite of signal at 5.28-5.38 ppm, acyl group)	
26	5.72-5.76	dt	=C*H*- (phenolic ring)	phenolic compounds
27	5.986		=C*H*- (phenolic ring)	phenolic compounds
28	6.551	dt	=C*H*- (phenolic ring)	phenolic compounds
29	6.607	dd	=C*H*- (C8'; phenolic ring)	dialdehyde of oleuropein lacking a carboxymethyl group aldehydic form of oleuropein
30	6.79-6.73	d	=C*H*- (C5', C7'; phenolic ring)	dialdehyde of secoiridoids (oleuropein, ligstroside) lacking a carboxymethyl group aldehydic form of secoiridoid (oleuropein, ligstroside)
31	7.05-7.00	dt	=C*H*- (C4', C8'; phenolic ring)	dialdehyde of ligstroside lacking a carboxymethyl group aldehydic form of ligstroside
32	7.562	s	=C*H*-O- (C3)	aldehydic form of secoiridoid (oleuropein, ligstroside)
33	8.14-8.06		>C(O*H*)OR	volatile compounds
34	9.215	d	-CHO (C1)	dialdehyde of secoiridoids (oleuropein, ligstroside) lacking a carboxymethyl group

#	Chemical shift (ppm)	Multiplicity[a]	Functional group	Attribution
34	9.215	d	-*CHO* (C1)	dialdehyde of secoiridoids (oleuropein, ligstroside) lacking a carboxymethyl group
35	9.51	d	-*CHO*	*E*-2-alkenals (*E*-2-hexenal)
36	9.626	dd	-*CHO* (C3)	dialdehyde of secoiridoids (oleuropein, ligstroside) lacking a carboxymethyl group
		dd	-*CHO* (C1)	aldehydic form of secoiridoids (oleuropein, ligstroside)

[a] Signal multiplicity: s, singlet; d, doublet: t, triplet; m, multiplet; q, quadruplet; dt, double triplet; dd, double doublet.

Adulteration of Olive Oil with Vegetable Oils

The proposed approach to detect blends of olive oils (VOOs or OOs) with other VOs and quantify the percentage of VO in the blend is based in the use of ^{1}H-NMR fingerprinting of the oil and a set of multivariate classification and regression models organized in the decision tree scheme shown in Figure 1. The analytical strategy proposed to detect the adulteration of olive oil with VOs was tested for the most common VOs used to be mixed with VOO and OO. The oils studied were genuine samples of virgin (VOO) and extra virgin olive oils (EVOO, n=176), olive oils (OO, n=3), refined conventional sunflower oil (normal type sunflower oil, NTSO, n=17), refined high oleic sunflower oil (HOSO, n=16), desterolized and deodorized high oleic sunflower oil (DOSO, n=1), refined hazelnut oil (HR, n=11), virgin hazelnut oil (HV, n=6), refined soybean oil (S, n=10), virgin avocado oil (EVAO, n=1), refined avocado oil (RAO, n=1), refined palm olein oil (RPOO, n=1) and refined corn oil (CO, n=1), which were used to prepare binary mixtures at different percentages (2%–90%) of VOs in VOOs or OOs (1007 blended samples). These samples were obtained in the framework of Autenfood project (ACCIÓ Programa Operatiu FEDER Catalunya 2014–2020) and OLEUM project (EC H2020 Programme 2014–2020), and were produced during two consecutive harvest years (2016/17 and 2017/18). Blends were prepared and preserved under controlled temperature conditions. All pure and blended oil samples were bottled with nitrogen headspace or minimal air headspace, kept at -20ºC and protected from light. The ^{1}H-NMR experiments and the univariate (ANOVA, Fisher index and Box & Whisker plots) and multivariate (Principal Component analysis (PCA), and supervised as partial least squares discriminant analysis (PLS-DA) and partial least squares regression (PLS-R)) data analysis of the ^{1}H-NMR fingerprints of the studied oils were performed according to the methodology described by Alonso-Salces et al. (2022).

Binary classification models were built by PLS-DA to (*i*) discriminate between oils containing VOOs and OOs, both pure oils and blends; (*ii*) distinguish oils containing or not a certain VO; and (*iii*) differentiate between 'legal' (containing NTSO or HOSO) and 'illegal' (containing other VOs) blends with VOO or OO, using sunflower oil as a vegetable oil model (Gómez-Coca, Pérez-Camino, Martínez-Rivas, et al., 2020; Monfreda et al., 2012). Moreover, regression models were built by PLS-R to determine the percentage of a VO in a blend with VOO or OO. The most influential variables on the models were not completely discriminant unless otherwise specified.

PLS-DA Models to Confirm the Presence of Virgin Olive Oil or Olive Oil in Blends with Vegetable Oils

The first stage of the decision tree (Figure 1) consists in identifying whether the oil sample contains VOO or OO using PLS-DA model 1, which presented recognition and prediction abilities of 97% and 98% of correct classification for the VOO and OO categories respectively (Table 2). The most influential NMR variables on the model were the methyl protons of the acyl group of oleic acid (#7b), the β-methylene protons of the acyl group of linoleic and linolenic acids (#10c), the methylene protons of the acyl group of oleic acid (#9b) and of saturated fatty acids (#9a), and the α-methylene protons of the acyl group of linolenic acid (#13d). The box and whiskers plot of these variables showed that the signal intensities were higher in VOO and their blends than in samples containing OO, even though the intensity ranges of both categories were partially overlapped. In contrast, the ^{1}H signals due to the allylic protons of the acyl group of linoleic acid (#12b) presented lower intensities and less variability in the VOO category than in the OO one; even though the box and whiskers plot of VOO intensity range completely overlapped the 1st, 2nd and 3rd quartiles of the OO intensity range. Moreover, the higher intensities of the glyceryl group of *sn*-1,3-diacylglycerides (#17) in the OO category also contributed to the discrimination of both types of oils. These observations agreed with previous studies reporting the differences in the composition of oleic, linolenic and saturated fatty acids and *sn*-1,3-

diacylglycerides between VOOs and OOs (Dais & Hatzakis, 2013; Guillén et al., 2003; Jiang et al., 2018; Pérez-Camino, Moreda, & Cert, 2001; Sacchi et al., 1997; Spyros, Philippidis, & Dais, 2004).

Table 2. PLS-DA models to discriminate between pure and blended oils containing olive oil or virgin olive oil and vegetable oils[1]

PLS-DA model	Data	PLS-comp	Boundary	Class[2]	Class code	n	p	%R	%P
1	Pure & blend VOO/OO	4	0.4079	VOO	0	838	0.70	97	97
				OO	1	356	0.30	98	98
2	Pure & blend VOO	6	0.3283	0-20% VOs in VOO	0	704	0.84	98	98
				25-90% VOs in VOO	1	132	0.16	97	97
3	0-20% VOs in VOO	5	0.2230	2-20% VOs in VOO	0	549	0.78	90	89
				Pure VOO	1	155	0.22	86	86
4	0-2% VOs in VOO	14	0.4264	2% VOs in VOO	0	204	0.57	90	90
				Pure VOO	1	155	0.43	93	92
5	2% VOs in VOO	19	0.4265	non-S	0	159	0.78	99	97
				S	1	45	0.22	100	100
29	Pure & blend OO	16	0.4388	0-20% VOs in OO	0	184	0.52	97	96
				30-80% VOs in OO	1	171	0.48	95	94

[1] Abbreviations: n, number of samples; centered data; PLS-comp, number of PLS components; p, prior probability; %R, % of recognition ability; %P, % of prediction ability in cross-validation; VOO, virgin olive oil; OO, olive oil; VO, vegetable oil; NTSO, refined conventional sunflower oil (normal type sunflower oil); HOSO, refined high oleic sunflower oil; DOSO, desterolized and deodorized high oleic sunflower oil; HR, refined hazelnut oil; HV, virgin hazelnut oil; S, refined soybean oil; EVAO, virgin avocado oil; RAO, refined avocado oil; RPOO, refined palm olein oil; CO, refined corn oil.

[2] Samples contained in each class: VOO, pure VOOs and blends of VOO with VOs (NTSO, HOSO, EVAO, HV, HR or S); OO, pure OOs and blends of OO with VOs (RPOO, CO, HOSO, NTSO, DOSO, RAO or HR); 0–20% VOs in VOO, pure VOOs and blends of VOO with 2–20% VOs (NTSO, HOSO, EVAO, HV, HR or S); 25–90% VOs in VOO, blends of VOO with 25–90% VOs (NTSO, HOSO, EVAO, HV, HR or S); 2–20% VOs in VOO, blends of VOO with 2–20% VOs (NTSO, HOSO, EVAO, HV, HR or S); Pure VOO, pure VOOs; 2% VOs in VOO, blends of VOO with 2% VOs (NTSO, HOSO, EVAO, HV, HR or S); non-S, blends of VOO with 2% VOs (NTSO, HOSO, EVAO, HV or HR); S, blends of VOO with 2% S; 0–20% VOs in OO, pure OOs and blends of OO with 2–20% VOs (RPOO, CO, HOSO, NTSO, DOSO, RAO or HR); 30–80% VOs in OO, blends of OO with 30–80% VOs (RPOO, CO, HOSO, NTSO, DOSO, RAO or HR).

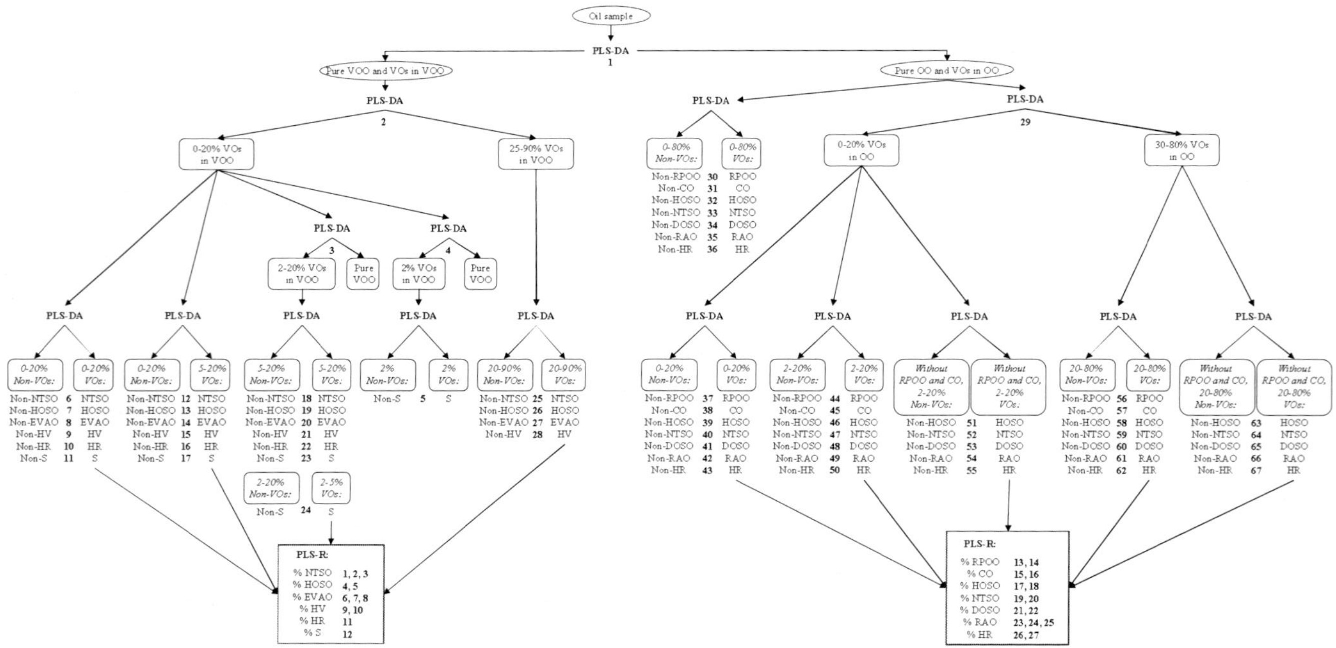

Figure 1. Decision tree constituted of PLS-DA classification and PLS-R regression models to determine the composition of binary mixtures of oils of the 'virgin olive oil' or 'olive oil' categories and other vegetable oils. *Abbreviations: VOO, virgin olive oil; OO, olive oil; VO, vegetable oil; NTSO, refined conventional sunflower oil (normal type sunflower oil); HOSO, refined high oleic sunflower oil; DOSO, desterolized and deodorized high oleic sunflower oil; HR, refined hazelnut oil; HV, virgin hazelnut oil; S, refined soybean oil; EVAO, virgin avocado oil; RAO, refined avocado oil; RPOO, refined palm olein oil; CO, refined corn oil.*

Once the oil sample is classified as containing VOO or OO according to the first step of decision tree scheme (Figure 1), further predictions are made using the binary classification models built separately for each type of olive oil, in order to elucidate whether the oil sample is mixed with VOs, in which proportion (low or high), and with which particular VO. The PLS-DA models achieved for samples with VOO, i.e., PLS-DA models 2–28 and 68–69, are shown in Tables 2–6 and 15; and for samples with OO, i.e., PLS-DA models 29–67 and 70–71, in Tables 2, 7–12 and 15. These PLS-DA models are described in the next sections.

PLS-DA Models to Discriminate Blends of Virgin Olive Oil with Vegetable Oils

In the second stage of the decision tree scheme (Figure 1), a PLS-DA model classifies the oil sample containing VOO according to their level of adulteration with VOs, i.e., low (VOOs with 0–20% VOs) and high (VOOs with 25–90% VOs), with correct prediction abilities of 98% and 97% for the low and high categories respectively (PLS-DA model 2 in Table 2). The most influential ^{1}H signals in this model were the methylene protons of oleic acid (#9b) and the methyl protons of squalene (#11), whose signal intensities were higher and presented less variability in the samples with low contents of VOs in VOO than in those VOO blends with more than 20% VOs, even though the intensity ranges of both categories were partially overlapping. This is in accordance with previous studies, since VOO is known to be one of the vegetable oils that presents the highest contents of oleic acid and squalene (Hayakawa, Yanagawa, Yamamoto, Aizawa, Taga, Mochizuki, et al., 2020; Jiang et al., 2018; Popescu et al., 2015; Tsimidou, 2010; Vigli et al., 2003). In addition, the α-methylene protons of the acyl group of linoleic (#13c) and linolenic acids (#13d) and the glyceryl protons of triacylglycerides (#18) also contributed to differentiate both categories; the box and whiskers plots of these signal intensities showed that the low category presented less variability but similar values to the samples in the 2nd and 3rd quartiles of the high category as expected since this category included blends with a larger variability of compositions.

In the next stage, pure VOOs are distinguished from blends with 2–20% VOs in VOO, identifying even 92% of the pure VOOs and 90% of the VOO blends with VO (PLS-DA models 3 and 4 in Table 2). The main characteristic

^{1}H signals involved in the distinction of both classes were the methyl (#7a) and the methylene (#9a) protons of the acyl group of saturated fatty acids, and the methylene protons at 1.38–1.42 ppm of the acyl group of unsaturated fatty acids (#9), which exhibited lower intensities and variabilities in blended VOOs overlapping with those of the 1st and 2nd quartiles of the pure VOOs. Other important variables, such as the methyl (#7c), allylic (#12b) and α-methylene (#13c) protons of linoleic acid, and vinyl protons of the acyl group of unsaturated fatty acids (#24) presented similar intensities in both categories. These observations are explained by the fact that saturated fatty acids are the 2nd major class of fatty acids in VOOs, being present in higher or similar concentrations than in the VOs studied, i.e., NTSO, HOSO, EVAO, HV, HR and S, whereas linoleic acid is contained in considerable lower amounts in VOOs than in the these VOs (Contiñas, Martínez, Carballo, & Franco, 2008; Guillén et al., 2003; Jabeur, Zribi, Makni, Rebai, Abdelhedi, & Bouaziz, 2014; Jiang et al., 2018; Jović, Smolić, Primožič, & Hrenar, 2016; Monfreda et al., 2012; Ranade & Thiagarajan, 2015; Souza, Ansolin, Batista, Meirelles, & Tubino, 2019; Yang et al., 2013).

Concerning the discrimination of blends of 2% VO in VOO, a satisfactory classification model was achieved only for refined soybean oil (S), which allowed to detect all blended samples containing 2% S in VOO, and predict correctly 97% of the VOO blends with any other VO at 2% (PLS-DA model 5 in Table 2). The most influential variables on this model were the methyl protons at 1.01 ppm (#7) and the β-methylene protons at 1.65 ppm (#10) of the acyl group of unsaturated fatty acids, which exhibited slightly lower intensities and less variability in VOO blends with 2% S, overlapping with the non-S category; and the vinylic protons of unsaturated fatty acids (#24) and the bis-allylic protons of the acyl group of linolenic acid (#15b), which showed higher intensities in VOO mixtures containing soybean oil, partially overlapping the intensity range of the non-S category. Soybean oil is one of the VOs with the highest contents of polyunsaturated fatty acids, in particular linolenic acid; and the highest IV, similar to sunflower oils but considerably higher than olive oil, hazelnut oil and avocado oil (Guillén et al., 2003; Souza et al., 2019; Vigli et al., 2003) (Contiñas et al., 2008; Christopoulou, Lazaraki, Komaitis, & Kaselimis, 2004; Jabeur et al., 2014; Jiang et al., 2018).

Table 3. PLS-DA models to detect the presence of a certain vegetable oil in a binary mixture of 2–20% vegetable oil in virgin olive oil[1]

PLS-DA model	Data	PLS-comp	Boundary	Class[2]	Class code	n	p	%R	%P	%P-EV
6	0-20% VOs in VOO	12	0.3623	non-NTSO	0	238	0.64	86	85	79
				NTSO	1	132	0.36	90	86	-
7	0-20% VOs in VOO	14	0.4713	non-HOSO	0	245	0.62	83	79	82
				HOSO	1	152	0.38	83	79	-
8	0-20% VOs in VOO	6	0.3791	non-EVAO	0	81	0.80	94	93	97
				EVAO	1	20	0.20	90	90	-
9	0-20% VOs in VOO	6	0.3815	non-HV	0	137	0.68	78	75	73
				HV	1	65	0.32	82	75	-
10	0-20% VOs in VOO	5	0.4011	non-HR	0	195	0.68	77	75	58
				HR	1	90	0.32	78	76	-
11	0-20% VOs in VOO	11	0.4248	non-S	0	208	0.70	98	96	95
				S	1	90	0.30	97	96	-

[1] See abbreviations in Table 2; %P-EV, % of prediction ability in external validation.

[2] Samples contained in each class: non-NTSO, pure VOOs and blends of VOO with 2–20% VOs (HOSO, EVAO, HV, HR or S); NTSO, blends of VOO with 2–20% NTSO; non-HOSO, pure VOOs and blends of VOO with 2–20% VOs (NTSO, EVAO, HV, HR or S); HOSO, blends of VOO with 2–20% HOSO; non-EVAO, pure VOOs and blends of VOO with 2–20% VOs (NTSO, HOSO, HV, HR or S); EVAO, blends of VOO with 2–20% EVAO; non-HV, pure VOOs and blends of VOO with 2–20% VOs (NTSO, HOSO, EVAO, HR or S); HV, blends of VOO with 2–20% HV; non-HR, pure VOOs and blends of VOO with 2–20% VOs (NTSO, HOSO, EVAO, HV or S); HR, blends of VOO with 2–10% HR; non-S, pure VOOs and blends of VOO with 2–20% VOs (NTSO, HOSO, EVAO, HV or HR); S, blends of VOO with 2–5% S.

Following the decision tree scheme (Figure 1), the ^{1}H-NMR fingerprint of an oil sample classified in the low category (0–20% VO in VOO) is then submitted to binary classification models developed for each individual VO in order to identify whether the oil sample contains a certain type of VO. Several PLS-DA models were built for different percentage ranges of VO (0–20% VO and/or 5–20% VO) in VOO (PLS-DA models 6–24 in Tables 3, 4 and 5). The performance of all these models supported and confirmed the predictions achieved. In general, the classification abilities of the PLS-DA models were better when the dataset contained only blended oils with 5–20% VO in VOO than when pure VOO and 2% VO data was also included (0–20% VO in

VOO). The prediction abilities ranged between 83% and 98% of hits depending on the VO blended with VOO.

Table 4. PLS-DA models to detect the presence of a certain vegetable oil in a binary mixture of 5–20% vegetable oil in virgin olive oil[1]

PLS-DA model	Data	PLS-comp	Boundary	Class[2]	Class code	n	p	%R	%P	%P-EV
12	0-20% non-NTSO in VOO	11	0.3508	non-NTSO	0	238	0.73	96	95	92
	5-20% NTSO in VOO			NTSO	1	87	0.27	94	90	-
13	0-20% non-HOSO in VOO	17	0.4098	non-HOSO	0	245	0.71	87	85	85
	5-20% HOSO in VOO			HOSO	1	102	0.29	90	86	-
14	0-20% non-EVAO in VOO	10	0.3805	non-EVAO	0	80	0.84	94	93	97
	5-20% EVAO in VOO			EVAO	1	15	0.16	100	93	-
15	0-20% non-HV in VOO	10	0.3675	non-HV	0	137	0.75	85	82	81
	5-20% HV in VOO			HV	1	45	0.25	80	78	
16	0-20% non-HR in VOO	14	0.3808	non-HR	0	195	0.76	85	79	72
	5-20% HR in VOO			HR	1	60	0.24	85	85	-
17	0-20% non-S in VOO	7	0.4156	non-S	0	208	0.82	98	98	97
	5-20% S in VOO			S	1	45	0.18	98	98	-

[1] See abbreviations in Table 3.

[2] Samples contained in each class: non-NTSO, pure VOOs and blends of VOO with 2–20% VOs (HOSO, EVAO, HV, HR or S); NTSO, blends of VOO with 5–20% NTSO; non-HOSO, pure VOOs and blends of VOO with 2–20% VOs (NTSO, EVAO, HV, HR or S); HOSO, blends of VOO with 5–20% HOSO; non-EVAO, pure VOOs and blends of VOO with 2–20% VOs (NTSO, HOSO, HV, HR or S); EVAO, blends of VOO with 5–20% EVAO; non-HV, pure VOOs and blends of VOO with 2–20% VOs (NTSO, HOSO, EVAO, HR or S); HV, blends of VOO with 5–20%HV; non-HR, pure VOOs and blends of VOO with 2–20% VOs (NTSO, HOSO, EVAO, HV or S); HR, blends of VOO with 5–10% HR; non-S, pure VOOs and blends of VOO with 2–20% VOs (NTSO, HOSO, EVAO, HV or HR); S, blends of VOO with 5% S.

Table 5. PLS-DA models to detect the presence of a certain vegetable oil in a binary mixture of 2–20% vegetable oil in virgin olive oil[1]

PLS-DA model	Data	PLS-comp	Boundary	Class[2,3]	Class code	n	p	%R	%P
18	5-20% non-NTSO in VOO	7	0.3029	non-NTSO	0	267	0.77	93	91
	5-20% NTSO in VOO			NTSO	1	78	0.23	94	90
19	5-20% non-HOSO in VOO	16	0.4039	non-HOSO	0	243	0.70	88	85
	5-20% HOSO in VOO			HOSO	1	102	0.30	92	88
20	5-20% non-EVAO in VOO	11	0.3002	non-EVAO	0	330	0.96	98	98
	5-20% EVAO in VOO			EVAO	1	15	0.04	93	93
21	5-20% non-HV in VOO	13	0.2335	non-HV	0	300	0.87	91	83
	5-20% HV in VOO			HV	1	45	0.13	91	87
22	5-20% non-HR in VOO	20	0.3291	non-HR	0	285	0.83	90	83
	5-20% HR in VOO			HR	1	60	0.17	93	88
23	5-20% non-S in VOO	7	0.3715	non-S	0	300	0.87	98	97
	5% S in VOO			S	1	45	0.13	98	98
24	2-20% non-S in VOO	13	0.4514	non-S	0	166	0.65	99	97
	2-5% S in VOO			S	1	90	0.35	98	97

[1] See abbreviations in Table 2.

[2] Samples contained in each class for PLS-DA models 18–23: non-NTSO, blends of VOO with 5–20% VOs (HOSO, EVAO, HV, HR or S); NTSO, blends of VOO with 5–20% NTSO; non-HOSO, blends of VOO with 5–20% VOs (NTSO, EVAO, HV, HR or S); HOSO, blends of VOO with 5–20% HOSO; non-EVAO, blends of VOO with 5–20% VOs (NTSO, HOSO, HV, HR or S); EVAO, blends of VOO with 5–20% EVAO; non-HV, blends of VOO with 5–20% VOs (NTSO, HOSO, EVAO, HR or S); HV, blends of VOO with 5–20%HV; non-HR, blends of VOO with 5–20% VOs (NTSO, HOSO, EVAO, HV or S); HR, blends of VOO with 5–10% HR; non-S, blends of VOO with 5–20% VOs (NTSO, HOSO, EVAO, HV or HR); S, blends of VOO with 5% S..

[3] Samples contained in each class for PLS-DA models 24: non-S, blends of VOO with 2-20% VOs (NTSO, HOSO, EVAO, HV or HR); S, blends of VOO with 2–5% S.

Among the VOs studied, sunflower oil is the oil that contains higher amounts of linoleic acid (Contiñas et al., 2008; Guillén et al., 2003; Jović et al., 2016; Ranade et al., 2015; Vigli et al., 2003). Indeed, the most important variables on models to discriminate the presence of NTSO in VOO were the signals due to the α-methylene (#13c), allylic (#12b) and methyl (#7c) protons of linoleic acid, and the vinylic protons of unsaturated fatty acids (#24), which presented higher intensities in 5-20% NTSO mixtures with VOO than in most of the other VO-VOO blends. However, the intensities ranges of both categories partially overlapped, since VOO blends with HOSO and S included in the non-NTSO category, also presents relative high contents of linoleic acid, even though lower than sunflower oil (Guillén et al., 2003; Vigli et al., 2003) (Fang et al., 2013; Fauhl et al., 2000; Popescu et al., 2015). In contrast, sunflower oil presents similar low amounts of linolenic acid as olive, hazelnut and avocado oils, therefore the ^{1}H signals of linolenic acid (#7d and #15b) presented larger variability in the non-NTSO category, as well as higher intensities due to VOs rich in this fatty acid, such as soybean oil (Vigli et al., 2003). This trend regarding linolenic signals (#7d, #15b, #12c) was also observed in the model built to detect mixtures of HOSO in VOO in low proportions. The methyl protons of oleic acid (#7b), the methylene protons of terpenic alcohols or sterols (#2), and the allylic protons of linoleic acid (#12b), which presented slightly higher intensities in the HOSO category than the non-HOSO category, were also important to discriminate blends containing low contents of HOSO (0–20% HOSO in VOO). Previous compositional studies support these remarks since HOSO contains both high contents of oleic acid and linoleic acid (Jović et al., 2016), and its mixture with VOO leads to an increase of the sterol content compared to pure olive oil (Al-Ismail, Alsaed, Ahmad, & Al-Dabbas, 2010).

Evaluating the most important variables on the binary classification models for the detection of EVAO-VOO blends in low proportions, it was observed that the ^{1}H-NMR spectra of the VOO mixtures of 5–20% EVAO showed higher signal intensities than the blends without EVAO for the β-methylene (#10a) and methyl (#7a, b) protons of saturated and oleic fatty acids, the allylic protons of oleic acid (#12a), and the methyl protons of squalene (#11) and β-sitosterol (#4). Meanwhile, the ^{1}H signals of the vinylic protons (#24), the methylene protons at 1.32–1.36 ppm of unsaturated fatty acids (#9) and the α-methylene protons of linolenic acid (#13d) presented lower intensities in the EVAO-VOO blends than in most of the other VO-VOO blends. This observation disclosed the presence of fatty acids with less degree of unsaturation in EVAO-VOO blends, as well as, a particular composition in fatty acids responsible for giving higher intensity ^{1}H signals at higher fields, i.e., 2.26–2.30 ppm (#13c, linoleic acid) and 1.96–2.00 ppm (#12a, oleic acid); and lower intensity signals at lower fields, i.e., 2.30–2.32 ppm (#13d, linolenic acid) and 2.02–2.06 ppm (#12b, c, linolenic acid), in comparison to most of the samples of other VO-VOO blends. Indeed, EVAO presents the highest contents of the saturated fatty acids, mainly palmitic acid, of all the VOs studied, similar medium amounts of oleic and linoleic acids as HOSO, and low quantities of linolenic acid as VOO, HV and HR (Guillén et al., 2003; Jabeur et al., 2014; Jović et al., 2016; Ranade et al., 2015).

Apropos of the most important variables in the models to discriminate blends with low percentages of HV in VOO from other VO blends with VOO, the signals of the allylic (#12a) and methyl (#7b) protons of oleic acid showed higher intensities in the blends with 5–20% HV than in those with other VOs, partially overlapping with the non-HV category. In contrast, the signal intensities of the methyl protons of squalene (#11), linolenic acid (#7d) and other unsaturated fatty acids at 1.00–1.02 ppm (#7), the vinylic protons of unsaturated fatty acids (#24), the allylic protons of linolenic acid (#12c) and the methylene protons at 1.38–1.40 ppm of unsaturated fatty acids (#9) in the HV-VOO blends were lower but overlapping with about 75% of the non-HV category range. In fact, HV is characterised by similar or slightly higher contents of oleic and linoleic acids and a lower content of linolenic acid respect to VOO (Christopoulou et al., 2004; Guillén et al., 2003; Vigli et al., 2003).

For the distinction of binary mixtures of VOO and HR from other VO-VOO mixtures in low proportions, the allylic protons of oleic and linolenic acids (#12a and c), the methyl protons of saturated fatty acids (#7a) and linolenic acid (#7d), and the methylene proton of terpenic alcohols or sterols (#2) exhibited lower signal intensities in blends of 5-20% HR in VOO than in the other VO-VOO blends, but partially overlapped with non-HR category. Regarding the other influential variables on this discrimination, i.e., ^{1}H signals of the acyl groups of linoleic (#13c, #15a) and other unsaturated fatty acids (#15, #24, #9 and #10) and the glyceryl group of triacylglycerides (#18), signal intensities presented less variability in the HR category, and intensity ranges in HR-VOO blends overlapping with 2nd and 3rd quartiles of the non-HR category. HR contains slightly lower amounts of oleic acid, saturated fatty acids and linolenic acid but higher amounts of linoleic acid than VOO (Guillén et al., 2003; Vigli et al., 2003). Moreover, VOO and the different VOs blended with VOO are characterised by different triacylglyceride profiles (Gómez-

Ariza, Arias-Borrego, García-Barrera, & Beltran, 2006; Parcerisa, Casals, Boatella, Codony, & Rafecas, 2000), which show characteristic resonances in the ^{1}H-NMR spectra.

The most discriminant variables in the PLS-DA models to detect the presence of soybean oil in VOO in low proportions were the characteristic ^{1}H signal of linolenic acid (#15b, #7d and #12c) and the vinylic protons of unsaturated fatty acids (#24), which presented significant higher intensities in S-VOO blends than in the other VO blends with VOO. Soybean oil is that with the highest contents of linolenic acid among the studied VOs (Contiñas et al., 2008; Christopoulou et al., 2004; Guillén et al., 2003; Jabeur et al., 2014; Vigli et al., 2003). The signal intensity of the glyceryl protons of *sn*-1,2-diacylglycerides (#16) was also slightly higher in the blends containing S, which can be explained by the fact that among the VOs studied, soybean oil is the one that contains the lowest amounts of *sn*-1,2-diacylglycerides, whereas VOO, the highest amounts (Vigli et al., 2003), and that in present study only samples with 2–5% S in VOO were available. Therefore, comparing VO-VOO blends with the same percentage of VO, the VOO blends with S presented the lowest contents of *sn*-1,2-diacylglycerides. As well, the signal intensities of the methyl protons oleic acid (#7b) and the α-methylene protons of linoleic acid (#13c), which were lower in the S category, overlapping mainly with the 1st and 2nd quartiles of the non-S category, contributed to the discrimination of both categories. This observation was explained by the fact that soybean oil presents significantly lower contents of oleic acid than VOO, and similar contents of linoleic acid as other VOs, such as sunflower oils (Guillén et al., 2003; Jović et al., 2016; Vigli et al., 2003).

When an oil sample is classified in the high category (25–90% VO in VOO) following the decision tree scheme (Figure 1), its ^{1}H-NMR fingerprint is submitted to the PLS-DA models developed for mixtures of 20–90% VO in VOO for each type of VO (PLS-DA models 25–28 in Table 6) in order to identify the VO contained in the blend. In the present study, only binary mixtures of NTSO, HOSO, EVAO or HV with VOO were available in this % VO range. The recognition and prediction abilities of the binary classification models built to determine whether the VOO blend contained NTSO, HV or EVAO were 99–100% for both categories, and 100% for the non-HOSO blends and 92% for the HOSO blends.

Table 6. PLS-DA models to detect the presence of a certain vegetable oil in a binary mixture of 20–90% vegetable oil in virgin olive oil[1]

PLS-DA model	Data	PLS-comp	Boundary	Class[2]	Class code	n	p	%R	%P
25	20-90% non-NTSO in VOO	4	0.4955	non-NTSO	0	73	0.47	100	100
	20-90% NTSO in VOO			NTSO	1	83	0.53	100	100
26	20-90% non-HOSO in VOO	4	0.4120	non-HOSO	0	130	0.83	100	100
	20-90% HOSO in VOO			HOSO	1	26	0.17	92	92
27	20-90% non-EVAO in VOO	4	0.3985	non-EVAO	0	131	0.84	100	99
	20-90% EVAO in VOO			EVAO	1	25	0.16	100	100
28	20-90% non-HV in VOO	3	0.3563	non-HV	0	134	0.86	100	100
	20-90% HV in VOO			HV	1	22	0.14	100	100

[1] See abbreviations in Table 2.

[2] Samples contained in each class: non-NTSO, blends of VOO with 20–80% VOs (HOSO, EVAO or HV); NTSO, blends of VOO with 20–90% NTSO; non-HOSO, blends of VOO with 20–90% VOs (NTSO, EVAO or HV); HOSO, blends of VOO with 20–80% HOSO; non-EVAO, blends of VOO with 20–90% VOs (NTSO, HOSO or HV); EVAO, blends of VOO with 20–80% EVAO; non-HV, blends of VOO with 20–90% VOs (NTSO, HOSO or EVAO); HV, blends of VOO with 20–80% HV.

The ^{1}H signal of the methylene protons of the acyl group of oleic acid (#9b) was completely discriminant between mixtures of VOOs with NTSO in high proportions from those with other VOs. The blends of 20–90% NTSO in VOO contained significantly lower amounts of oleic acid than those VOO blends with 20–90% HOSO, EVAO or HV. It is well-documented that virgin hazelnut oil, high oleic sunflower oil and virgin avocado oil present significant higher contents of oleic acid than sunflower oil (Contiñas et al., 2008; Guillén et al., 2003; Jabeur et al., 2014; Jović et al., 2016; Ranade et al., 2015; Vigli et al., 2003; Yang et al., 2013). Other ^{1}H signals that are significantly different between both categories but not completely discriminant were the α-methylene (#13d) and β-methylene (#10c) protons of linolenic acid, and the allylic protons of oleic acid (#12a), which presented lower intensities in the NTSO-VOO blends, as expected according to the contents of these compounds in these oils reported in the aforementioned literature. In contrast, sunflower oils contains the highest amounts of linoleic acid of the VOs studied, which explained that the allylic (#12b), α-methylene (#13c), methylene (#9c) and methyl (#7c) protons of the acyl group of linoleic acid showed higher intensities in the NTSO category. This trend was also observed for other important signal, such as the glyceryl protons of triacylglycerides (#18). Indeed, VOO, NTSO and the different VOs studied present characteristic profiles of triacylglicerides (Andrikopoulos, Giannakis, & Tzamtzis, 2001; Gómez-Ariza et al., 2006), which is reflected in their NMR spectra and can contribute to the differentiation between them and their mixtures. As a result, NTSO presents a characteristic composition of fatty acids and triacylglycerides that allows the complete discrimination of NTSO from HOSO, EVAO and HV, and their binary mixtures with VOO at proportions higher than 20% VO in the blend.

With regard to the model for discriminating blends of high proportions of HOSO in VOO, several influential ^{1}H signals were disclosed. The signal intensities of the α-methylene (#13d) and allylic (#12c) protons of linolenic acid, and the vinylic protons at 5.30–5.32 ppm of unsaturated fatty acids (#24) were lower in the mixtures of HOSO and VOO; whereas those of the α-methylene (#13c), allylic (#12b), methylene (#9c) of linoleic acid, and the vinylic protons at 5.32–5.34 ppm of unsaturated fatty acids (#24) were higher in HOSO-VOO mixtures. Other important variables in the model were the methyl protons of oleic acid (#7b); HOSO-VOO blends exhibited signal intensity ranges overlapping the 1st and 2nd quartiles of the non-HOSO category. The trends observed for these ^{1}H signals agree with the chemical data of the oils studied already published (Guillén et al., 2003; Jović et al., 2016; Ranade et al., 2015): HOSO presents higher concentrations of linoleic acid than VOO, HV and EVAO and lower than NTSO; HOSO contains lower amounts of linolenic acid than NTSO, VOO and EVAO, and similar to HV; and HOSO has lower concentrations of oleic acid than HV and VOO, similar to EVAO, and higher than NTSO.

The optimised PLS-DA model for the identification of VOO blends containing EVAO in high proportions presented high B regression coefficients (in absolute value) for some of the ^{1}H signals underlined in blends with low proportions of EVAO, and followed the same trends as well. On the one hand, the intensities of the vinylic protons (#24) and the methylene protons at 1.32–1.34 ppm (#9) of unsaturated fatty acids and the allylic protons of linolenic acid (#12c) were lower in the EVAO category. On the other hand, the allylic (#12b), α-methylene (#13c) and β-methylene protons (#10c) protons of linoleic acid exhibited higher intensities in the EVAO category.

However, the most influential variables were the signals of the methylene protons of oleic acid (#9b) and saturated fatty acids (#9a), which presented higher intensities in the EVAO category, overlapping the 3rd and 4th quartiles of the non-EVAO category; as well as the methylene protons of linolenic acid (#9c), with lower intensities in the EVAO category, overlapping with the 2nd and 3rd quartiles of the non-EVAO category. These observations agreed with the data of the fatty acid composition of these oils previously published (Guillén et al., 2003; Jović et al., 2016; Ranade et al., 2015).

The most important variables that contribute to distinguish blends with high proportions of HV in VOO were the methyl (#7b), methylene (#9b) and allylic (#12a) protons of oleic acid, whose intensities were significantly higher in the HV category, partially overlapping with the 4th quartile of the non-HV category. Indeed, HV presents similar or slightly higher contents of oleic acid than VOO, and considerably higher amounts respect to the other VOs studied (Guillén et al., 2003). The opposite trend was shown by the signals of the methyl protons of linoleic acid (#7c) and the allylic protons of linolenic acid (#12c), which displayed lower intensity values and variability in the HV blends than in the non-HV category. Certainly, the concentrations of linoleic acid in HV are lower than in the other VOs and slightly higher than in VOO, whereas linolenic acid is present in HV in similar amounts than in HOSO, and in lower amounts than in NTSO, VOO and EVAO (Christopoulou et al., 2004; Jović et al., 2016; Vigli et al., 2003).

PLS-DA Models to Discriminate Blends of Olive Oil with Vegetable Oils

The procedure to distinguish OO blends is similar to the one described for VOO blends in previous section. In contrast with the VOO classification models obtained, satisfactory PLS-DA models for all the VOs (RPOO, CO, HOSO, NTSO, DOSO, RAO and HR) were achieved using the whole percentage range of VO in the OO mixture, i.e., 0–80% VO in OO (PLS-DA models 30–36 in Table 7). Prediction abilities of the binary classification models developed to discriminate between OO blends with RPOO, CO or HOSO were 95–100% for both categories; with NTSO, DOSO or RAO, 84–87% for the OO blends with the specific VO, and 91-97% for the OO blends that did not contain the specific VO; and with HR, 97% for HR-OO category, and 89% for the non-HR category.

Table 7. PLS-DA models to detect the presence of a certain vegetable oil in a binary mixture of 2–80% vegetable oil in olive oil[1]

PLS-DA model	Data	PLS-comp	Boundary	Class[2]	Class code	n	p	%R	%P
30	0-80% VOs in OO	2	0.1815	non-RPOO	0	315	0.88	100	100
				RPOO	1	41	0.12	95	95
31	0-80% VOs in OO	7	0.3545	non-CO	0	310	0.87	96	95
				CO	1	46	0.13	100	100
32	0-80% VOs in OO	7	0.3662	non-HOSO	0	319	0.90	98	97
				HOSO	1	37	0.10	95	95
33	0-80% VOs in OO	12	0.2809	non-NTSO	0	268	0.75	98	97
				NTSO	1	88	0.25	85	85
34	0-80% VOs in OO	5	0.1652	non-DOSO	0	319	0.90	91	91
				DOSO	1	37	0.10	84	84
35	0-80% VOs in OO	11	0.2354	non-RAO	0	318	0.89	96	92
				RAO	1	38	0.11	95	87
36	0-80% VOs in OO	15	0.2270	non-HR	0	319	0.90	93	89
				HR	1	37	0.10	100	97

[1] See abbreviations in Table 3; OO, olive oil; DOSO, desterolized and deodorized high oleic sunflower oil; RAO, refined avocado oil; RPOO, refined palm olein oil; CO, refined corn oil.

[2] Samples contained in each class: non-RPOO, pure OOs and blends of OO with 2–80% VOs (CO, HOSO, NTSO, DOSO, RAO or HR); RPOO, blends of OO with 2–80% RPOO; non-CO, pure OOs and blends of OO with 2–80% VOs (RPOO, HOSO, NTSO, DOSO, RAO or HR); CO, blends of OO with 2–80% CO; non-HOSO, pure OOs and blends of OO with 2–80% VOs (RPOO, CO, NTSO, DOSO, RAO or HR); HOSO, blends of OO with 2–80% HOSO; non-NTSO, pure OOs and blends of OO with 2–80% VOs (RPOO, CO, HOSO, DOSO, RAO or HR); NTSO, blends of OO with 2–80% NTSO; non-DOSO, pure OOs and blends of OO with 2–80% VOs (RPOO, CO, HOSO, NTSO, RAO or HR); DOSO, blends of OO with 2–80% DOSO; non-RAO, pure OOs and blends of OO with 2–80% VOs (RPOO, CO, HOSO, NTSO, DOSO or HR); RAO, blends of OO with 2–80% RAO; non-HR, pure OOs and blends of OO with 2–80% VOs (RPOO, CO, HOSO, NTSO, DOSO or RAO); HR, blends of OO with 2–80% HR.

The main ^{1}H signals responsible for the identification of OO blends containing RPOO were the methylene protons of saturated fatty acids (#9a), which presented significantly higher intensities in RPOO-OO blends, even though were not completely discriminant; whereas the methylene (#9c) and allylic (#12b) protons of linoleic acid showed lower intensities in the RPOO category. Palm oil is, among the VOs studied, the one that contains the highest amounts of saturated fatty acids (Vigli et al., 2003). Indeed, palmitic acid is the major saturated fatty acid in palm oil, being present in similar amounts as oleic acid. Meanwhile, linoleic acid is a minor compound in palm oil, present in similar concentrations as in OO, and in lower amounts than in the other VOs

(Mancini, Imperlini, Nigro, Montagnese, Daniele, Orrù, et al., 2015; Montoya, Cochard, Flori, Cros, Lopes, Cuellar, et al., 2014).

The blends of CO in OO were distinguished from the other VOs in OO as a result of its characteristic profiles of fatty acids and triacylglycerides (Christopoulou et al., 2004; Gómez-Ariza et al., 2006; Jabeur et al., 2014; Yang et al., 2013). The intensity of the most influential signals, i.e., the methyl protons of the acyl groups of linoleic acid (#7c) and saturated fatty acids (#7a), the bis-allylic protons of linolenic acid (#15b) and the glyceryl protons at 4.30–4.32 ppm of triacylglycerides (#18), were higher in the blends containing CO; while the glyceryl protons at 4.28–4.30 ppm of triacylglycerides (#18) showed the opposite trend. In fact, corn oil present similar linoleic contents to sunflower oil, and significantly higher ones than refined avocado, refined hazelnut, palm and olive oil; slightly higher amounts of linolenic acid than the other oils studied; and a level of saturated fatty acids lower than palm oil but similar or slightly higher than the rest of the oils considered in the model (Guillén et al., 2003; Jiang et al., 2018; Monfreda et al., 2012; Vigli et al., 2003).

The major contribution to the binary classification models built to determine the presence of HOSO in OO was due to the methylene protons oleic acid (#9b), the allylic protons of linoleic acid (#12b) and the vinylic protons of unsaturated fatty acids (#24), which exhibited higher intensities in the blends with HOSO, partially overlapping with the non-HOSO category. These observations were in accordance with the fact that HOSO contains higher amounts of oleic acid than sunflower, corn and palm oils, similar to avocado oil, and lower than hazelnut and olive oils. Linoleic acid is present in larger amounts in HOSO than in palm, olive, hazelnut and avocado oils and lower than in sunflower and corn oils. The allylic protons of linolenic acid (#12c) of HOSO-OO blends displayed intensities overlapping with the 1st and 2nd quartiles of the non-HOSO category. Actually, linolenic contents of HOSO are similar to those of HR and slightly lower than the other studied oils. Furthermore, the methylene protons of saturated fatty acids (#9a) were also influent in the model, displaying intermediate-high values for HOSO-OO blends overlapping the 2nd, 3rd and 4th quartiles of the non-HOSO category, but far from RPOO-OO blends, which exhibit the largest contents. These observations are supported by previously descriptions of the composition of the pure oils (Hilary S. Green & Wang, 2020; Guillén et al., 2003; Jović et al., 2016; Vigli et al., 2003).

Among the most influential variables in the binary classification models achieved for the detection of NTSO in OO, the methyl (#7c) and bis-allylic

(#15a) protons of linoleic acid, and the vinylic protons (#24), the methyl protons at 1.00–1.02 ppm (#7) and the methylene protons at 1.32–1.34 ppm (#9) of unsaturated fatty acids, displayed higher intensities in the OO blends with NTSO. The opposite trends were observed for the allylic (#12a) and methyl (#7b) protons of oleic acid. This behaviour agreed with composition of sunflower oil, which is characterised by the largest contents of linoleic acid and unsaturated fatty acids, and the lowest contents of oleic acid respect to the other oils studied (Christopoulou et al., 2004; Guillén et al., 2003; Jabeur et al., 2014; Jović et al., 2016; Monfreda et al., 2012; Orsavova, Misurcova, Ambrozova, Vicha, & Mlcek, 2015; Yang et al., 2013).

The PLS-DA model built to distinguish OO blends containing DOSO presented the highest absolute B regression coefficients for the allylic protons of oleic (#12a) and linoleic (#12b) acids and the methylene protons of oleic acid (#9b), whose intensities were higher in the DOSO category. During the desterolization process, the dehydration of sterols and the elimination of the acid group of sterol esters take place by bleaching, producing olefinic degradation products and di-steryl ethers; however the profiles of triacylglycerides and fatty acids are practically unaltered (Grob, Biedermann, Bronz, & Giuffré, 1994). Instead, the deodorization process may affect the composition of triglycerides, diglycerides, fatty acids and minor components of the unsaponifiable fraction, depending mainly on the temperature and time of the process (Aparicio et al., 2013). Thus, DOSO is characterized by relatively high contents of oleic and linoleic acids as HOSO. Nevertheless, the intensities of the signal due to oleic acid (#12a and #9b) in DOSO-OO blends presented a large variability and were significantly higher than in the non-DOSO category, even though not completely discriminant between both categories. In contrast, linoleic acid signal (#12b) intensities of DOSO-OO blends exhibited less variability and overlapped with the 1st and 2nd quartiles of the non-DOSO category. The methyl protons of linoleic acid (#7c) in DOSO-OO blends exhibited a narrow range of intensities close to the 2nd quartiles of the non-DOSO category. The vinylic protons of unsaturated fatty acids (#24) displayed lower intensities and high variability in the blends with DOSO overlapping the 1st, 2nd and 3rd quartiles of the non-DOSO category.

Refined avocado oil presents intermediate compositions regarding its fatty acid profiles compared to the other oils studied (Guillén et al., 2003; Jabeur et al., 2014; Jović et al., 2016; Vigli et al., 2003; Yang et al., 2013), as well as its sterol contents, in particular β-sitosterol (Al-Ismail et al., 2010; Fernandes, Gómez-Coca, Pérez-Camino, Moreda, & Barrera-Arellano, 2017; Hilary S. Green et al., 2020; Karabulut, Topcu, Yorulmaz, Tekin, & Ozay, 2005;

Parcerisa et al., 2000). In fact, the intermediate composition of RAO was reflected in the NMR spectra, and in particular in the most important variables in the binary classification model to detect RAO in OO. Thus, the methyl (#7c), allylic (#12b) and α-methylene (#13c) protons of linoleic acid and the methyl protons of β-sitosterol (#4) exhibited signals for the RAO-OO blends with intensity ranges overlapping with 2nd, 3rd and 4th quartiles of the non-RAO category. Whereas, the α-methylene protons of linolenic acid (#13d) and the methylene protons at 1.32–1.34 ppm of unsaturated fatty acids (#9) of RAO-OO blends presented intensity ranges overlapping with 1st, 2nd and 3rd quartiles of the non-RAO category, and a median intensity value lower for RAO-OO mixtures than for the other types of mixtures. The signal intensities of the methylene protons at 1.20–1.22 ppm of the acyl group of saturated fatty acids (#9) in RAO-OO blend overlapped the 2nd and 3rd quartiles of the non-RAO category.

The most contributing variables for the identification of HR in OO were the methyl (#7b) and allylic (#12a) protons of oleic acid, presenting higher intensities in the HR-OO blends; and the methyl (#7d), bis-allylic (#15b) and allylic (#12c) protons of linolenic acid, the vinylic protons of unsaturated fatty acids (#24), the β-methylene (#10a) and methyl (#7a) protons of saturated fatty acids and the methyl protons of terpenic alcohols or sterols (#2), showing lower intensities in the mixtures of HR with OO than in the other VO-OO blends. Refined hazelnut oil contains the highest amounts of oleic acid among the VOs studied, and comparable contents to those of olive oil. HR presents the lowest linolenic acid contents similar to HOSO, therefore the linolenic signal intensities overlapped with 1st and 2nd quartiles of the non-HR category (Hilary S. Green et al., 2020; Guillén et al., 2003; Jović et al., 2016; Parcerisa et al., 2000; Vigli et al., 2003). As well, each oil type presents characteristic profiles of sterols and terpenic alcohols (Al-Ismail et al., 2010; Aparicio et al., 2013; Fernandes et al., 2017; Karabulut et al., 2005; Parcerisa et al., 2000), which in the present model contributed to the distinction of OO blends with and without HR.

Similar to the procedure described for VOO blends, once the oil sample is classified as containing OO in the first stage of the decision tree scheme (Figure 1), a PLS-DA model classifies the sample according to their level of VO in the OO, i.e., low (OOs with 0–20% VOs) and high (OOs with 30–80% VOs); the prediction abilities being of 96% and 94% for the low and high categories respectively (PLS-DA model 29 in Table 2). The most influential variables on the model were the methyl protons of saturated fatty acids (07a) and β-sitosterol (#4); the allylic (#12b), bis-allylic (#15a) and α-methylene (#13c) of linoleic acid; and the vinylic protons (#24), the methyl protons at 1.00–1.02 ppm (#7), and the methylene protons at 1.32–1.33 ppm (#9) of unsaturated fatty acids, which exhibited lower intensities in the low category, overlapping with the 1st and 2nd quartiles of the high category. In contrast, the ^{1}H signals of linolenic acid (#7d and #15b) and oleic acid (#12a) displayed higher intensities in the low category, overlapping with the 3rd and 4th quartiles of the high category. The chemical composition of the blends that constituted each category justified these observations, since the low category contained samples with highest proportions of OO, which is the oil that contains the highest concentrations of oleic acid together with HR; whereas in the high category, blends with high percentages of VOs characterised by high linoleic and β-sitosterol contents were included (Al-Ismail et al., 2010; Aparicio et al., 2013; Fernandes et al., 2017; Hilary S. Green et al., 2020; Guillén et al., 2003; Jović et al., 2016; Karabulut et al., 2005; Parcerisa et al., 2000; Vigli et al., 2003).

No models to discriminate pure OOs from blends of VO with OOs were built due to the low number of pure OO samples available for modelling. In next stage, an oil sample classified in the low category (oils with 0-20% VOs in OO) according to PLS-DA model 29 (Table 2) is predicted using individual VO binary classification models in order to identify whether the oil contains or not a specific VO (PLS-DA models 37–43 in Table 8). Besides, PLS-DA models were built using a data matrix without pure OO samples (PLS-DA models 44–50 in Table 9). The recognition and prediction abilities achieved by the models developed with and without pure OO samples were similar, being higher than 95% of hits in the models for detecting RPOO, CO and HOSO in OO blends; c.a. 90% for NTSO, DOSO and HR in OO blends; and c.a. 80–85% for RAO in OO blends.

The most influential variables in the binary classification models for OO blends with 0–20% RPOO, 0–20% NTSO, 0–20% DOSO and 0–20% HR (except for the saturated fatty acid signals) were the same as in the corresponding models built for the full percentage range of VO in OO. In the case of the models for mixtures of CO and OO, the methyl protons of the acyl group of linoleic acid (#7c) was the main responsible for the distinction of OO mixtures containing CO from those without CO. The identification of low proportions of HOSO in OO were mainly due to the higher intensity of the ^{1}H signals of the methylene protons of saturated fatty acids (#9a) and unsaturated fatty acids with signals at 1.30–1.34 ppm (#9), which presented higher intensities in the HOSO category. Referring to the models for distinguishing low percentages of RAO in OO, the most influent signals were those due to the polyunsaturated fatty acids. Thus, the intensities of the methyl protons of linoleic acid (#7c) and the β-methylene protons of of linoleic acid (#10c) displayed intensities with similar median values in both categories but different variabilities. Besides, the signal intensities of the methylene protons of linolenic acid (#9c) and the methylene protons at 1.31–1.34 ppm (#9) and vinylic protons (#24) of unsaturated fatty acids were lower in the OO blends containing RAO, overlapping with the 1st and 2nd quartiles of the non-RAO category.

Table 8. PLS-DA models to detect the presence of a certain vegetable oil in a binary mixture of 2–20% vegetable oil in olive oil[1]

PLS-DA model	Data	PLS-comp	Boundary	Class[2]	Class code	n	p	%R	%P
37	0-20% VOs in OO	2	0.2399	non-RPOO	0	162	0.89	98	98
				RPOO	1	21	0.11	95	95
38	0-20% VOs in OO	12	0.3522	non-CO	0	164	0.89	97	95
				CO	1	20	0.11	100	100
39	0-20% VOs in OO	4	0.3039	non-HOSO	0	172	0.93	96	96
				HOSO	1	12	0.07	100	100
40	0-20% VOs in OO	11	0.2770	non-NTSO	0	143	0.79	93	90
				NTSO	1	38	0.21	97	89
41	0-20% VOs in OO	8	0.1904	non-DOSO	0	164	0.89	88	89
				DOSO	1	20	0.11	95	90
42	0-20% VOs in OO	7	0.2110	non-RAO	0	163	0.89	82	80
				RAO	1	21	0.11	90	81
43	0-20% VOs in OO	14	0.2809	non-HR	0	162	0.90	94	90
				HR	1	19	0.10	95	95

1 See abbreviations in Tables 3 and 7.

2 Samples contained in each class: non-RPOO, pure OOs and blends of OO with 2–20% VOs (CO, HOSO, NTSO, DOSO, RAO or HR); RPOO, blends of OO with 2–20% RPOO; non-CO, pure OOs and blends of OO with 2–20% VOs (RPOO, HOSO, NTSO, DOSO, RAO or HR); CO, blends of OO with 2–20% CO; non-HOSO, pure OOs and blends of OO with 2–20% VOs (RPOO, CO, NTSO, DOSO, RAO or HR); HOSO, blends of OO with 2–20% HOSO; non-NTSO, pure OOs and blends of OO with 2–20% VOs (RPOO, CO, HOSO, DOSO, RAO or HR); NTSO, blends of OO with 2–20% NTSO; non-DOSO, pure OOs and blends of OO with 2–20% VOs (RPOO, CO, HOSO, NTSO, RAO or HR); DOSO, blends of OO with 2–20% DOSO; non-RAO, pure OOs and blends of OO with 2–20% VOs (RPOO, CO, HOSO, NTSO, DOSO or HR); RAO, blends of OO with 2–20% RAO; non-HR, pure OOs and blends of OO with 2–20% VOs (RPOO, CO, HOSO, NTSO, DOSO or RAO); HR, blends of OO with 2–20% HR.

Table 9. PLS-DA models to detect the presence of a certain vegetable oil in a binary mixture of 2–20% vegetable oil in olive oil[1]

PLS-DA model	Data	PLS-comp	Boundary	Class[2]	Class code	n	p	%R	%P
44	2-20% VOs in OO	2	0.2604	non-RPOO	0	130	0.86	98	97
				RPOO	1	21	0.14	95	95
45	2-20% VOs in OO	7	0.3987	non-CO	0	132	0.87	96	96
				CO	1	20	0.13	100	100
46	2-20% VOs in OO	3	0.3359	non-HOSO	0	140	0.92	98	98
				HOSO	1	12	0.08	100	100
47	2-20% VOs in OO	12	0.3176	non-NTSO	0	114	0.75	96	89
				NTSO	1	38	0.25	97	89
48	2-20% VOs in OO	8	0.2189	non-DOSO	0	131	0.87	92	85
				DOSO	1	20	0.13	95	95
49	2-20% VOs in OO	6	0.2633	non-RAO	0	131	0.86	83	82
				RAO	1	21	0.14	90	90
50	2-20% VOs in OO	14	0.3408	non-HR	0	131	0.87	97	92
				HR	1	19	0.13	100	95

[1] See abbreviations in Table 2.

[2] Samples contained in each class: non-RPOO, blends of OO with 2–20% VOs (CO, HOSO, NTSO, DOSO, RAO or HR); RPOO, blends of OO with 2–20% RPOO; non-CO, blends of OO with 2–20% VOs (RPOO, HOSO, NTSO, DOSO, RAO or HR); CO, blends of OO with 2–20% CO; non-HOSO, blends of OO with 2–20% VOs (RPOO, CO, NTSO, DOSO, RAO or HR); HOSO, blends of OO with 2–20% HOSO; non-NTSO, blends of OO with 2–20% VOs (RPOO, CO, HOSO, DOSO, RAO or HR); NTSO, blends of OO with 2–20% NTSO; non-DOSO, blends of OO with 2–20% VOs (RPOO, CO, HOSO, NTSO, RAO or HR); DOSO, blends of OO with 2–20% DOSO; non-RAO, blends of OO with 2–20% VOs (RPOO, CO, HOSO, NTSO, DOSO or HR); RAO, blends of OO with 2–20% RAO; non-HR, blends of OO with 2–20% VOs (RPOO, CO, HOSO, NTSO, DOSO or RAO); HR, blends of OO with 2–20% HR.

Taking into account that with the above models, all CO-OO blends and 95% of the RPOO-OO blends were identified, as well as at least 95% of the OO blends not containing CO or RPOO, further classification models were built with datasets without the ^{1}H-NMR spectral data of RPOO-OO and CO-OO mixtures. The PLS-DA models achieved (PLS-DA models 51–55 in Table 10) afforded better classification abilities to detect NTSO and RAO in OO-blends, and similar results to resolve the presence of HOSO, DOSO or HR in OO-blends. Hence, once it is discarded that the OO sample contains RPOO or CO using the PLS-DA models 37, 38, 44 and 45 (Tables 8 and 9), the presence of HOSO, NTSO, DOSO, RAO or HR in an OO blend is predicted by PLS-DA model 51–55. These classification models support previous models for OO blends with HOSO, DOSO or HR, and confirm the classifications achieved. Regarding the model for the determination of low proportions of

NTSO in OO (PLS-DA model 52), a part from the ^{1}H signals that were important in previous models (PLS-DA models 33, 40 and 47), the methylene protons of oleic acid (#9b), which presented lower intensities in the NTSO category; and the glyceryl protons of triacylglycerides (#18), which showed higher intensities in the NTSO category, contributed to improve the classification abilities of previous models. Concerning the most influential variables in the models to distinguish low percentages of RAO in OO (PLS-DA model 54), besides the important variables in previous models (PLS-DA models 35, 42 and 49), the signal intensities of the methyl protons of oleic acid (#7b) and the glyceryl protons of triacylglycerides (#23) in the OO blends containing RAO were overlapping with the 1st, 2nd and 3rd quartiles of the non-RAO category; and the α-methylene of linoleic (#13c) and linolenic acids (#13d), with the 2nd, 3rd and 4th quartiles of the non-RAO category. This is in agreement with the characteristic intermediate composition of fatty acids that RAO presents, as referred above.

PLS-DA models were developed for the blends with 20–80% VOs in OO in order to identify the VO contained in the mixture (PLS-DA models 56–62 in Table 11). The recognition and prediction abilities of the binary classification models achieved for OO blends with RPOO, CO, DOSO or HR were 98–100% for both categories; with NTSO and RAO, ≥91% for both categories; and with HOSO, 86% for the HOSO category and 99% for the non-HOSO category. Since all blends were correctly classified by the RPOO-OO and CO-OO models, further PLS-DA models to detect 20–80% VO in OO were built using a dataset without the ^{1}H-NMR spectral data of RPOO and CO blends with OO (PLS-DA models 63–67 in Table 12), obtaining models with better recognition and prediction abilities except for HR-OO blend. Indeed, the models for the identification of OO mixtures with and without NTSO or HOSO allowed the correct classification of all samples of both categories; the model for OO mixtures with and without RAO identified all samples containing RAO and 92% of samples in the non-RAO category; and the model for OO mixtures with and without DOSO gave the same classification results as the model built including RPOO and CO spectral data. The model to differentiate OO mixtures with and without HR provided slightly worse classifications than previous model but over 90% of hits for both categories. In fact, the detection of the adulteration of OO with HR is one of the main challenge in fraud detection due to the close composition of both refined oils (Agiomyrgianaki et al., 2010; García-González et al., 2004; Gómez-Ariza et al., 2006; Mannina et al., 2009).

Table 10. PLS-DA models to detect the presence of a certain vegetable oil in a binary mixture of 2–20% vegetable oil in olive oil, once the presence of RPOO or CO is discarded[1]

PLS-DA model	Data	PLS-comp	Boundary	Class[2]	Class code	n	p	%R	%P
51	2-20% VOs in OO *without RPOO and CO data*	2	0.3689	non-HOSO	0	98	0.89	98	97
				HOSO	1	12	0.11	100	100
52	2-20% VOs in OO *without RPOO and CO data*	7	0.3706	non-NTSO	0	72	0.65	100	99
				NTSO	1	38	0.35	95	92
53	2-20% VOs in OO *without RPOO and CO data*	8	0.2569	non-DOSO	0	89	0.82	91	85
				DOSO	1	20	0.18	100	95
54	2-20% VOs in OO *without RPOO and CO data*	10	0.3905	non-RAO	0	87	0.81	91	87
				RAO	1	20	0.19	100	95
55	2-20% VOs in OO *without RPOO and CO data*	15	0.3948	non-HR	0	89	0.82	97	92
				HR	1	19	0.18	100	95

[1] See abbreviations in Tables 3 and 7.

[2] Samples contained in each class: non-HOSO, blends of OO with 2–20% VOs (NTSO, DOSO, RAO or HR); HOSO, blends of OO with 2–20% HOSO; non-NTSO, blends of OO with 2–20% VOs (HOSO, DOSO, RAO or HR); NTSO, blends of OO with 2–20% NTSO; non-DOSO, blends of OO with 2–20% VOs (HOSO, NTSO, RAO or HR); DOSO, blends of OO with 2–20% DOSO; non-RAO, blends of OO with 2–20% VOs (HOSO, NTSO, DOSO or HR); RAO, blends of OO with 2–20% RAO; non-HR, blends of OO with 2–20% VOs (HOSO, NTSO, DOSO or RAO); HR, blends of OO with 2–20% HR.

The PLS-DA model for OO blends with RPOO (PLS-DA model 56) disclosed that the signals of the methylene protons of saturated fatty acids (#9a), mainly due to palmitic acid, and the methylene protons of linoleic acid (#9c) were completely discriminant between blends containing ≥20% RPOO in OO and blends with the other VOs studied. As a result, the measurement of just one of these two variables would be enough to establish whether an OO is mixed with RPOO in percentages ≥20%. The trends observed for these signals were the same as in the OO blends with low proportions of VOs.

Table 11. PLS-DA models to detect the presence of a certain vegetable oil in a binary mixture of 20–80% vegetable oil in olive oil[1]

PLS-DA model	Data	PLS-comp	Boundary	Class[2]	Class code	n	p	%R	%P
56	20-80% VOs in OO	1	0.3445	non-RPOO	0	185	0.88	100	100
				RPOO	1	25	0.12	100	100
57	20-80% VOs in OO	7	0.4410	non-CO	0	178	0.85	100	100
				CO	1	31	0.15	100	100
58	20-80% VOs in OO	5	0.4063	non-HOSO	0	182	0.87	99	99
				HOSO	1	28	0.13	86	86
59	20-80% VOs in OO	6	0.3650	non-NTSO	0	151	0.72	100	99
				NTSO	1	59	0.28	93	92
60	20-80% VOs in OO	4	0.3127	non-DOSO	0	188	0.90	100	99
				DOSO	1	20	0.10	100	100
61	20-80% VOs in OO	5	0.3195	non-RAO	0	187	0.89	95	94
				RAO	1	23	0.11	91	91
62	20-80% VOs in OO	9	0.3083	non-HR	0	187	0.91	99	98
				HR	1	19	0.09	100	100

[1] See abbreviations in Table 2.

[2] Samples contained in each class: non-RPOO, blends of OO with 20–80% VOs (CO, HOSO, NTSO, DOSO, RAO or HR); RPOO, blends of OO with 20–80% RPOO; non-CO, blends of OO with 20–80% VOs (RPOO, HOSO, NTSO, DOSO, RAO or HR); CO, blends of OO with 20–80% CO; non-HOSO, blends of OO with 20–80% VOs (RPOO, CO, NTSO, DOSO, RAO or HR); HOSO, blends of OO with 20–80% HOSO; non-NTSO, blends of OO with 20–80% VOs (RPOO, CO, HOSO, DOSO, RAO or HR); NTSO, blends of OO with 20–80% NTSO; non-DOSO, blends of OO with 20–80% VOs (RPOO, CO, HOSO, NTSO, RAO or HR); DOSO, blends of OO with 20–80% DOSO; non-RAO, blends of OO with 20–80% VOs (RPOO, CO, HOSO, NTSO, DOSO or HR); RAO, blends of OO with 20–80% RAO; non-HR, blends of OO with 20–80% VOs (RPOO, CO, HOSO, NTSO, DOSO or RAO); HR, blends of OO with 20–80% HR.

With regard to the binary classification model obtained to detect OO blends with ≥20% CO (PLS-DA model 57), the most significant variables were the same as those in the full range model, and showed similar trends; even though in the model for the high range (≥20%), the methylene protons of oleic acid (#9b) and the methyl protons of linolenic acid (#7d) and β-sitosterol (#4) were also preponderant. Oleic acid signal presented lower intensities in the CO category and overlapped with the 1st and 2nd quartiles of the non-CO category, since corn oil together with sunflower oil are the oils with the lowest contents of this fatty acid. While, the signals of linolenic acid and β-sitosterol displayed higher intensities in the CO category, in agreement with the known higher amounts of both compounds in corn oil compared to the other VOs blended with OO (Aparicio et al., 2013; Guillén et al., 2003; Monfreda et al., 2012; Vigli et al., 2003).

Table 12. PLS-DA models to detect the presence of a certain vegetable oil in a binary mixture of 20–80% vegetable oil in olive oil, once the presence of RPOO or CO is discarded[1]

PLS-DA model	Data	PLS-comp	Boundary	Class[2]	Class code	n	p	%R	%P
63	20-80% VOs in OO *without RPOO and CO data*	3	0.4447	non-HOSO	0	125	0.82	100	100
				HOSO	1	27	0.18	100	100
64	20-80% VOs in OO *without RPOO and CO data*	3	0.4443	non-NTSO	0	95	0.62	100	100
				NTSO	1	59	0.38	100	100
65	20-80% VOs in OO *without RPOO and CO data*	4	0.2963	non-DOSO	0	131	0.87	99	99
				DOSO	1	20	0.13	100	100
66	20-80% VOs in OO *without RPOO and CO data*	2	0.3560	non-RAO	0	131	0.85	92	92
				RAO	1	23	0.15	100	100
67	20-80% VOs in OO *without RPOO and CO data*	8	0.2858	non-HR	0	132	0.86	97	95
				HR	1	22	0.14	91	91

[1] See abbreviations in Tables 3 and 7.

[2] Samples contained in each class: non-HOSO, blends of OO with 20–80% VOs (NTSO, DOSO, RAO or HR); HOSO, blends of OO with 20–80% HOSO; non-NTSO, blends of OO with 20–80% VOs (HOSO, DOSO, RAO or HR); NTSO, blends of OO with 20–80% NTSO; non-DOSO, blends of OO with 20–80% VOs (HOSO, NTSO, RAO or HR); DOSO, blends of OO with 20–80% DOSO; non-RAO, blends of OO with 20–80% VOs (HOSO, NTSO, DOSO or HR); RAO, blends of OO with 20–80% RAO; non-HR, blends of OO with 20–80% VOs (HOSO, NTSO, DOSO or RAO); HR, blends of OO with 20–80% HR.

The models achieved to identify the presence of HOSO in OO in high proportions (PLS-DA models 58 and 63) presented the highest absolute B regression coefficients for the same variables than in the full range model (PLS-DA model 32), and followed the same trend. In PLS-DA model 63, additional signals due to oleic acid (#12a) and linoleic acid (#9c) contributed to the discrimination. The signal due to saturated fatty acids (#9a) presented a

B regression coefficient three times higher than the rest of the signals in this model than in the full range model.

The binary classification models built to detect OO blends with ≥20% NTSO (PLS-DA model 59 and 64) presented the same influential variables as the previous NTSO-OO models described above (the full percentage and low content models), complying with the same tendencies. However, once the presence of RPOO and CO in the OO blend was discarded by the PLS-DA models 56 and 57 respectively, the signals of the bis-allylic (#15a) and allylic (#12b) protons of linoleic acid, the methylene protons at 1.34–1.36 ppm (#9) and the vinylic protons (#24) of unsaturated fatty acids, and the methylene protons of oleic acid (#9b) were completely discriminant between both categories, thus any of this variable can be used as markers to determine whether an OO blend contains NTSO at concentration ≥20%.

The most discriminant variables in the binary classification models for OO blends with ≥20% DOSO (PLS-DA model 60 and 65) were the same as in the corresponding model achieved for the full percentage range. With respect to the PLS-DA models for determination of ≥20% RAO in OO (PLS-DA models 61 and 66), less ^{1}H signals were influent in these models compared to the full percentage range and the low content models. Thus, the signals of the methylene protons of oleic acid (#9b) and the allylic protons of linoleic acid (#12b) in the RAO category overlapped with the 3rd and 4th quartiles of the non-RAO category, as a result of RAO characteristic intermediate composition of these fatty acids among the VOs studied. Meanwhile, the signal intensities of methylene protons of linolenic acids (#9c) and the vinylic protons of unsaturated fatty acids (#24) showed lower intensities in the RAO-OO blends, overlapping with the 1st quartile of the non-RAO category.

As for the full percentage range and the low range models built to detect HR in OO, the most contributing variables to PLS-DA models 62 and 67 were the allylic protons of linoleic (#12b) and oleic (#12a) acids and the methyl protons of oleic acid (#7b), but also the methylene protons of oleic acid (#9b) were preponderant; all of them presenting higher intensities in the HR-OO blends. The trends of the signal intensities of the methyl protons of saturated fatty acids (#7a) and linolenic acid (#7d), the vinylic protons of unsaturated fatty acids (#24) were similar to those observed in previous models, presenting lower intensities in the HR-OO blends. Besides, following the same trend, the allylic (#12c) and α- methylene (#13d) protons of linolenic acids were also significant for the differentiation of the OO blends with and without HR.

PLS-R Models to Determine the Percentage of Vegetable Oil in the Blend with Virgin Olive Oil or Olive Oil

After the VO contained in the blend with VOO or OO is identified using the classification models described in previous sections according to the decision tree scheme (Figure 1), PLS regression (PLS-R) models are used to determine the percentage of VO contained in the mixture with VOO or OO. Thus, PLS-R models for binary mixtures of NTSO, HOSO, EVAO, HV, HR or S with VOO (PLS-R models 1–12 in Table 13) and for binary mixtures of RPOO, CO, HOSO, NTSO, DOSO, RAO or HR with OO (PLS-R models 13–27 in Table 14) were successfully built. The full range of the % VO in the blend was divided in smaller sub-ranges and a PLS-R model was developed for each one. These regression models provided more accurate predictions than the PLS-R models constructed for the full range of % VO. The most influential variables on these models corresponded to those observed for the classification models, and therefore, were explained by the characteristic composition in fatty acids, triacylglycerides and squalene of the oils present in the blend. The coherence in the interpretation of the B regression coefficients supported the robustness of the models attained.

The regression models assembled for VOO blends yielded R^2 values >0.97, indicating excellent precisions, except for the low range models of VOO mixtures with NTSO, HOSO, HR and S. The models for VOO blends with NTSO, HOSO and S in the low percentage range presented R^2 values <0.70, indicating that the equation can only be used for screening purposes, which enables to distinguish between low, medium and high values of % VO. The model for HR-VOO blends in the low percentage range showed R^2 values <0.50, so the equation only discriminates between high and low values (Priego Capote et al., 2007); in the same way that PLS-DA model 73 distinguished 2–5% HR and 10% HR in VOO (Table 15).

The PLS-R models allowed to determine the % VO in VOO with accuracies under 5% R-RMSEP for contents of ≥10% NTSO, ≥34% EVAO, ≥39% HOSO and ≥45% HV; 5–10% R-RMSEP for contents of 13–45% HV; 5–15% R-RMSEP for contents of 8–10% NTSO, 7–34% EVAO, 20–39% HOSO and 10–26% HV; 15–20% R-RMSEP for contents of 6–8% NTSO, 5–7% EVAO, 17–20% HOSO and 5% S; and with accuracy of 28% R-RMSEP for contents of 10% HR.

Table 13. PLS-R models to determine the percentage of a certain vegetable oil in a binary mixture with virgin olive oil[1]

PLS-R model	Data[2]	n	PLS-comp	R-cal	R-val	R^2-val	RMSEP (% VO)
1	2-10% NTSO in VOO[3]	113	6	0.86	0.83	0.68	1.2
2	10-20% NTSO in VOO[3]	24	6	0.9995	0.9946	0.989	0.49
3	20-90% NTSO in VOO[3]	76	1	0.9990	0.9989	0.998	0.96
4	2-20% HOSO in VOO[3]	100	7	0.75	0.71	0.50	3.4
5	20-80% HOSO in VOO[4]	21	5	0.998	0.994	0.987	1.9
6	2-20% EVAO in VOO[4]	20	6	0.998	0.988	0.98	1.0
7	20-45% EVAO in VOO[4]	14	3	0.995	0.987	0.97	1.7
8	45-80% EVAO in VOO[4]	12	3	0.998	0.996	0.992	1.3
9	10-30% HV in VOO[4]	25	7	0.995	0.986	0.97	1.3
10	30-80% HV in VOO[4]	16	1	0.994	0.993	0.986	2.3
11	2-10% HR in VOO[3]	84	3	0.58	0.55	0.30	2.8
12	2-5% S in VOO[3]	86	9	0.87	0.78	0.61	0.95

[1] Abbreviations: n, number of samples; centered data; PLS-comp, number of PLS components; R-cal, correlation coefficient in calibration; R-val, correlation coefficient in validation; R^2-val, coefficient of determination in validation; RMSEP, root mean square error in the prediction (% VO).

[2] Samples used to build each model.

[3] 3-fold cross validation. [4] Leave-one-out cross validation.

Table 14. PLS-R models to determine the percentage of a certain vegetable oil in a binary mixture with olive oil[1]

PLS-R model	Data[2]	n	PLS-comp	R-cal	R-val	R2-val	RMSEP (% VO)
13	2-20% RPOO in OO[4]	20	4	0.9997	0.9993	0.9986	0.25
14	20-80% RPOO in OO[3]	25	1	0.9993	0.9992	0.998	0.80
15	2-10% CO in OO[4]	12	1	0.997	0.996	0.992	0.32
16	10-80% CO in OO[3]	32	1	0.99992	0.99990	0.9998	0.32
17	2-20% HOSO in OO[4]	10	2	0.994	0.983	0.97	1.0
18	10-80% HOSO in OO[3]	25	3	0.9994	0.9992	0.998	0.80
19	2-20% NTSO in OO[3]	34	4	0.9989	0.9978	0.996	0.45
20	20-80% NTSO in OO[3]	54	1	0.997	0.994	0.989	1.4
21	2-20% DOSO in OO[4]	19	6	0.998	0.994	0.987	0.78
22	20-80% DOSO in OO[4]	18	2	0.997	0.996	0.991	2.0
23	2-10% RAO in OO[4]	11	5	0.997	0.963	0.93	0.76
24	2-20% RAO in OO[4]	17	9	0.9994	0.9812	0.963	1.3
25	20-80% RAO in OO[4]	17	4	0.9991	0.9974	0.995	1.5
26	2-20% HR in OO[4]	14	3	0.9988	0.9977	0.995	0.49
27	20-80% HR in OO[3]	21	3	0.9997	0.9995	0.9990	0.64

[1] See abbreviations in Table 13.

[2] Samples used to build each model.

[3] 3-fold crossvalidation. [4] Leave-one-out crossvalidation.

All regression models disclosed for OO blends presented excellent precisions; most of the models yielded R^2 values >0.99, except for the low % range models of HOSO and RAO with R^2 values of 0.97 and 0.93–0.96 respectively. The models permitted the quantification of the % VO in OO with accuracies under 5% R-RMSEP for contents of ≥5% RPOO, ≥6% CO, ≥10% HR, ≥16% DOSO, ≥16% HOSO, ≥9% NTSO and ≥31% RAO; 5–15% R-RMSEP for contents of 2–5% RPOO, 2–6% CO, 3–10% HR, 5–16% DOSO, 7–16% HOSO, 3–9% NTSO and 5–31% RAO; and 15–20% R-RMSEP for contents of 2–3% HR, 4–5% DOSO, 5–7% HOSO, 2–3% NTSO and 4–5% RAO.

The limits of detection for a VO in VOO or OO of the present strategy based in ^{1}H-NMR fingerprinting coupled to multivariate data analysis were 2–5%. As seen in previous sections, the classification abilities of the models to identify HV, HR, HOSO and NTSO in VOO improved considerably when the 2% VO blends and pure VOO were removed from the dataset to develop the models, indicating that these samples were close to the boundary and therefore misclassified. As a result, the detection limit for the VOO blends with HV, HR, HOSO and NTSO was found to be between 2% and 5% VO in VOO. The difficulty in the detection of lower amounts than 5% of these oils in VOO lies in the fact that HV and HR present very close contents of oleic acid, saturated fatty acids and linoleic acid to VOO; HOSO and NTSO contains similar amounts of saturated fatty acids to VOO; and HOSO also has relative high concentrations of oleic acid. In the same way, the classification abilities of the PLS-DA models to identify RAO-OO blends at low RAO percentages became better when pure OO samples were excluded of the dataset in the modelling step, implying that the samples of 2% RAO in OO and pure OO were close to the boundary and hence, misclassified. Thus, the limit of detection was established as in the range between 2% and 5% RAO in OO. Indeed, RAO and OO contains similar amounts of saturated fatty acids, linoleic and linolenic acids, and RAO presents relative high contents of oleic acid respect to most of the other VOs blended with OO. The detection limits of the models to verify the presence of EVAO or S in VOO, and RPOO, CO, HOSO, NTSO, DOSO and HR in OO were under 2% VO in the blend. These results are alike or outperform those reported in literature using NMR or other analytical techniques. In previous NMR studies on VOO adulteration, the detection of 5% of hazelnut oil was possible using ^{13}C-NMR and discriminant data analysis (Zamora et al., 2001). ^{1}H and ^{31}P-NMR together with discriminant analysis allowed the detection of adulterations as low as 5% of hazelnut, corn, sunflower and soybean oils in VOO (Vigli et al., 2003). The determination of the contents of oleic, linoleic, linolenic and saturated fatty acids and squalene by ^{1}H-NMR enabled to detect 4.5% of soybean in VOO (Jiang et al., 2018). Low field NMR with different pattern recognition techniques (PLS-R, super vector machine) managed to discriminate VOO blends with corn, soybean, hazelnut, sunflower and high oleic sunflower oils at levels of 10–13% (Parker et al., 2014; Wang et al., 2020). High field ^{1}H-NMR detected the adulteration of refined hazelnut oil in olive oil at a proportion of 10% using linear discriminant analysis (Mannina et al., 2009), and even at 8% using ^{1}H and ^{13}C-NMR and artificial neural networks (García-González et al., 2004) or as low as 1% using ^{1}H and ^{31}P-NMR and canonical discriminant analysis or classification binary trees (Agiomyrgianaki et al., 2010). ^{13}C-NMR and discriminant data analysis distinguished several VOs such as sunflower, avocado, soybean and hazelnut oils at 5% in OO (Guyader et al., 2018). Other approaches that detected contents of 1–4% sunflower, soybean and corn oils in olive oil were based on voltammetric fingerprinting (Tsopelas, Konstantopoulos, & Kakoulidou, 2018), mass spectrometry fingerprinting (Sánchez-Hernández, Nozal, Marina, & Crego, 2012), fluorescence spectroscopy (Tan et al., 2018), Raman spectroscopy (Philippidis, Poulakis, Papadaki, & Velegrakis, 2017), or the determination of the composition of fatty acids, sterols, triglycerides and different chemical parameters and ratios (Contiñas et al., 2008; Christopoulou et al., 2004; Jabeur et al., 2014; Monfreda et al., 2012; Yang et al., 2013). The volatile profile analysis managed to disclose hazelnut oil in olive oil in level as low as 5% (Mildner-Szkudlarz & Jeleń, 2008). FTIR or triacylglyceride fingerprints allowed the detection of high oleic sunflower oil in levels as low as 10% (De La Mata-Espinosa et al., 2011; Jović et al., 2016). Chromatographic techniques afforded the lowest limits of detection for sunflower, soybean corn and palm oils in VOO. In fact, sunflower, soybean and corn oils were able to be detected in VOO even at 0.5% adulteration level by UHPLC-QTOF/MS (Luo et al., 2020), and at 0.5%, 0.2% and 0.1% by GC (Jabeur et al., 2016), as well as 1% palm oil in VOO by GC (Jabeur et al., 2016). The analysis of sterols and their degradation products were proposed to determine the adulteration of edible oils with desterolized sunflower oil, but the limit of detection of these approaches were not determined (Biedermann, Grob, Mariani, & Schmidt, 1996; Grob et al., 1994). To the authors' knowledge, no data was found in literature regarding the adulteration of VOO or OO with avocado oil, neither virgin nor refined.

PLS-DA Models to Discriminate Between 'Legal' and 'Illegal' Blends of Virgin Olive Oil or Olive Oil with Vegetable Oils

Taking into account the aforementioned possibilities that Reg. (EU) 29/2012 and its amendments offer to the Member States, a case study was included as an example of the potential of the present approach. Being normal type and high oleic sunflower oils the most representatives vegetable oils used for blending with olive oil, sunflower oil was considered as a model vegetable oil of 'legal' blends in a supposed situation that a Member State would allow the production and commercialization of these blends in its territory, the marketing of these blends from other countries or the production of these blends for exportation; while the blends of olive oil with the other VOs studied would not be permitted in some or all of the situations specified in the regulation. In this section, it is described the binary classification models developed to distinguish 'legal' and 'illegal' blends, and to differentiate which of the two types of sunflower oils, i.e., NTSO and HOSO, is blended with VOO or OO (Figure 2). Afterwards, the percentage of NTSO or HOSO in the blend can be determined by the PLS regression (PLS-R) models (Tables 13 and 14) in order to verify the percentage of NTSO or HOSO in the declared blend or determine the level of adulteration of the VOO or OO.

The binary classification models discriminating between 'legal' blends, i.e., VOO or OO blends containing NTSO or HOSO, and 'illegal' blends, i.e., VOO blends with EVAO, HV, HR or S, and OO blends with RPOO, CO, DOSO, RAO or HR, provided prediction abilities of 77% for both 'legal' and 'illegal' VOO blends (PLS-DA models 68 in Table 15), and 98% for 'illegal' OO blends and 86% for 'legal' OO blends (PLS-DA models 70 in Table 15). In both PLS-DA models, the bis-allylic (#15b) and methyl (#7d) protons of linolenic acid, presenting higher intensities and/or variability in the 'illegal' category, were among the most influential variables. The methyl (#7c) and bis-allylic (#15a) protons of linoleic acid and the vinylic protons of unsaturated fatty acids (#24), showing higher intensities and/or variability in the 'legal' category, were also decisive in both models. Other relevant variables in the model for OO blends were the allylic (#12a) and methyl (#7b) protons of oleic acid, which exhibit lower intensities and high variability in the 'legal' category; and the α-methylene protons of linoleic acid (#13c), the methyl protons of β-sitosterol (#4) and the methyl protons of terpenic alcohols or sterols (#2), with higher intensities in the 'legal' blends. These observations agreed with the known differences in the composition of fatty acids, sterols and terpenic alcohols of NTSO and HOSO respect to the VOs included in the 'illegal' category, VOO and OO, already mentioned above.

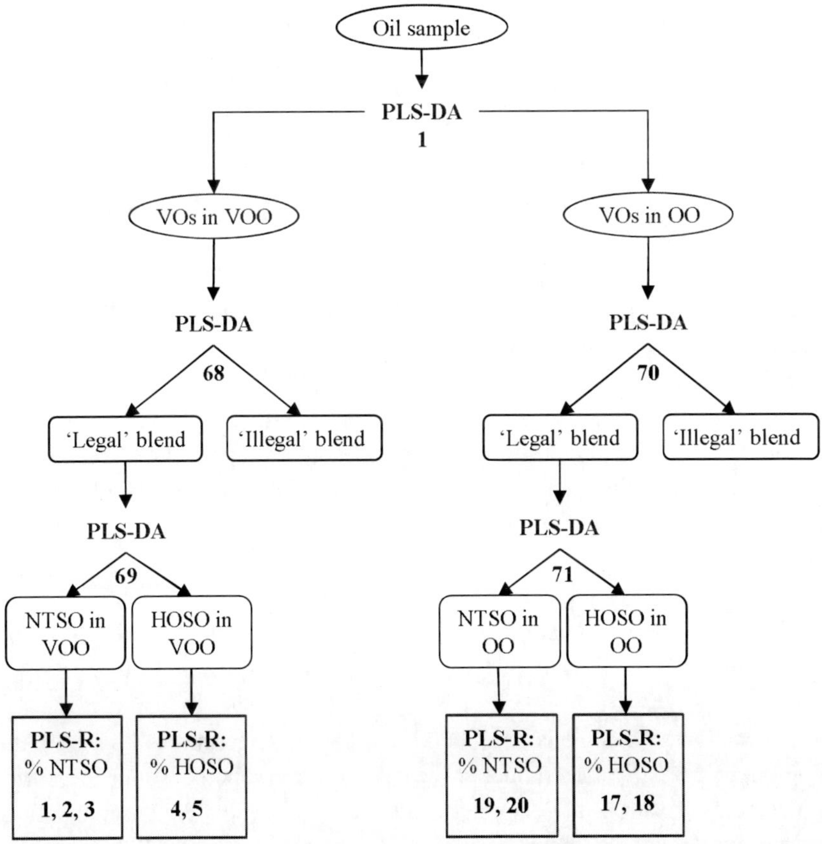

Figure 2. Decision tree constituted of PLS-DA classification and PLS-R regression models for a case-study: Discrimination between 'legal' (containing NTSO or HOSO) and 'illegal' (not containing NTSO or HOSO) blends, and determination of % NTSO or HOSO in binary mixtures with oils of the 'virgin olive oil' or 'olive oil' categories. *Abbreviations: VOO, virgin olive oil; OO, olive oil; VO, vegetable oil; NTSO, refined conventional sunflower oil (normal type sunflower oil); HOSO, refined high oleic sunflower oil.*

Moreover, binary classification models were constructed to distinguish 'legal' blends containing NTSO from those with HOSO, affording prediction abilities of 83–85% for both catergories in VOO blends (PLS-DA models 69 in Table 15) and 97% for both categories in OO blends (PLS-DA models 71 in Table 15). In both models, the signal intensities of the α-methylene protons of oleic acid (#13b) and the vinylic protons at 5.32–5.34 ppm of unsaturated fatty acids (#24) were higher in HOSO blends than in NTSO blends. In contrast, the α-methylene (#13d) and allylic (#12c) of linolenic acid, the methylene protons at 1.32–1.36 ppm of unsaturated fatty acids (#9) and the methyl protons of linoleic (#7c) exhibit higher intensities and variability in NTSO blends. In the PLS-DA model for VOO blends, the methyl (#7b) and allylic (#12a) protons of oleic acid also contributed to the discrimination between both categories, showing higher intensities for the HOSO category. In the particular case of the model for OO blends, the most discriminant variables were the methylene protons of oleic acid (#9b) and the glyceryl protons of triacylglycerides (#18), which presented higher intensities and lower variability in HOSO category. The vinylic protons at 5.30–5.32 ppm of unsaturated fatty acids (#24), which exhibited lower intensities and variabilities in HOSO category, were also of significant importance in the discrimination of both types of sunflower oils. In fact, HOSO contains higher amounts of oleic acid and lower concentrations of linoleic and linolenic acids (polyunsaturated fatty acids) than NTSO, as reported in literature (Jović et al., 2016).

PLS-DA Models to Discriminate between Blends of Virgin Olive Oil or Olive Oil with Different Compositions of Vegetable Oils

Whenever an oil sample is classified as containing a certain VO by more than one binary classification model of the decision tree scheme (Figure 1), further binary classification models can be built using datasets containing only the information related to those specific VOs. For instance, PLS-DA models were built to distinguish OO mixtures containing DOSO or HR (PLS-DA model 74 in Table 15), RAO or HR (PLS-DA model 75 in Table 15), RAO or DOSO (PLS-DA model 76 in Table 15) and DOSO or HOSO (PLS-DA model 77 in Table 15).

The prediction abilities of the model to differentiate between mixtures with DOSO and HR were 84% for the DOSO-OO blends and 95% for the HR-OO blends. The major contributing signals to this model were the allylic protons of oleic acid (#12a) and the methyl protons of saturated fatty acids (#9a) and oleic acid (#9b), which presented lower intensities in the HR-OO blends overlapping with the 1st, 2nd and 3rd quartiles of the DOSO category; and the allylic protons (#12b), bis-allylic (#15a), methyl (#7c) and methylene

(#9c) protons of linoleic acid, which showed lower intensities in the DOSO-OO blends. The allylic and bis-allylic signal intensity ranges of the DOSO category partially overlapped with the 1st and 2nd quartiles of the HR category, whereas the methyl signal of the DOSO category completely overlapped with the 1st quartile, and the methylene signal, with the 1st, 2nd and 3rd quartiles of the HR category.

Concerning the distinction of RAO-OO and HR-OO blends, the prediction abilities achieved by the corresponding model were 82% for RAO category and 84% for HR category. The most influential variables in the model were the oleic acid signals due to the methylene (#9b), methyl (#7b) and allylic (#12a) protons of oleic acid and the methylene protons of linoleic acid (#9c), which presented higher intensities in HR-OO blends, overlapping with 3rd and 4th quartiles of the RAO category. In contrast, the intensity range of the methylene protons of saturated fatty acids (#9a) in the RAO category overlapped with the 3rd and 4th quartiles of the HR category. Refined hazelnut oils are known to present higher contents of oleic acid, similar concentrations of linoleic acid and lower amounts of saturated fatty acids than refined avocado oil (Hilary S. Green et al., 2020; Parcerisa et al., 2000).

With respect to the model to differentiate RAO-OO and DOSO-OO blends, the prediction abilities afforded by the model were 95% for the RAO category and 97% for the DOSO category. Among the most dominant variables in the model, the linoleic signals due to the methyl (#7c) and allylic (#12b) protons, and the methyl proton signal of squalene (#11) displayed higher intensities in RAO-OO mixtures; whereas the oleic signals due to the allylic (#12a), methylene (#9b) and methyl (#7b) protons and the signals of the methylene (#9c) and β-methylene (#10c) protons of linolenic acid showed higher intensities in DOSO-OO mixtures.

The binary classification model for the differentiation of DOSO-OO and HOSO-OO mixtures achieved prediction abilities of 95% and 100% respectively. The main signals responsible for the discrimination were the methylene protons of saturated fatty acids (#9a), which presented similar intensity values for both categories but higher variability in the HOSO-OO blends; the allylic (#12b) and methyl (#7c) protons of linoleic acid, and the vinylic protons at 5.32–5.34 ppm of unsaturated fatty acids (#24), which exhibited higher intensities in the OO blends with HOSO; and the allylic (#12a) and methylene (#9b) protons of oleic acid, and the vinylic protons at 5.35–5.38 ppm of unsaturated fatty acids (#24), which presented higher intensities and variabilities in OO blends with DOSO.

Table 15. PLS-DA models to discriminate between 'legal' and 'illegal' blends of olive oil and vegetable oils, 'legal' blends of VOO or OO with NTSO and HOSO, VOO blends with 2% S and 5% S, VOO blends with 2–5% HR and 10% HR, OO blends with DOSO and HR, OO blends with RAO and HR, OO blends with RAO and DOSO, and OO blends of with DOSO and HOSO[1]

PLS-DA model	Data	PLS-comp	Boundary	Class[2,3,4,5]	Class code	n	p	%R	%P
68	2-90% VOs in VOO	10	0.5290	'Illegal' blend	0	302	0.44	78	77
				'Legal' blend	1	381	0.56	81	77
69	2-90% NTSO in VOO 2-80% HOSO in VOO	9	0.5543	NTSO	0	207	0.54	85	83
				HOSO	1	174	0.46	88	85
70	2-80% VOs in OO	13	0.3960	'Illegal' blend	0	199	0.61	99	98
				'Legal' blend	1	125	0.39	87	86
71	2-80% NTSO in OO 2-80% HOSO in OO	5	0.3979	NTSO	0	88	0.70	98	97
				HOSO	1	37	0.30	97	97
72	2-5% S in VOO	9	0.4643	2% S in VOO	0	44	0.50	95	93
				5% S in VOO	1	44	0.50	93	93
73	2-10% HR in VOO	6	0.4429	2–5% HR	0	59	0.66	83	80
				10% HR	1	30	0.34	80	77
74	2-80% DOSO in OO 2-80% HR in OO	3	0.4805	DOSO	0	37	0.50	86	84
				HR	1	37	0.50	97	95
75	2-80% RAO in OO 2-80% HR in OO	3	0.5011	RAO	0	38	0.51	79	82
				HR	1	37	0.49	86	84
76	2-80% RAO in OO 2-80% DOSO in OO	6	0.4723	RAO	0	38	0.51	95	95
				DOSO	1	37	0.49	100	97
77	2-80% DOSO in OO 2-80% HOSO in OO	3	0.4280	DOSO	0	37	0.50	95	95
				HOSO	1	37	0.50	100	100

[1] See abbreviations in Table 2.

[2] Samples contained in each class for PLS-DA models 68–69: 'Illegal' blend, blends of VOO with 2–80% VOs (EVAO, HV, HR or S); 'Legal' blend, blends of VOO with 2-90% VOs (NTSO or HOSO); NTSO, blends of VOO with 2–90% NTSO; HOSO, blends of VOO with 2-80% HOSO. [3] Samples contained in each class for PLS-DA models 70–71: 'Illegal' blends, blends of OO with 2–80% VOs (RPOO, CO, DOSO, RAO or HR); 'Legal' blends, blends of OO with 2–80% VOs (HOSO or NTSO); NTSO, blends of OO with 2–80% NTSO; HOSO, blends of OO with 2–80% HOSO. [4] Samples contained in each class PLS-DA models 72–73: 2% S in VOO, blends of VOO with 2% S; 5% S in VOO, blends of VOO with 5% S; 2–5% HR in VOO, blends of VOO with 2–5% HR; 10% HR in VOO, blends of VOO with 10% HR. [5] Samples contained in each class PLS-DA models 74–77: DOSO, blends of OO with 2–80% DOSO; HR, blends of OO with 2–80% HR; RAO, blends of OO with 2–80% RAO; HOSO, blends of OO with 2–80% HOSO.

Taking into account the trends observed for the most influential variables in the models achieved for the discrimination of DOSO-OO blends from other VO-OO mixtures, DOSO-OO blends presented higher concentrations of oleic acid than OO blends with HR, RAO and HOSO, which are the VOs that presented the highest contents according to literature (Hilary S. Green et al., 2020; Guillén et al., 2003; Jović et al., 2016; Parcerisa et al., 2000); and contained lower amounts of linoleic acid than the OO blends with these VOs. DOSO, which is an oil obtained from the desterolization and deodorization of high oleic sunflower oil was expected to present higher contents of linoleic acid, close to HOSO. However, DOSO blends contained even lower amounts of linoleic acid than the blends with HR and RAO. These results evidenced that during the desodorization and/or desterolization process the fatty acid profile of the oil was altered resulting in lower linoleic acid contents and higher oleic acid contents. It is known that during this raffination processes drastic conditions are used that leads to olefinic degradation of sterols, the isomerization of squalene and linoleic and linolenic acids, among other changes in the chemical composition of the oil (Aparicio et al., 2013; Grob, Artho, & Mariani, 1992; Grob et al., 1994).

For blends that contain VO at lower levels than 10% in VOO or OO, in the event that the determination of the VO content is not enough accurate by the regression models achieved (Tables 13 and 14), it could be interesting to discriminate between blends with different % VO. For instance, a PLS-DA model was developed to distinguish between VOO blends with 2% S and those with 5% S, affording 93% of correct predictions (PLS-DA model 72 in Table 15). In this model, the most influential variables were the bis-allylic (#15b) and methyl (#7d) protons of linolenic acid, which exhibited higher intensities in the blends with 5% S in VOO; and the α-methylene (#13b) and methyl (#7b) protons of oleic acid, which displayed higher intensities in the blends with 2% S in VOO. These remarks are in accordance with the relative compositions of S and VOO; the major fatty acids of these oils being linolenic and oleic acids respectively (Vigli et al., 2003).

Moreover, a binary classification model was built to discriminate between VOO blends with 2–5% HR and 10% HR (PLS-DA model 73 in Table 15), achieving predictions of 80% and 77% of hits respectively. The most significant variables in the model were the β-methylene (#10c), allylic (#12c) and bis-allylic (#15b) protons of linolenic acid and the methyl protons of squalene (#11), which presented overlapping intensity ranges for both categories but lower median of intensity in blends with 10% HR; and the methyl protons of linoleic acid (#7c), which presented a slightly higher median of intensities in blends with 10% HR. Both observations agreed with the known composition of VOO and HR, since the former contains higher amounts of squalene and linolenic acid than the latter, and the opposite occurs for linoleic acid (Guillén et al., 2003; Vigli et al., 2003; Vingering, Oseredczuk, du Chaffaut, Ireland, & Ledoux, 2010).

Prediction of Blends of Olive Oil with other Vegetable Oils

Blind oil samples provided within the OLEUM project and eight commercial oils in the Swedish market were analysed. The label of the commercial oils described that they were mixtures of VOO or OO, and other VO such as rapeseed oil, sunflower oil, or a non-identified vegetable oil. The composition of these samples was predicted by the classification and regression models developed for blends of olive oil with other vegetable oils following the decision trees shown in Figures 1 and 2. Table 16 shows: *i*) the PLS-DA models used according to the decision tree and the corresponding classification predictions; *ii*) the % VO in the blend calculated by the PLS-R model built for the blend (VO-VOO or VO-OO) previously predicted by the classification model; and *iii*) the predictions of the complementary PLS-DA models (PLS-DA models 74-77 in Table 15).

Most of the samples were predicted satisfactory according to the description provided. Indeed, the type of olive oil, i.e., VOO or OO, the particular VO and the percentage of VO in the oil sample were accurately determined. All 'legal' mixtures of VOO or OO with 40–60% NTSO or HOSO; all RPOO-OO and HV-VOO blends (5–30% VO); and EVAO-VOO and HR-OO blends with ≥10% VO, were correctly identified and the percentage of VO accurately determined; only blind samples 16, 17 and 19 were predicted to present slightly higher % VO than those given in the description, and blind sample 26, scarcely lower % HR in OO than in the description. The DOSO-OO blends were satisfactory determined by the classification and regression models achieved for these mixtures; just for sample 36 the %DOSO in OO was barely lower than predicted. However, 10% DOSO in OO was confused with mixtures of 2–11% of HOSO in OO.

Table 16. Prediction of the composition of blind oil samples using the classification and regressions models in the decision trees and the complementary PLS-DA models[1,2,3]

	PLS-DA		PLS-R					
Blind sample	Models applied	Predictions	Predicting model	Blend	% VO			Description
1	1, 2, 25-28, 68, 69	'Legal' NTSO in VOO	3	NTSO-VOO	39.6	±	1.9	*EVOO + NTSO, 60:40*
2	1, 2, 25-28, 68, 69	'Legal' NTSO in VOO	3	NTSO-VOO	50.8	±	1.9	*EVOO + NTSO, 50:50*
3	1, 2, 25-28, 68, 69	'Legal' NTSO in VOO	3	NTSO-VOO	61.4	±	1.9	*EVOO + NTSO, 40:60*
4	1, 2, 25-28, 68, 69	'Legal' HOSO in VOO	5	HOSO-VOO	40.0	±	3.9	*EVOO + HOSO, 60:40*
5	1, 2, 25-28, 68, 69	'Legal' HOSO in VOO	5	HOSO-VOO	50.1	±	3.9	*EVOO + HOSO, 50:50*
6	1, 2, 25-28, 68, 69	'Legal' HOSO in VOO	5	HOSO-VOO	60.3	±	3.9	*EVOO + HOSO, 40:60*
7	1, 30-36, 29, 56-67, 70, 71	'Legal' NTSO in OO	20	NTSO-OO	41.7	±	2.8	*OO + NTSO, 60:40*
8	1, 30-36, 29, 56-67, 70, 71	'Legal' NTSO in OO	20	NTSO-OO	51.2	±	2.8	*OO + NTSO, 50:50*
9	1, 30-36, 29, 56-67, 70, 71	'Legal' NTSO in OO	20	NTSO-OO	62.1	±	2.8	*OO + NTSO, 40:60*
10	1, 30-36, 29, 56-67, 70, 71	'Legal' HOSO in OO	18	HOSO-OO	39.9	±	1.6	*OO + HOSO, 60:40*
11	1, 30-36, 29, 56-67, 70, 71	'Legal' HOSO in OO	18	HOSO-OO	49.9	±	1.6	*OO + HOSO, 50:50*
12	1, 30-36, 29, 56-67, 70, 71	'Legal' HOSO in OO	18	HOSO-OO	60.3	±	1.6	*OO + HOSO, 40:60*
13	1, 2, 3-24, 68, 69	VOO; low; non-VO; 'illegal'	6	EVAO-VOO	6.5	±	2.1	*EVOO + EVAO, 95:5*
			4	HOSO-VOO	3.9	±	6.8	
	73	2-5% HR in VOO	11	HR-VOO	3.9	±	5.6	
14	1, 2, 3-24, 68, 69	VOO; low; EVAO; 'illegal'	6	EVAO-VOO	12.9	±	2.1	*EVOO + EVAO, 90:10*
15	1, 2, 3-24, 68, 69	VOO; low; EVAO; 'illegal'	6	EVAO-VOO	23.9	±	2.1	*EVOO + EVAO, 80:20*
16	1, 2, 25-28, 68, 69	VOO; high; EVAO; 'illegal'	7	EVAO-VOO	42.6	±	3.4	*EVOO + EVAO, 70:30*

Table 16. (Continued)

	PLS-DA		PLS-R					
Blind sample	Models applied	Predictions	Predicting model	Blend	% VO			Description
17	1, 2, 3-24, 68, 69	VOO; low; HV; 'illegal'	9	HV-VOO	9.5	±	2.6	*EVOO + HV, 95:5*
18	1, 2, 3-24, 68, 69	VOO; low; HV; 'illegal'	9	HV-VOO	10.9	±	2.6	*EVOO + HV, 90:10*
19	1, 2, 3-24, 68, 69	VOO; low; HV; 'illegal'	9	HV-VOO	26.0	±	2.6	*EVOO + HV, 80:20*
20	1, 2, 25-28, 68, 69	VOO; high; HV; 'illegal'	9	HV-VOO	27.4	±	2.6	*EVOO + HV, 70:30*
21	1, 30-36, 29, 37-67, 70, 71 76	OO; low; RAO, DOSO; 'illegal' RAO in OO	21 23	DOSO-OO RAO-OO	1.4 0.0	± ±	1.6 1.5	*OO + RAO, 95:5*
22	1, 30-36, 29, 37-67, 70, 71 76	OO; low; RAO, DOSO; 'illegal' DOSO in OO	21 23	DOSO-OO RAO-OO	4.4 9.0	± ±	1.6 1.5	*OO + RAO, 90:10*
23	1, 30-36, 29, 37-67, 70, 71 76	OO; low; RAO, DOSO; 'illegal' RAO in OO	21 24	DOSO-OO RAO-OO	13.2 22.3	± ±	1.6 2.7	*OO + RAO, 80:20*
24	1, 30-36, 29, 37-67, 70, 71 76	OO; low; RAO, DOSO; 'illegal' RAO in OO	21 24	DOSO-OO RAO-OO	19.2 22.6	± ±	1.6 2.7	*OO + RAO, 70:30*
25	1, 30-36, 29, 37-55, 70, 71	OO; low; RAO; 'illegal'	24	RAO-OO	12.7	±	2.7	*OO + HR, 95:5*
26	1, 30-36, 29, 37-67, 70, 71 75	OO; low; HR, RAO; 'illegal' HR in OO	25 26	RAO-OO HR-OO	36.2 6.4	± ±	3.1 1.0	*OO + HR, 90:10*
27	1, 30-36, 29, 37-55, 70, 71	OO; low; HR; 'illegal'	26 27	HR-OO HR-OO	15.0 20.3	± ±	1.0 1.3	*OO + HR, 80:20*
28	1, 30-36, 29, 37-55, 70, 71	OO; low; HR; 'illegal'	27	HR-OO	28.3	±	1.3	*OO + HR, 70:30*
29	1, 30-36, 29, 37-67, 70, 71	OO; low; RPOO, RAO, DOSO; 'illegal'	13	RPOO-OO	5.2	±	0.5	*OO + RPOO, 95:5*
30	1, 30-36, 29, 37-67, 70, 71	OO; low; RPOO, RAO, DOSO; 'illegal'	13	RPOO-OO	10.1	±	0.5	*OO + RPOO, 90:10*
31	1, 30-36, 29, 37-67, 70, 71	OO; low; RPOO; 'illegal'	13 14	RPOO-OO RPOO-OO	19.8 20.4	± ±	0.5 1.6	*OO + RPOO, 80:20*
32	1, 30-36, 29, 37-67, 70, 71	OO; low; RPOO; 'illegal'	14	RPOO-OO	30.7	±	1.6	*OO + RPOO, 70:30*

	PLS-DA		PLS-R					
Blind sample	Models applied	Predictions	Predicting model	Blend	% VO			Description
33	1, 30-36, 29, 37-55, 70, 71	OO; low; DOSO; 'illegal'	21	DOSO-OO	4.8	±	1.6	*OO + DOSO, 95:5*
34	1, 30-36, 29, 37-55, 70, 71	OO; low; DOSO/HOSO; legal-HOSO	17	HOSO-OO	2.0	±	2.1	*OO + DOSO, 90:10*
	77	HOSO in OO	18	HOSO-OO	11.2	±	1.6	
			21	DOSO-OO	12.4	±	1.6	
35	1, 30-36, 29, 37-55, 70, 71	OO; low; DOSO; 'illegal'	21	DOSO-OO	21.0	±	1.6	*OO + DOSO, 80:20*
			22	DOSO-OO	20.1	±	4.0	
36	1, 30-36, 29, 37-55, 70, 71	OO; low; DOSO/HR; 'illegal'	22	DOSO-OO	35.1	±	4.0	*OO + DOSO, 70:30*
	74	DOSO in OO	27	HR-OO	29.4	±	1.3	
37	1, 2, 25-28, 68, 69	VOO; high; NTSO; legal-NTSO	3	NTSO-VOO	99.4*	±	1.9	*Label: 65% NTSO + 35% EVOO*[4]
38	1, 2, 25-28, 68, 69	VOO; high; NTSO; legal-NTSO	3	NTSO-VOO	104.9*	±	1.9	*Label: Vegetable oil + VOO*[4]
39	1, 30-36, 29, 37-67, 70, 71	OO; low; CO, RAO, HR; 'illegal'	16	CO-OO	56.4	±	0.6	*Label: Rapeseed oil +*
	75	HR in OO	27	HR-OO	107.3*	±	1.3	*EVOO*[4,5]
40	1, 30-36, 29, 56-67, 70, 71	OO; high; NTSO; legal-NTSO	20	NTSO-OO	93.2*	±	2.8	*Label: 80% Rapeseed oil + 20% VOO*[4,5]
41	1, 30-36, 29, 37-67, 70, 71	OO; low; CO, RAO, HR; 'illegal'	16	CO-OO	52.0	±	0.6	*Label: 75% Rapeseed*
	75	HR in OO	27	HR-OO	106.9*	±	1.3	*oil + 25% EVOO*[4,5]
42	1, 30-36, 29, 37-67, 70, 71	OO; low; CO, RAO, HR; 'illegal'	16	CO-OO	41.6	±	0.6	*Label: 75% Rapeseed*
	75	HR in OO	27	HR-OO	95.5*	±	1.3	*oil + 25% EVOO*[4,5]

Table 16. (Continued)

	PLS-DA		PLS-R					
Blind sample	Models applied	Predictions	Predicting model	Blend	% VO			Description
43	1, 30-36, 29, 37-67, 70, 71	OO; low; CO, RAO, HR, DOSO; 'illegal'	16	CO-OO	51.2	±	0.6	*Label: 80% Rapeseed*
	75	HR in OO	27	HR-OO	106.9*	±	1.3	*oil + 20% EVOO*[4,5]
44	1, 30-36, 29, 56-67, 70, 71	OO; high; NTSO; legal-NTSO	20	NTSO-OO	93.3*	±	2.8	*Label: 80% Vegetable oil + 20% VOO*[4]

[1] See abbreviations in Tables 3 and 7.

[2] Decision trees in Figures 1 and 2.

[3] Complementary PLS-DA models: PLS-DA models 72–77 in Table 15.

[4] The label did not comply with the Reg. (EU) 29/2012 and its amendments, since the commercial blend did not contain at least 50% of olive oil, and therefore, the presence of olive oil on the label is forbidden.

[5] From the predictions achieved, it could be infer that samples (39, 41–43) did not contain NTSO or HOSO, and presented close composition to pure HR or blends of 50% CO in OO. Sample 40 was identified by all classification models as a NTSO-OO blend.

* Extrapolated results (outside the calibration range of the regression model).

The low content mixtures of 5% EVAO in VOO and 5% HR in OO were not properly identified. For the 5% EVAO-VOO blend, the VO contained was not recognised by any of the classification models, but the percentage of VO was inside the calibration range of the regression models made for EVAO-VOO, HOSO-VOO and HR-VOO blends. Indeed, the EVAO-VOO model predicted correctly the percentage of EVAO in the mixture. The blend of 5% HR in OO was not detected by any of the HR-OO classification models but it was by the RAO-OO models. The RAO-OO blends were identified and the VO percentage determined by the classification and regression models constructed for both RAO-OO and DOSO-OO, but the PLS-DA model 76, which distinguishes these two mixtures, predicted satisfactorily that the blends contained RAO, except for the mixture of 10% RAO in OO. Despite this, the RAO-OO regression model calculated accurately the content of RAO in the blends of 10% and 20% RAO in OO. The blends with 5% and 30% RAO in OO were not correctly determined by the corresponding regression models.

Regarding, the commercial oils analysed, it was declared in the label to be mixtures of vegetable oils or NTSO with EVOO or VOO. Samples 37 and 38 were confirmed to contain VOO, whereas sample 44 was classified as an OO blend. Moreover, the three of them were predicted to contain NTSO at percentages >90%. All the other commercial oil samples (39–43) were labelled as mixtures of VOO or EVOO with rapeseed oil, however all of them were classified as blends with OO. These results are not conclusive since no blends of rapeseed oil with VOO or OO were available to be included in the modelling step of present study. From the predictions of the classification and regression models of the decision tree scheme (Figures 1 and 2), it could be infer that most of these samples (39, 41–43) presented close composition to blends of 50% CO in OO. Sample 40 was identified by all classification models as NTSO-OO blends, and predicted to have >90% NTSO. In order to predict blends of rapeseed oil with VOO and OO, these type of blends must be included in the dataset used to build the models of the decision tree scheme, as well as any other VO that could be of interest so as to certify the percentage of VOs in VOO or OO blends or detect olive oil adulterations.

Conclusion

¹H-NMR fingerprinting of olive oils contain useful information for detecting blends between olive oils and VOs, and quantifying the percentage of each oil

in the mixture. Pattern recognition techniques applied to ^{1}H-NMR spectral data provide chemical tools to (*i*) detect and distinguish the presence of VOO or OO in an oil sample; (*ii*) to discriminate between pure olive oils and their blends with VOs to a certain extent, given by the detection limit disclosed for each VO in the mixture; (*iii*) to identify the VO in the blend with VOO or OO; (*iv*) to differentiate between blends made with different VOs in VOO or OO, or (*v*) with the same VO at different concentrations in the mixture; and (*vi*) to determine the percentage of the VO blended with VOO or OO. In this sense, ^{1}H-NMR spectral data of pure virgin olive oils and their mixtures with sunflower, high oleic sunflower, virgin avocado, virgin and refined hazelnut and soybean oils; and pure olive oils and their mixtures with refined palm olein oil, corn oil, sunflower oil, high oleic sunflower oil, desterolized high oleic sunflower oil, refined avocado oil and refined hazelnut oil, was used to optimize and validate binary classification models built by PLS-DA and regression models by PLS-R, which were arranged in a decision tree scheme in order to determine the composition of an oil sample. The performance of the proposed strategy is endorsed by the satisfactory classification and regression models developed. The binary classification models achieved present classification abilities of 90–100% for most of the models; and the fact that the recognition abilities are the same or greater and close to the prediction abilities means that the models are robust and stable, and that the classification results are not dependent on the samples included in the training set in the modelling step. The regression models obtained present excellent precisions and acceptable accuracies for the determination of the percentage of VOs in the VOO or OO mixtures studied. The percentage of the VO in the blend is determined with accuracies under the 20% of R-RMSEP for contents as low as 5% EVAO or S, 6% NTSO, 10% HV and 17% HOSO in VOO, and 2% RPOO, CO, NTSO or HR, 4% DOSO or RAO and 5% HOSO in OO. The limits of the detection were <2% EVAO or S and 2–5% NTSO, HOSO, HV or HR in VOO, and <2% RPOO, CO, HOSO, NTSO, DOSO or HR and 2–4% RAO in OO. Moreover, the chemical interpretation of the most influential variables in the models corroborates their reliability. Besides, the effectiveness of the proposed strategy was tested with blind samples, which confirmed its feasibility to support Reg. (EU) 29/2012.

The present research is a proof of concept of a stepwise strategy based on ^{1}H-NMR fingerprinting together with pattern recognition techniques in order to determine the VO in olive oil blends for authentication and fraud detection purposes. Further studies should be carried out with larger balanced sample sets that cover the variability of olive oils and the VOs of interest. The different possible sources of variability, i.e., varieties of each botanical species, agronomical and climatic conditions, harvests and geographical origins among others should be considered. The implementation of this approach requires the development of a databank of ^{1}H-NMR fingerprints of oils legally blended or submitted to adulteration, as well as of the adulterants and their blends, representative of the natural oil variability and compositional differences, in order to guarantee robust models for both authentication and fraud detection. It is worth to note that this requirement is feasible taking into account that one of the aims of the OLEUM project is to develop the OLEUM databank, an online integrated quality assurance database of olive oil analytical methods and chemical data, as well as a OLEUM Network of a worldwide community of proficient analytical laboratories involved in the analysis of olive oil, which can also contribute to feed and update the databank over time.

Acknowledgments

This work was developed in the framework of the project OLEUM "Advanced solutions for assuring authenticity and quality of olive oil at global scale" funded by the European Commission within the Horizon 2020 Programme (2014–2020), grant agreement No. 635690; and the project AUTENFOOD funded by ACCI´O-Generalitat de Catalunya and the European Union through the Programa Operatiu FEDER Catalunya 2014-2020 (Ref COMRDI-15-1-0035). The information contained in this chapter reflects the authors' views; the European Commission is not liable for any use of the information contained herein. The authors would like to thank all producers that supplied the olive oils, virgin olive oils and vegetable oils for this study, and the technical and staff support provided by SGIker (UPV/EHU, MICINN, GV/EJ, ESF).

References

Agiomyrgianaki, A., Petrakis, P. V., & Dais, P. (2010). Detection of refined olive oil adulteration with refined hazelnut oil by employing NMR spectroscopy and multivariate statistical analysis. *Talanta, 80*(5), 2165-2171.

Al-Ismail, K. M., Alsaed, A. K., Ahmad, R., & Al-Dabbas, M. (2010). Detection of olive oil adulteration with some plant oils by GLC analysis of sterols using polar column. *Food Chemistry, 121*(4), 1255-1259.

Alonso-Salces, R. M. , Berrueta, Luis Ángel Beatriz Quintanilla-Casas, Alba Tres, S.V. et al. (2022). Stepwise strategy based on ^{1}H-NMR fingerprinting in combination with chemometrics to determine the content of vegetable oils in olive oil mixtures. *Food Chemistry, 366*, 130588.

Alonso-Salces, R. M., Héberger, K., Holland, M. V., Moreno-Rojas, J. M., Mariani, C., Bellan, G., Reniero, F., & Guillou, C. (2010). Multivariate analysis of NMR fingerprint of the unsaponifiable fraction of virgin olive oils for authentication purposes. *Food Chemistry, 118*(4), 956-965.

Alonso-Salces, R. M., Holland, M. V., & Guillou, C. (2011). ^{1}H-NMR fingerprinting to evaluate the stability of olive oil. *Food Control, 22*(12), 2041-2046.

Alonso-Salces, R. M., Moreno-Rojas, J. M., Holland, M. V., Reniero, F., Guillou, C., & Heberger, K. (2010). Virgin olive oil authentication by multivariate analyses of ^{1}H NMR fingerprints and δ^{13}C and δ^{2}H data. *Journal of Agricultural and Food Chemistry, 58*(9), 5586-5596.

Alonso-Salces, R. M., Segebarth, N., Garmón-Lobato, S., Holland, M. V., Moreno-Rojas, J. M., Fernández-Pierna, J. A., Baeten, V., Fuselli, S. R., Gallo, B., Berrueta, L. A., Reniero, F., Guillou, C., & Héberger, K. (2015). ^{1}H-NMR and isotopic fingerprinting of olive oil and its unsaponifiable fraction: Geographical origin of virgin olive oils by pattern recognition. *European Journal of Lipid Science and Technology, 117*(12), 1991-2006.

Andrikopoulos, N. K., Giannakis, I. G., & Tzamtzis, V. (2001). Analysis of olive oil and seed oil triglycerides by capillary gas chromatography as a tool for the detection of the adulteration of olive oil. *Journal of Chromatographic Science, 39*(4), 137-145.

Aparicio, R., & Harwood, J. (2013). *Handbook of Olive Oil: Analysis and Properties* (2nd ed.). USA: Springer US.

Azizian, H., Mossoba, M. M., Fardin-Kia, A. R., Karunathilaka, S. R., & Kramer, J. K. G. (2016). Developing FT-NIR and PLS1 Methodology for Predicting Adulteration in Representative Varieties/Blends of Extra Virgin Olive Oils. *Lipids, 51*(11), 1309-1321.

Biedermann, M., Grob, K., Mariani, C., & Schmidt, J. P. (1996). Detection of desterolized sunflower oil in olive oil through isomerized Δ7-sterols. *Zeitschrift fur Lebensmittel -Untersuchung und -Forschung, 202*(3), 199-204.

Caponio, F., Summo, C., Bilancia, M. T., Paradiso, V. M., Sikorska, E., & Gomes, T. (2011). High performance size-exclusion chromatography analysis of polar compounds applied to refined, mild deodorized, extra virgin olive oils and their blends: An approach to their differentiation. *LWT - Food Science and Technology, 44*(8), 1726-1730.

Carranco, N., Farrés-Cebrián, M., Saurina, J., & Núñez, O. (2018). Authentication and quantitation of fraud in extra virgin olive oils based on HPLC-UV fingerprinting and multivariate calibration. *Foods, 7*(4).

Conte, L., Bendini, A., Valli, E., Lucci, P., Moret, S. et al. (2019). Olive oil quality and authenticity: A review of current EU legislation, standards, relevant methods of analyses, their drawbacks and recommendations for the future. *Trends in Food Science & Technology*, in press.

Contiñas, A., Martínez, S., Carballo, J., & Franco, I. (2008). Detection of contaminations and/or adulterations of the extra virgin olive oil with seeds oils (sunflower and soybean) and olive pomace oil. *Grasas y Aceites, 59*(2), 97-103.

Christophoridou, S., & Dais, P. (2009). Detection and quantification of phenolic compounds in olive oil by high resolution ^{1}H nuclear magnetic resonance spectroscopy. *Analytica Chimica Acta, 633*(2), 283-292.

Christopoulou, E., Lazaraki, M., Komaitis, M., & Kaselimis, K. (2004). Effectiveness of determinations of fatty acids and triglycerides for the detection of adulteration of olive oils with vegetable oils. *Food Chemistry, 84*(3), 463-474.

Dais, P., & Hatzakis, E. (2013). Quality assessment and authentication of virgin olive oil by NMR spectroscopy: A critical review. *Analytica Chimica Acta, 765*, 1-27.

Dais, P., & Spyros, A. (2007). ^{31}P NMR spectroscopy in the quality control and authentication of extra-virgin olive oil: A review of recent progress. *Magnetic Resonance in Chemistry, 45*(5), 367-377.

De La Mata-Espinosa, P., Bosque-Sendra, J. M., Bro, R., & Cuadros-Rodríguez, L. (2011). Olive oil quantification of edible vegetable oil blends using triacylglycerols chromatographic fingerprints and chemometric tools. *Talanta, 85*(1), 177-182.

De la Mata-Espinosa, P., Dominguez-Vidal, A., Bosque-Sendra, J. M., Ruiz-Medina, A., Cuadros-Rodríguez, L., & Ayora-Cañada, M. J. (2012). Olive oil assessment in edible oil blends by means of ATR-FTIR and chemometrics. *Food Control, 23*(2), 449-455.

Dimperio, M., Mannina, L., Capitani, D., Bidet, O., Rossi, E., Bucarelli, F., Quaglia, G., & Segre, A. (2007). NMR and statistical study of olive oils from Lazio: A geographical, ecological and agronomic characterization. *Food Chemistry, 105*(3), 1256-1267.

European Commission. (2012). European Commission Regulation 29/2012 on marketing standards for olive oil. *Official Journal of the European Union, January 13, 2012*, 12-21.

European Commission. (2020). European Commission website. *Food Fraud.* https://ec.europa.eu/knowledge4policy/food-fraud-quality/food-fraud-data-bases_en, Accessed: December 1, 2020.

European Parliament. (2014). Committee on the Environment, Public Health and Food Safety. Resolution of 14 January 2014 on the food crisis, fraud in the food chain and the control thereof (2013/2091 (INI)).

Fang, G., Goh, J. Y., Tay, M., Lau, H. F., & Li, S. F. Y. (2013). Characterization of oils and fats by ^{1}H NMR and GC/MS fingerprinting: Classification, prediction and detection of adulteration. *Food Chemistry, 138*(2-3), 1461-1469.

Fauhl, C., Reniero, F., & Guillou, C. (2000). ^{1}H NMR as a tool for the analysis of mixtures of virgin olive oil with oils of different botanical origin. *Magnetic Resonance in Chemistry, 38*(6), 436-443.

Fernandes, G. D., Gómez-Coca, R. B., Pérez-Camino, M. d. C., Moreda, W., & Barrera-Arellano, D. (2017). Chemical Characterization of Major and Minor Compounds of Nut Oils: Almond, Hazelnut, and Pecan Nut. *Journal of Chemistry, 2017*, Article 2609549.

García-González, D. L., Mannina, L., D'Imperio, M., Segre, A. L., & Aparicio, R. (2004). Using ^{1}H and ^{13}C NMR techniques and artificial neural networks to detect the adulteration of olive oil with hazelnut oil. *European Food Research and Technology, 219*(5), 545-548.

Gómez-Ariza, J. L., Arias-Borrego, A., García-Barrera, T., & Beltran, R. (2006). Comparative study of electrospray and photospray ionization sources coupled to quadrupole time-of-flight mass spectrometer for olive oil authentication. *Talanta, 70*(4), 859-869.

Gómez-Coca, R. B., Pérez-Camino, M. d. C., Bendini, A., Gallina Toschi, T., & Moreda, W. (2020). Olive oil mixtures. Part two: Detection of soft deodorized oil in extra virgin olive oil through diacylglycerol determination. Relationship with free acidity. *Food Chemistry, 330*, Article 127226.

Gómez-Coca, R. B., Pérez-Camino, M. d. C., Martínez-Rivas, J. M., Bendini, A., Gallina Toschi, T., & Moreda, W. (2020). Olive oil mixtures. Part one: Decisional trees or how to verify the olive oil percentage in declared blends. *Food Chemistry, 315*, Article 126235.

Gouilleux, B., Marchand, J., Charrier, B., Remaud, G. S., & Giraudeau, P. (2018). High-throughput authentication of edible oils with benchtop Ultrafast 2D NMR. *Food Chemistry, 244*, 153-158.

Green, H. S., Li, X., De Pra, M., Lovejoy, K. S., Steiner, F., Acworth, I. N., & Wang, S. C. (2020). A rapid method for the detection of extra virgin olive oil adulteration using UHPLC-CAD profiling of triacylglycerols and PCA. *Food Control, 107*, Article 106773.

Green, H. S., & Wang, S. C. (2020). First report on quality and purity evaluations of avocado oil sold in the US. *Food Control, 116*, Article 107328.

Grob, K., Artho, A., & Mariani, C. (1992). Determination of Raffination of Edible Oils and Fats by Olefinic Degradation Products of Sterols and Squalene, Using Coupled LC-GC. *Lipid/Fett, 94*(10), 394-400.

Grob, K., Biedermann, M., Bronz, M., & Giuffré, A. M. (1994). The Detection of Adulteration with Desterolized Oils. *Lipid/Fett, 96*(9), 341-345.

Grošelj, N., Vračko, M., Pierna, J. A. F., Baeten, V., & Novič, M. (2008). The use of FT-MIR spectroscopy and counter-propagation artificial neural networks for tracing the adulteration of olive oil. *Acta Chimica Slovenica, 55*(4), 935-941.

Guillén, M. D., & Ruiz, A. (2001). High resolution ^{1}H nuclear magnetic resonance in the study of edible oils and fats. *Trends in Food Science and Technology, 12*(9), 328-338.

Guillén, M. D., & Ruiz, A. (2003). Rapid simultaneous determination by proton NMR of unsaturation and composition of acyl groups in vegetable oils. *European Journal of Lipid Science and Technology, 105*(11), 688-696.

Guyader, S., Thomas, F., Portaluri, V., Jamin, E., Akoka, S., Silvestre, V., & Remaud, G. (2018). Authentication of edible fats and oils by non-targeted ^{13}C INEPT NMR spectroscopy. *Food Control, 91*, 216-224.

Hayakawa, T., Yanagawa, M., Yamamoto, A., Aizawa, S., Taga, A., Mochizuki, N., Itabashi, Y., Uchida, H., Ishihara, Y., & Kodama, S. (2020). A simple screening method for extra virgin olive oil adulteration by determining squalene and tyrosol. *Journal of Oleo Science, 69*(7), 677-684.

International Olive Council.. (2019). *Economic Affairs & Promotion Unit.* World Olive Oil Figures. https://www.internationaloliveoil.org/what-we-do/economic-affairs-promotion-unit/#figures, Accessed: December 1, 2020.

Izquierdo, M., Lastra-Mejías, M., González-Flores, E., Cancilla, J. C., Aroca-Santos, R., & Torrecilla, J. S. (2020). Deep thermal imaging to compute the adulteration state of extra virgin olive oil. *Computers and Electronics in Agriculture, 171.*

Jabeur, H., Zribi, A., & Bouaziz, M. (2016). Extra-virgin olive oil and cheap vegetable ols: Distinction and dtection of aulteration as dtermined by GC and chemometrics. *Food Analytical Methods, 9*(3), 712-723.

Jabeur, H., Zribi, A., Makni, J., Rebai, A., Abdelhedi, R., & Bouaziz, M. (2014). Detection of chemlali extra-virgin olive oil adulteration mixed with soybean oil, corn oil, and sunflower oil by using GC and HPLC. *Journal of Agricultural and Food Chemistry, 62*(21), 4893-4904.

Jergović, A. M., Peršurić, Ž., Saftić, L., & Kraljević Pavelić, S. (2017). Evaluation of MALDI-TOF/MS Technology in Olive Oil Adulteration. *Journal of the American Oil Chemists' Society, 94*(6), 749-757.

Jiang, X. Y., Li, C., Chen, Q. Q., & Weng, X. C. (2018). Comparison of ^{19}F and ^{1}H NMR spectroscopy with conventional methods for the detection of extra virgin olive oil adulteration. *Grasas y Aceites, 69*(2), Article e249.

Jiménez-Carvelo, A. M., González-Casado, A., & Cuadros-Rodríguez, L. (2017). A new analytical method for quantification of olive and palm oil in blends with other vegetable edible oils based on the chromatographic fingerprints from the methyl-transesterified fraction. *Talanta, 164*, 540-547.

Jiménez-Carvelo, A. M., Osorio, M. T., Koidis, A., González-Casado, A., & Cuadros-Rodríguez, L. (2017). Chemometric classification and quantification of olive oil in blends with any edible vegetable oils using FTIR-ATR and Raman spectroscopy. *LWT - Food Science and Technology, 86*, 174-184.

Jović, O., Smolić, T., Primožič, I., & Hrenar, T. (2016). Spectroscopic and chemometric analysis of binary and ternary edible oil mixtures: Qualitative and quantitative study. *Analytical Chemistry, 88*(8), 4516-4524.

Karabulut, I., Topcu, A., Yorulmaz, A., Tekin, A., & Ozay, D. (2005). Effects of industrial refining process on some properties of hazelnut oil. *European Journal of Lipid Science and Technology, 107*, 476-480.

Lioupi, A., Nenadis, N., & Theodoridis, G. (2020). Virgin olive oil metabolomics: A review. *Journal of Chromatography B: Analytical Technologies in the Biomedical and Life Sciences, 1150*, Article 122161.

Luo, Y., Gao, B., Zhang, Y., & Yu, L. (2020). Detection of olive oil adulteration with vegetable oils by ultra-performance convergence chromatography-quadrupole time-of-flight mass spectrometry (UPC^{2}-QTOF MS) coupled with multivariate data analysis based on the differences of triacylglycerol compositions. *Food Science and Nutrition, 8*, 3759-3767.

Maggio, R. M., Cerretani, L., Chiavaro, E., Kaufman, T. S., & Bendini, A. (2010). A novel chemometric strategy for the estimation of extra virgin olive oil adulteration with edible oils. *Food Control, 21*(6), 890-895.

Mancini, A., Imperlini, E., Nigro, E., Montagnese, C., Daniele, A., Orrù, S., & Buono, P. (2015). Biological and Nutritional Properties of Palm Oil and Palmitic Acid: Effects on Health. *Molecules, 20*(9), 17339-17361.

Mannina, L., D'Imperio, M., Capitani, D., Rezzi, S., Guillou, C., Mavromoustakos, T., Vilchez, M. D. M., Fernández, A. H., Thomas, F., & Aparicio, R. (2009). ^{1}H NMR-based protocol for the detection of adulterations of refined olive oil with refined hazelnut oil. *Journal of Agricultural and Food Chemistry, 57*(24), 11550-11556.

Mannina, L., Patumi, M., Fiordiponti, P., Emanuele, M. C., & Segre, A. L. (1999). Olive and hazelnut oils: A study by high-field ^{1}H NMR and gas chromatography. *Italian Journal of Food Science, 11*(2), 139-149.

Mannina, L., & Segre, A. (2002). High resolution nuclear magnetic resonance: From chemical structure to food authenticity. *Grasas y Aceites, 53*(1), 22-33.

Mannina, L., Sobolev, A. P., & Segre, A. (2003). Olive oil as seen by NMR and chemometrics. *Spectroscopy Europe, 15*(3), 6-14.

Meenu, M., Cai, Q., & Xu, B. (2019). A critical review on analytical techniques to detect adulteration of extra virgin olive oil. *Trends in Food Science and Technology, 91*, 391-408.

Mignani, A. G., Ciaccheri, L., Ottevaere, H., Thienpont, H., Conte, L., Marega, M., Cichelli, A., Attilio, C., & Cimato, A. (2011). Visible and near-infrared absorption spectroscopy by an integrating sphere and optical fibers for quantifying and discriminating the adulteration of extra virgin olive oil from Tuscany. *Analytical and Bioanalytical Chemistry, 399*(3), 1315-1324.

Mildner-Szkudlarz, S., & Jeleń, H. H. (2008). The potential of different techniques for volatile compounds analysis coupled with PCA for the detection of the adulteration of olive oil with hazelnut oil. *Food Chemistry, 110*(3), 751-761.

Monfreda, M., Gobbi, L., & Grippa, A. (2012). Blends of olive oil and sunflower oil: Characterisation and olive oil quantification using fatty acid composition and chemometric tools. *Food Chemistry, 134*(4), 2283-2290.

Monfreda, M., Gobbi, L., & Grippa, A. (2014). Blends of olive oil and seeds oils: Characterisation and olive oil quantification using fatty acids composition and chemometric tools. Part II. *Food Chemistry, 145*, 584-592.

Montoya, C., Cochard, B., Flori, A., Cros, D., Lopes, R., Cuellar, T., Espeout, S., Syaputra, I., Villeneuve, P., Pina, M., Ritter, E., Leroy, T., & Billotte, N. (2014). Genetic architecture of palm oil fatty acid composition in cultivated oil palm (*Elaeis guineensis* Jacq.) compared to its wild relative *E. oleifera* (H.B.K) Cortés. *PLoS ONE, 9*(5), Article e95412.

OLEUM Project. (2016). *About OLEUM: Aims and Objectives*. http://www.oleum project.eu/about-oleum/aims-and-objectives, Accessed date: December 1, 2020.

Orsavova, J., Misurcova, L., Ambrozova, J. V., Vicha, R., & Mlcek, J. (2015). Fatty Acids Composition of Vegetable Oils and Its Contribution to Dietary Energy Intake and Dependence of Cardiovascular Mortality on Dietary Intake of Fatty Acids. *International Journal of Molecular Sciences, 16*(6), 12871-12890.

Osorio, M. T., Haughey, S. A., Elliott, C. T., & Koidis, A. (2014). Evaluation of methodologies to determine vegetable oil species present in oil mixtures: Proposition of an approach to meet the EU legislation demands for correct vegetable oils labelling. *Food Research International, 60*, 66-75.

Owen, R. W., Mier, W., Giacosa, A., Hull, W. E., Spiegelhalder, B., & Bartsch, H. (2000). Identification of lignans as major components in the phenolic fraction of olive oil. *Clinical Chemistry, 46*(7), 976-988.

Ozcan-Sinir, G. (2020). Detection of adulteration in extra virgin olive oil by selected ion flow tube mass spectrometry (SIFT-MS) and chemometrics. *Food Control, 118*.

Papadia, P., Del Coco, L., Muzzalupo, I., Rizzi, M., Perri, E., Cesari, G., Simeone, V., Mondelli, D., Schena, F. P., & Fanizzi, F. P. (2011). Multivariate analysis of ^{1}H-NMR spectra of genetically characterized extra virgin olive oils and growth soil correlations. *Journal of the American Oil Chemists' Society, 88*(10), 1463-1475.

Parcerisa, J., Casals, I., Boatella, J., Codony, R., & Rafecas, M. (2000). Analysis of olive and hazelnut oil mixtures by high-performance liquid chromatography-atmospheric pressure chemical ionisation mass spectrometry of triacylglycerols and gas-liquid chromatography of non-saponifiable compounds (tocopherols and sterols). *Journal of Chromatography A, 881*(1-2), 149-158.

Parker, T., Limer, E., Watson, A. D., Defernez, M., Williamson, D., & Kemsley, E. K. (2014). 60MHz ^{1}H NMR spectroscopy for the analysis of edible oils. *TrAC - Trends in Analytical Chemistry, 57*, 147-158.

Pérez-Camino, M. C., Moreda, W., & Cert, A. (2001). Effects of Olive Fruit Quality and Oil Storage Practices on the Diacylglycerol Content of Virgin Olive Oils. *Journal of Agricultural and Food Chemistry, 49*(2), 699-704.

Pérez-Trujillo, M., Gómez-Caravaca, A. M., Segura-Carretero, A., Fernández-Gutiérrez, A., & Parella, T. (2010). Separation and identification of phenolic compounds of extra virgin olive oil from *Olea europaea* L. by HPLC-DAD-SPE-NMR/MS. Identification of a new diastereoisomer of the aldehydic form of oleuropein aglycone. *Journal of Agricultural and Food Chemistry, 58*(16), 9129-9136.

Philippidis, A., Poulakis, E., Papadaki, A., & Velegrakis, M. (2017). Comparative Study using Raman and Visible Spectroscopy of Cretan Extra Virgin Olive Oil Adulteration with Sunflower Oil. *Analytical Letters, 50*(7), 1182-1195.

Popescu, R., Costinel, D., Dinca, O. R., Marinescu, A., Stefanescu, I., & Ionete, R. E. (2015). Discrimination of vegetable oils using NMR spectroscopy and chemometrics. *Food Control, 48*, 84-90.

Priego Capote, F., Ruiz Jiménez, J., & Luque De Castro, M. D. (2007). Sequential (step-by-step) detection, identification and quantitation of extra virgin olive oil adulteration by chemometric treatment of chromatographic profiles. *Analytical and Bioanalytical Chemistry, 388*(8), 1859-1865.

Ranade, S., & Thiagarajan, P. (2015). A review on *Persea Americana* Mill. (Avocado) - Its fruit and oil. *International Journal of PharmTech Research, 8*(6), 72-77.

Rotondo, A., Salvo, A., Giuffrida, D., Dugo, G., & Rotondo, E. (2011). NMR analysis of aldehydes in sicilian extra-virgin olive oils by DPFGSE techniques. *AAPP Atti della Accademia Peloritana dei Pericolanti, Classe di Scienze Fisiche, Matematiche e Naturali, 89*(1).

Sacchi, R., Addeo, F., & Paolillo, L. (1997). ^{1}H and ^{13}C NMR of virgin olive oil. An overview. *Magnetic Resonance in Chemistry, 35*, S133-S145.

Sacchi, R., Patumi, M., Fontanazza, G., Barone, P., Fiordiponti, P., Mannina, L., Rossi, E., & Segre, A. L. (1996). A high-field ^{1}H nuclear magnetic resonance study of the minor components in virgin olive oils. *Journal of the American Oil Chemists' Society, 73*(6), 747-758.

Sacco, A., Brescia, M. A., Liuzzi, V., Reniero, F., Guillou, G., Ghelli, S., & van der Meer, P. (2000). Characterization of Italian olive oils based on analytical and nuclear magnetic resonance determinations. *Journal of the American Oil Chemists' Society, 77*(6), 619-625.

San Martín, E., Avenoza, A., Peregrina, J. M., & Busto, J. H. (2020). Solvent-based strategy improves the direct determination of key parameters in edible fats and oils by ^{1}H NMR. *Journal of the Science of Food and Agriculture, 100*(4), 1726-1734.

Sánchez-Hernández, L., Nozal, L., Marina, M. L., & Crego, A. L. (2012). Determination of nonprotein amino acids and betaines in vegetable oils by flow injection triple-quadrupole tandem mass spectrometry: A screening method for the detection of adulterations of olive oils. *Journal of Agricultural and Food Chemistry, 60*(4), 896-903.

Santos, P. M., Kock, F. V. C., Santos, M. S., Lobo, C. M. S., Carvalho, A. S., & Colnago, L. A. (2017). Non-invasive detection of adulterated olive oil in full bottles using time-domain NMR relaxometry. *Journal of the Brazilian Chemical Society, 28*(2), 385-390.

Shi, T., Zhu, M., Zhou, X., Huo, X., Long, Y., Zeng, X., & Chen, Y. (2019). ^{1}H NMR combined with PLS for the rapid determination of squalene and sterols in vegetable oils. *Food Chemistry, 287*, 46-54.

Souza, P. T., Ansolin, M., Batista, E. A. C., Meirelles, A. J. A., & Tubino, M. (2019). Identification of extra virgin olive oils modified by the addition of soybean oil, using ion chromatography. *Journal of the Brazilian Chemical Society, 30*(5), 1055-1062.

Spyros, A., Philippidis, A., & Dais, P. (2004). Kinetics of diglyceride formation and isomerization in virgin olive oils by employing ^{31}P NMR spectroscopy. Formulation of a quantitative measure to assess olive oil storage history. *Journal of Agricultural and Food Chemistry, 52*(2), 157-164.

Tan, J., Li, R., Jiang, Z.-T., Shi, M., Xiao, Y.-Q., Jia, B., Lu, T.-X., & Wang, H. (2018). Detection of extra virgin olive oil adulteration with edible oils using front-face fluorescence and visible spectroscopies. *Journal of the American Oil Chemists' Society, 95*(5), 535-546.

Torrecilla, J. S., Cancilla, J. C., Matute, G., Díaz-Rodríguez, P., & Flores, A. I. (2013). Self-organizing maps based on chaotic parameters to detect adulterations of extra virgin olive oil with inferior edible oils. *Journal of Food Engineering, 118*(4), 400-405.

Torrecilla, J. S., García, J., García, S., & Rodríguez, F. (2011). Application of lag-k autocorrelation coefficient and the TGA signals approach to detecting and quantifying adulterations of extra virgin olive oil with inferior edible oils. *Analytica Chimica Acta, 688*(2), 140-145.

Tsimidou, M. Z. (2010). Squalene and tocopherols in olive oil: Importance and methods of analysis. In *Olives and Olive Oil in Health and Disease Prevention*, (pp. 561-567): Elsevier Inc.

Tsopelas, F., Konstantopoulos, D., & Kakoulidou, A. T. (2018). Voltammetric fingerprinting of oils and its combination with chemometrics for the detection of extra virgin olive oil adulteration. *Analytica Chimica Acta, 1015*, 8-19.

Ün, İ., & Ok, S. (2018). Analysis of olive oil for authentication and shelf life determination. *Journal of Food Science and Technology, 55*(7), 2476-2487.

Vigli, G., Philippidis, A., Spyros, A., & Dais, P. (2003). Classification of edible oils by employing ^{31}P and ^{1}H NMR spectroscopy in combination with multivariate statistical analysis. A proposal for the detection of seed oil adulteration in virgin olive oils. *Journal of Agricultural and Food Chemistry, 51*(19), 5715-5722.

Vingering, N., Oseredczuk, M., du Chaffaut, L., Ireland, J., & Ledoux, M. (2010). Fatty acid composition of commercial vegetable oils from the French market analysed using a long highly polar column. *OCL, 17*(3), 185-192.

Vlahov, G. (1997). Quantitative 13C NMR method using the DEPT pulse sequence for the detection of olive oil adulteration with soybean oil. *Magnetic Resonance in Chemistry, 35*, S8-S12.

Vlahov, G. (2009). ^{13}C nuclear magnetic resonance spectroscopic detection of the adulteration of extra virgin olive oils extracted from different cultivars with cold-pressed hazelnut oil. *Journal of AOAC International, 92*(6), 1747-1754.

Wang, X., Wang, G., Hou, X., & Nie, S. (2020). A rapid screening approach for authentication of olive oil and classification of binary blends of olive oils using low-field nuclear magnetic resonance spectra and support vector machine. *Food Analytical Methods, 13*, 1894–1905.

Yang, Y., Ferro, M. D., Cavaco, I., & Liang, Y. (2013). Detection and identification of extra virgin olive oil adulteration by GC-MS combined with chemometrics. *Journal of Agricultural and Food Chemistry, 61*(15), 3693-3702.

Zamora, R., Alba, V., & Hidalgo, F. J. (2001). Use of high-resolution ^{13}C nuclear magnetic resonance spectroscopy for the screening of virgin olive oils. *Journal of the American Oil Chemists' Society, 78*(1), 89-94.

Zhang, X. F., Zou, M. Q., Qi, X. H., Liu, F., Zhang, C., & Yin, F. (2011). Quantitative detection of adulterated olive oil by Raman spectroscopy and chemometrics. *Journal of Raman Spectroscopy, 42*(9), 1784-1788.

Chapter 3

A Metabolomics Approach Based on ^{1}H-NMR Fingerprinting and Multivariate Data Analysis for Virgin Olive Oil Stability Assessment

Gabriela E. Viacava[1], Blanca Gallo[2], Luis A. Berrueta[2] and Rosa M. Alonso-Salces[3,*]

[1]Consejo Nacional de Investigaciones Científicas y Técnicas (CONICET), Grupo de Investigación en Ingeniería en Alimentos (GIIA), INCyTAA, Department of Chemical and Food Engineering, Faculty of Engineering, Universidad Nacional de Mar del Plata (UNMdP), Mar del Plata, Argentina

[2] Department of Analytical Chemistry, Faculty of Science and Technology, Universidad del País Vasco/Euskal Herriko Unibertsitatea (UPV/EHU), Bilbao, Spain

[3] Consejo Nacional de Investigaciones Científicas y Técnicas (CONICET), CIAS-IIPROSAM, Faculty of Exact and Natural Sciences, Universidad Nacional de Mar del Plata (UNMdP), Mar del Plata, Argentina

Abstract

The quality of virgin olive oil (VOO) is related to its oxidative stability, storage, and sensory and nutritional properties. VOO oxidation leads to the formation of off-flavour substances, degradation of its bioactive antioxidants and accumulation of degradation compounds, causing the loss of its sensory and health-promoting qualities, its commercial value and consumer acceptance. VOO resistance to oxidation depends on its chemical composition and its exposure to pro-oxidant factors such as oxygen, light and temperature. VOO stability during storage has been studied by an untargeted metabolomics approach based on ^{1}H-NMR fingerprinting and pattern recognition. In a preliminary study, the

[*] Corresponding Author's Email: rosamaria.alonsosalces@gmail.com.

In: Chemometrics
Editor: Larry D. Crenshaw
ISBN: 979-8-88697-274-0
© 2022 Nova Science Publishers, Inc.

stability of a VOO was studied under normal storage conditions, i.e., in dark and at room temperature, over a period of 43 months to evaluate the approach for virgin olive oil stability assessment. Principal component analysis (PCA) of the ^{1}H-NMR fingerprints allowed to visualize the evolution of VOO with time. In a further study, shelf life conditions during commercialization of VOOs were simulated, and a representative set of VOOs covering the full range of possible chemical compositions were exposed to these conditions: light (500 lux for 12h/day) at 25°C for 12 months or stored in the dark at 25 °C, 30°C and 35°C for 24 months. Supervised multivariate data analysis of the ^{1}H-NMR spectra of VOOs provided classification models to evaluate VOO freshness and verify the light exposure of the VOO during storage, and regression models to determine VOO storage time and tentatively the best before date of a fresh VOO. The different pattern recognition techniques disclosed VOO compositional changes due to hydrolytic and oxidative degradation taking place during storage, and confirmed that light and increasing temperature enhance these processes. The presence of characteristic resonances of hydroperoxides (primary oxidation products) and the decrease of ^{1}H signals assigned to phenolic compounds, fatty acids, squalene and native (*E*)-2-hexenal present in fresh VOO revealed its oxidative degradation. The emergence of low intensity ^{1}H signals of saturated aldehydes with increasing temperature in darkness, as well as in light exposure, meant that the secondary oxidation process has started at a low rate and yield in the stored oils. The decrease of the ^{1}H signals of triacylglycerides and *sn*-1,2-diacylglycerides, and increase of *sn*-1,3-diacylglycerides indicated that VOO hydrolytic degradation and diacylglyceride isomerisation was occurring. ^{1}H-NMR fingerprinting of VOO together with pattern recognition techniques afford relevant information to assess VOO quality regarding legal, sensory and health-promoting aspects.

Keywords: nuclear magnetic resonance, chemometrics, stability, oxidative degradation, hydrolytic degradation, olive oil

Introduction

Chemical Composition and Quality of Virgin Olive Oil

Virgin olive oil (VOO) is a high added value agricultural product of the Mediterranean diet from nutritional, health, sensory and commercial points of view. The exclusive nutritional, health-promoting and sensory properties of

VOO are due to its particular chemical composition. VOO is a complex mixture of triacylglycerides (more than 98%) and a wide range of minor components (Aparicio & Harwood, 2013). Triacylglycerides in VOO are rich in monounsaturated fatty acids (such as oleic acid), which confer nutritional value to this edible oil. Besides, its natural antioxidants, mainly phenolics and tocopherols, contribute to protect human health from several diseases (López-Miranda, Pérez-Jiménez, Ros, De Caterina, Badimón, Covas, et al., 2010). Mostly secoiridoid derivatives, but also simple phenols (hydroxytyrosol and tyrosol), flavonols and lignans are the families of phenolic compounds found in VOO. Phenolic compounds also attribute the positive sensory characteristics of pungency and bitterness to VOOs (Dais & Hatzakis, 2013; Servili, Esposto, Fabiani, Urbani, Taticchi, Mariucci, et al., 2009). However, VOO organoleptic properties are principally influenced by their volatile compounds, e.g., aldehydes, aliphatic alcohols and esters (García-González, Vivancos, & Aparicio, 2011; Morales, Aparicio-Ruiz, & Aparicio, 2013). VOO is a plant source of the hydrocarbon squalene, which is considered a valuable functional ingredient (Tsimidou, 2010). Other important minor constituents of VOO are β-sitosterol and α-tocopherol, being the most abundant phytosterol and tocopherol respectively; carotenoids, chlorophylls, pheophytins, terpenic alcohols, free fatty acids, diacylglycerides and monoacylglycerides (Aparicio et al., 2013; Dais et al., 2013).

VOO chemical composition depends on the olive cultivar, agronomical, environmental and pedoclimatic factors, time of olive harvest, olive storage after harvest, extraction technology and storage conditions (Aparicio et al., 2013; Servili, Esposto, Taticchi, Urbani, Di Maio, Sordini, et al., 2014). In particular, the profiles of phenolic compounds, volatile compounds and monounsaturated fatty acids in VOOs are highly related to the olive cultivar and strongly influenced by agronomical and technological factors; the triacylglycerides and fatty acid profiles of VOO are mostly determined by the olive cultivar genotype, and less influenced by environmental factors and the ripeness stage of olives (Servili et al., 2014; Velasco & Dobarganes, 2002); VOO composition in some sterols, such as campesterol, is affected by the olive cultivar genetics and pedoclimatic conditions (Servili et al., 2014). As a result, VOO can present a wide assortment of compositions (Servili, Esposto, Taticchi, Urbani, Di Maio, Veneziani, et al., 2015). Therefore the assessment of quality and traceability of VOO is still a challenging task for the olive oil industry and the food control laboratories due to the complexity and variability of this multicomponent food matrix.

Stability of Virgin Olive Oil

Despite VOO is recognised by its high stability, hydrolytic, autoxidation and photo-oxidation processes can occur during its production and storage. VOO oxidation leads to the formation of off-flavour substances, the degradation of its bioactive antioxidants and the accumulation of degradation compounds, which causes the loss of its sensory and health-promoting qualities as well as its commercial value and consumer acceptance. During the VOO oxidation process, oxygen is incorporated into the triglyceride structure by a free radical mechanism, and very reactive species are generated causing the breakdown of the acyl group chains and producing primary oxidation compounds called hydroperoxides (Guillen & Ruiz, 2001, 2006). These primary oxidation products may degrade into secondary oxidation products such as aldehydes, ketones, lactones, alcohols, acids, etc. Moreover, reactions between different acyl group chains may occur resulting in oligomeric and polymeric systems. Furthermore, VOO can also undergo hydrolytic degradation. During this process, triacylglycerides hydrolyse, which increases the contents of free fatty acids and therefore the acidity of the oil, which implies a deterioration of its quality. The oxidation of edible oils is a matter of major concern also from a safety point of view because some oxidation products such as aldehydes are toxic (Guillen et al., 2001, 2006). Regarding the sensory relevance of oxidation products, several saturated and unsaturated aldehydes were found to be responsible for rancid sensory defect in VOO (Morales, Luna, & Aparicio, 2005), as well as for off-odours (Kalua, Allen, Bedgood, Bishop, Prenzler, & Robards, 2007), altering its organoleptic characteristics.

The resistance of a VOO to oxidation depends on its chemical composition and its exposure to pro-oxidant factors such as oxygen, light, temperature and activators (chlorophylls and transition metals) during storage (Esposto, Taticchi, Di Maio, Urbani, Veneziani, Selvaggini, et al., 2015). With regard to chemical composition, VOO oxidative stability is mainly due to its relatively low degree of fatty acid unsaturation, its relatively high contents of oleic acid (a monounsaturated fatty acid), and to the antioxidant activity of some of its unsaponifiable components, mainly phenolic compounds (Esposto, Taticchi, Urbani, Selvaggini, Veneziani, Di Maio, et al., 2017; Frankel, 2010; Fregapane, Gómez-Rico, Inarejos, & Salvador, 2013; Krichene, Salvador, & Fregapane, 2015). α-Tocopherol also exhibits notorious antioxidant activity (Esposto et al., 2015; Esposto et al., 2017), and even shows a synergistic effect in association with some phenolic compounds (Deiana, Rosa, Cao, Pirisi, Bandino, & Dessi, 2002). Phenolics are hydrophilic antioxidants that inhibit oxidation by singlet oxygen quenching, free radical scavenging and metal chelating mechanisms; whereas α-tocopherol is a lipophilic antioxidant that acts scavenging lipid peroxy radicals and also quenching singlet oxygen (Esposto et al., 2017).

The ratio between monounsaturated and polyunsaturated fatty acids is considered an appropriate index of olive oil tendency to undergo autoxidation; the higher the ratio, the higher oxidative stability of the VOO (Farhoosh & Hoseini-Yazdi, 2013; Krichene, Allalout, Mancebo-Campos, Salvador, Zarrouk, & Fregapane, 2010). Indeed, VOO autoxidation rate depends to a great extent on the formation rate of alkyl radicals of fatty acids or acylglycerides, and this rate, in turn, on the types of these constituents in VOO (Choe & Min, 2006). Regarding minor components in VOO, some accelerates VOO oxidation performing as pro-oxidant such as free fatty acids, mono- and diacylglycerides and metals; and others act as antioxidant, such as phenolic compounds, carotenoids, sterols, tocotrienols, phospholipids and squalene. Meanwhile, tocopherols present both functions, and chlorophylls is pro-oxidant in the presence of light and antioxidant in the dark (Aparicio et al., 2013; Choe et al., 2006; Velasco et al., 2002). Esposto et al. (2017) reported that the initial contents of antioxidant compounds, particularly oleuropein derivatives and α-tocopherol, highly influence the loss of VOO quality from the health-promoting and sensory points of view. Oleuropein derivatives are involved in the primary inhibition of oxidation processes in VOO, and thus VOOs with high initial contents of these compounds experiment lower losses of secoiridoid derivatives and α-tocopherol, and develop fewer amounts of volatile C_7-C_{11} compounds responsible for the rancid defect during storage (Esposto et al., 2017). High phenolic contents in VOOs seem to limit the involvement of α-tocopherol in the oxidation processes and preserve its antioxidant effect (Esposto et al., 2017). Whereas in VOOs with low phenolic contents, α-tocopherol plays an important role in the inhibition of the oxidation phenomena, and a larger decrease of α-tocopherol contents is observed (Esposto et al., 2015). Squalene presents a moderate antioxidant activity, which is concentration dependent in VOO stored at low temperatures and in the dark (Psomiadou & Tsimidou, 1999), and appears to protect α-tocopherol during the photo-oxidation of VOO (Psomiadou & Tsimidou, 2002). During light exposure, VOO stability is also influenced by their pigment contents, which is characteristic of the olive cultivar (Aparicio et al., 2013), since VOO susceptibility to photo-oxidation rather depends on its chlorophyll content (Psomiadou et al., 2002; Tena, Aparicio, & García-González, 2017). The diacylglyceride isomerisation is not affected by storage conditions of temperature and light exposure but by the olive cultivar and storage time (Caponio, Paradiso, Bilancia, Summo, Pasqualone, & Gomes, 2013).

Quality Assessment of Virgin Olive Oil

The quality of VOO is related to its oxidative stability, storage history, sensory characteristics and nutritional properties (Dais et al., 2013), and is currently determined by chemical-physical measurements together with sensory assessment carried out by panel test. The oxidative stability of VOO has been evaluated by several methods reviewed elsewhere (Rosa M. Alonso-Salces, Holland, & Guillou, 2011; Frankel, 2010): Oil stability index (OSI), also known as Rancimat test, peroxide value, conjugated dienes, conjugated trienes, K_{232} and K_{270} UV indexes, acidity, total phenol content, TBARS test, DPPH assay, ORAC assay, analysis of individual components (phenolic compounds, tocopherols, squalene) by liquid or gas chromatography, electron paramagnetic resonance analyses, among others. Most of the studies on the oxidative stability of VOO previous to 2011 employed questionably high-temperatures, which contributed to the understanding of the autoxidation process under thermal stressing, but unfortunately were not conclusive to predict stability of olive oil under usual conditions during its transport and storage, as well as the role of its natural antioxidants during oxidation process taking place along its shelf life. The evolution of VOO oxidation is a slow process that usually requires from 9 to 24 months (Dais et al., 2013; Esposto et al., 2017). Another of the major drawbacks of previous approaches to evaluate VOO stability is that they focused on a single metabolite or a single family of metabolites, whereas the stability of VOO is a complex phenomenon and depends on the different constituents in the oil, as said above. Moreover, synergisms among oil components in relation to oxidation have been observed. This means that to have a global picture of VOO stability, the quantification of different metabolites may require different analytical methods and techniques making even more complicate the evaluation of the oil stability. Besides, legal parameters and analytical methods used to determine the quality of VOO and described by the International Olive Council (IOC) as well as by the Codex Alimentarius, do not include the evaluation of the quality decay of VOO during storage considering the complexity of alteration processes and their effect in both health-promoting and sensory properties. Indeed, the evaluation of the quality of a VOO should include not only the legal aspects, but also those related to human health and sensory ones as part of a whole (Servili et al., 2015).

According to the European Regulation 1335/2013, the storage conditions of olive oil should be clearly indicated on the label of the product to ensure that the consumer is properly informed about the best conditions for its preservation. However, the labelling of the harvest year, which is an indicator related to the freshness of the oil, is not compulsory. Moreover, this regulation establishes neither a storage time nor the control of VOO deterioration during storage. In this regard, the Directorate General of the European Commission for Agriculture and Rural Development (DG-AGRI) expressed the imperative of studying the storage period, i.e., the best-before date, for olive oil (DG-AGRI, 2012). In the framework of this need, ^{1}H-NMR fingerprinting was used to evaluate the stability of VOO under usual commercial storage conditions. ^{1}H-NMR is a useful analytical tool to contribute to this aim, since this technique can provide information related to legal, health-promoting and sensory features of VOO in the same analysis, require little or no sample pre-treatment, use small amounts of organic solvents or reagents, and are typically less time-consuming, allowing high and fast throughput analysis (Rosa M. Alonso-Salces et al., 2011; Dais et al., 2013). Indeed, major and several minor components in VOO, as well as its degradation products can be determined by ^{1}H-NMR. The ^{1}H-NMR spectrum of a VOO is considered its fingerprint, which can be further analysed by pattern recognition techniques to extract the information required for each subject in question.

Next sections of this chapter gathers the results achieved by the untargeted metabolomics approach based on the multivariate data analysis of the ^{1}H-NMR spectral data of VOO aliquots stored under normal (room temperature and dark for 43 months) and simulated commercial conditions (light (500 lux for 12h/day) at 25ºC for 12 months or dark at 25ºC, 30ºC and 35ºC for 24 months) for quality and stability assessment of VOO. This study allowed to: (*i*) disclose the chemical compounds responsible for the compositional changes taking place in VOO due to its hydrolytic and/or oxidative degradation (Rosa María Alonso-Salces, Gallo, Collado, Sasía-Arriba, Viacava, García-González, et al., 2021; Rosa M. Alonso-Salces et al., 2011); (*ii*) build classification models to discriminate between fresh and non-fresh VOO for the control of VOO freshness, and to verify the light exposure of the oil during storage; and (*iii*) develop regression models to determine the age of a VOO, i.e., the VOO storage time, and tentatively, the best before date of a fresh VOO (Rosa María Alonso-Salces et al., 2021).

^{1}H-NMR Spectrum of Fresh Virgin Olive Oil

Signals of major and some minor compounds typically observed in the ^{1}H-NMR spectra of VOOs analysed at time zero (fresh samples), along with their chemical shifts and their assignments to protons of the different functional groups are gathered in Table 1 (Rosa María Alonso-Salces et al., 2021; Rosa M. Alonso-Salces et al., 2011). The ^{1}H signals of triacylglycerides, the major components of olive oil, are well-known (Mannina & Segre, 2002; Sacco, Brescia, Liuzzi, Reniero, Guillou, Ghelli, et al., 2000). Triacylglycerides present different substitution patterns depending on the length, degree and kind of unsaturation of the acyl groups (Aparicio et al., 2013).

Minor components are only observed by ^{1}H-NMR when their signals are not overlapped with those of the main components and their concentrations are high enough to be detected; this is the case for the signals of cycloartenol at 0.318 ppm and 0.543 ppm; β-sitosterol at 0.669 ppm; stigmasterol at 0.687 ppm; squalene at 1.662 ppm; *sn*-1,2 diacylglyceryl at 3.71 ppm and 5.10 ppm; three unknown terpenes at 4.571 ppm; 4.648 ppm and 4.699 ppm; phenolics at 5.73 ppm, 5.99 ppm, 6.55 ppm, 6.61 ppm, 6.75 ppm and 7.03 ppm; aldehydic protons of secoiridoids at 9.215 ppm and 9.626 ppm, and of (*E*)-2-hexenal at 9.51 ppm; and hydroxyl protons of volatile compounds in the region 8.06–8.14 ppm (Rosa María Alonso-Salces et al., 2021; R. M. Alonso-Salces, Héberger, Holland, Moreno-Rojas, Mariani, Bellan, et al., 2010; Rosa M. Alonso-Salces et al., 2011; Rosa M. Alonso-Salces, Moreno-Rojas, Holland, Reniero, Guillou, & Heberger, 2010; R. M. Alonso-Salces, Segebarth, Garmón-Lobato, Holland, Moreno-Rojas, Fernández-Pierna, et al., 2015; Christophoridou & Dais, 2009; D'Imperio, Mannina, Capitani, Bidet, Rossi, Bucarelli, et al., 2007; Guillén & Ruiz, 2001; Mannina et al., 2002; Mannina, Sobolev, & Segre, 2003; Owen, Mier, Giacosa, Hull, Spiegelhalder, & Bartsch, 2000; Papadia, Del Coco, Muzzalupo, Rizzi, Perri, Cesari, et al., 2011; Pérez-Trujillo, Gómez-Caravaca, Segura-Carretero, Fernández-Gutiérrez, & Parella, 2010; Rotondo, Salvo, Giuffrida, Dugo, & Rotondo, 2011; Sacchi, Patumi, Fontanazza, Barone, Fiordiponti, Mannina, et al., 1996a; Shi, Zhu, Zhou, Huo, Long, Zeng, et al., 2019; Ün & Ok, 2018).

Table 1. Chemical shift assignments of the ^{1}H-NMR signals of the main components of VOOs

#	Chemical shift (ppm)	Multiplic.[a]	Functional group	Attribution
01	0.318	d	-CH_2- (cyclopropanic ring)	cycloartenol
02	0.527	s	-CH_2-	alcohol, sterol
03	0.543	d	-CH_2- (cyclopropanic ring)	cycloartenol
04	0.669	s	-CH_3 (C18-steroid group)	β-sitosterol
05	0.687	s	-CH_3 (C18-steroid group)	stigmasterol
06	0.740	t	-CH_3 (^{13}C satellite of signal at 0.87 ppm, acyl group)	
07	0.866	t	-CH_3 (acyl group)	saturated, oleic (or ω-9) and linoleic (or ω-6)
08	0.960	t	-CH_3 (acyl group)	linolenic (or ω-3)
09	0.987	t	-CH_3 (^{13}C satellite of signal at 0.87 ppm, acyl group)	
010	1.19-1.37		-$(CH_2)_n$- (acyl group)	
011	1.243		-$(CH_2)_n$- (acyl group)	saturated (palmitic, stearic)
012	1.256		-$(CH_2)_n$- (acyl group)	oleic
013	1.288		-$(CH_2)_n$- (acyl group)	linoleic and linolenic
014	1.51-1.65		-OCO-CH_2-CH_2- (acyl group)	
015	1.662	s	-CH_3	squalene
016	1.96-2.07		-CH_2-CH=CH- (acyl group)	
017	2.26-2.32	m	-OCO-CH_2- (acyl group)	
018	2.40-2.45	m	-OCO-CH_2- (^{13}C satellite of signal at 2.26-2.32 ppm, acyl group)	
019	2.72-2.82		=CH-CH_2-CH= (acyl group)	
020	2.754	t	=CH-CH_2-CH= (acyl group)	linoleic
021	2.789	t	=CH-CH_2-CH= (acyl group)	linolenic
022	3.69-3.73	d	-CH_2OH (glyceryl group)	*sn*-1,2-diacylglycerides
023	4.09-4.32		-CH_2OCOR (glyceryl group)	triacylglycerides
024	4.571	d		terpene
025	4.648	s		terpene
026	4.699	s		terpene

Table 1. (Continued)

#	Chemical shift (ppm)	Multiplic.[a]	Functional group	Attribution
027	5.05-5.15	m	>*CH*OCOR (glyceryl group)	*sn*-1,2-diacylglycerides
028	5.22-5.28	m	>*CH*OCOR (glyceryl group)	triacylglycerides
029	5.28-5.38	m	-*CH*=*CH*- (acyl group)	
030	5.52-5.43	m	-*CH*=*CH*- (^{13}C satellite of signal at 5.28-5.38 ppm, acyl group)	
031	5.72-5.76	dt	=*CH*- (phenolic ring)	phenolic compounds
032	5.986		=*CH*- (phenolic ring)	phenolic compounds
033	6.551	dt	=*CH*- (phenolic ring)	phenolic compounds
034	6.607	dd	=*CH*- (C8'; phenolic ring)	dialdehyde of oleuropein lacking a carboxymethyl group aldehydic form of oleuropein
035	6.79-6.73	d	=*CH*- (C5', C7'; phenolic ring)	dialdehyde of secoiridoids (oleuropein, ligstroside) lacking a carboxymethyl group aldehydic form of secoiridoid (oleuropein, ligstroside)
036	7.05-7.00	dt	=*CH*- (C4', C8'; phenolic ring)	dialdehyde of ligstroside lacking a carboxymethyl group aldehydic form of ligstroside
037	7.562	s	=*CH*-O- (C3)	aldehydic form of secoiridoid (oleuropein, ligstroside)
038	8.14-8.06		>C(O*H*)OR	volatile compounds
039	9.215	d	-*CHO* (C1)	dialdehyde of secoiridoids (oleuropein, ligstroside) lacking a carboxymethyl group
040	9.51	d	-*CHO*	*E*-2-alkenals (*E*-2-hexenal)
041	9.626	dd	-*CHO* (C3)	dialdehyde of secoiridoids (oleuropein, ligstroside) lacking a carboxymethyl group
		dd	-*CHO* (C1)	aldehydic form of secoiridoids (oleuropein, ligstroside)

[a] Signal multiplicity: s, singlet; d, doublet; t, triplet; m, multiplet; dd, double-doublet; dt, double-triplet; R, alkyl chain.

Stability of VOO Stored at Room Temperature in the Dark

The untargeted metabolomics approach based on ^{1}H-NMR fingerprinting and multivariate data analysis to assess VOO stability was evaluated in a first study. Thus, VOO stability under normal storage conditions, i.e., in dark and at room temperature was studied. About a liter of VOO was divided into aliquots contained in dark glass 40 mL-vials completely filled and kept at –20ºC in a freezer. Each month, over a period of more than 3 and half years (samples for the months 20th, 32nd, 38th and 42nd were missing), one vial was taken from the freezer and stored at room temperature in a closed box. A preliminary supposition was made; this considered that the degradation of VOO at –20ºC is not significant and thus the last aliquot taken out of the freezer was time 0. All aliquots were analyzed by ^{1}H-NMR once the last sample was taken from the freezer, and the multivariate data analysis of the ^{1}H-NMR fingerprints were carried out as described by Alonso-Salces et al., (2011).

In this preliminary study, ^{1}H-NMR fingerprinting combined with pattern recognition was used to study VOO stability under normal storage conditions, i.e., in dark and at room temperature, over a period of 43 months (Rosa M. Alonso-Salces et al., 2011). PCA of the ^{1}H-NMR fingerprints allowed to visualize the evolution of VOO with time. Figure 1 shows the PCA score plot generated for the first two dimensions (Rosa M. Alonso-Salces et al., 2011). The samples represented in this PCA score plot were distributed in the direction of the first principal component (PC1) in clusters, partially overlapped, according to the time they had been at room temperature. Moreover, the most influential features on PC1 corresponded to ^{1}H-NMR signals of compounds involved in the hydrolytic and oxidative degradation of VOO. So, PC1, accounting for 15% of the total variability of the system, contained information related to the change and evolution of the chemical composition of VOO during storage in the dark at room temperature, and hence, about the degradation of olive oil under these conditions.

^{1}H-NMR signals of some minor compounds resulting from the oxidative degradation of VOO were observed. The presence of hydroperoxides was confirmed by the ^{1}H-NMR signals at 8.09-8.19 ppm due to hydroperoxide protons; 6.51-6.57 ppm, 5.95-6.01 ppm and 5.55-5.57 ppm due to protons of the conjugated diene systems; and 4.35-4.37 ppm due to the methine proton of the hydroperoxide group, as reported by other authors (Guillen et al., 2001).

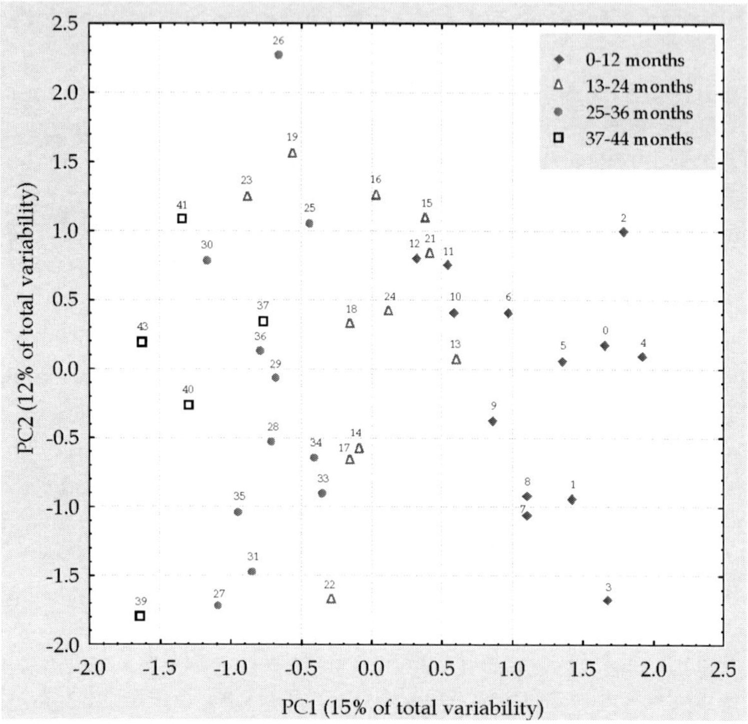

Figure 1. PCA score plot of the samples used to study the stability of VOO on the space defined by the two first principal components. Samples are numbered according to the time (months) that they had been at room temperature in the dark before analysis (Rosa M. Alonso-Salces et al., 2011).

Several resonances also indicated the occurrence of hydrolytic degradation. In this sense, slight changes in the intensity of the tryglyceride signals at 5.25 ppm, 4.45 ppm and 4.27 ppm and the α-methylene protons of the acyl group at 2.15-2.21 ppm were observed. Moreover, a slight decrease in the intensity of the signals at 2.75-2.79 ppm of the diallylic protons and at 2.03 ppm of the allylic protons of linoleic and linolenic acyl groups, and at 1.25-1.29 ppm of the methylene proton signal of oleic, linoleic and linolenic acyl groups, during storage at room temperature, revealed that tryglycerides were degrading. The increase in the intensity of the signal at 4.05-4.09 ppm, assigned to the proton of the glyceryl group of sn-1,3-diglycerides, was indicative of the loss of quality and freshness of the VOO (Guillen et al., 2001). Young, good quality olive oils contain mainly native sn-1,2-diacylglyceride and only small amounts of sn-1,3-diacylglyceride. The increase in the latter was observed after one year of storage at room temperature, which seems to be caused by intermolecular transposition and/or lipolytic phenomena (Sacchi, Patumi, Fontanazza, Barone, Fiordiponti, Mannina, et al., 1996b). Moreover, in the samples stored for longer than 18 months, a broad signal also appeared in the region of saturated alcohols at 3.85-3.89 ppm, which can arise from lipolysis (Sacchi et al., 1996b). All these signals presented very small intensities in comparison with characteristic VOO signals, indicating that the oxidative degradation was taking place at a very low rate and yield. This was also supported by the fact that characteristic resonances of aldehydes (9.3-9.9 ppm), the main secondary oxidation products, were not detected in the VOO over the 3 and half years of storage at room temperature, so the secondary oxidation process had not yet occurred. These facts agree with the recognized high oxidative stability of VOO. Some minor signals at 6.97 ppm and 6.75 ppm were assigned to phenolic compounds (Owen, Haubner, Mier, Giacosa, Hull, Spiegelhalder, et al., 2003). The decrease or disappearance, respectively, of these signals during storage at room temperature was in agreement with the role that these substances play as antioxidants during the oxidative degradation process of VOO.

Stability of VOO Exposed to Light at 25°C for 12 Months or Stored in the Dark at 25°C, 30°C and 35°C for 24 Months

In a second study, twenty three extra virgin olive oils (EVOOs) from the harvest 2016/2017 were supplied by EU and extra-EU providers. The legal control parameters, i.e., free acidity content (g of oleic acid/100 g of oil), peroxide value (PV, amount of hydroperoxides expressed as milli-equivalents of O_2/kg of oil), K_{232} and K_{270} extinction coefficients, ΔK and the fatty acid profile of the EVOOs were determined according to the official methods of the Commission Regulation (EEC) No 2568/91 and the Commission Delegated Regulation (EU) 2015/1830. Their α-tocopherol contents were determined according to Esposto et al., (2015); and their phenolic compounds were extracted as reported by Montedoro et al., (1992) and analysed according to Selvaggini et al., (2006). The legal quality parameters and the antioxidant compositions (α-tocopherol and phenolic compounds) of the 23 EVOOs were used to select 6 of them according to the following criteria: (*i*) free acidity < 0.5% and PV < 12 meq O_2/kg oil; and (*ii*) cover the concentration range of oleic acid, total phenols and α-tocopherols found in the market. These 6

EVOOs ("mother" samples) were mixed virtually according to different percentage ratios issued by an experimental design, resulting in 51 virtual EVOO samples characterised by different contents of oleic acid, hydrophilic phenols, and α-tocopherol. According to the D-optimal design using the Most Descriptive Compound (MDC) algorithm, the most descriptive 20 EVOOs among the 51 virtual blends (0%, 50%, 75%, 100%) were selected from the Principal Component Analysis (PCA) space, in order to cover the experimental domain with low, medium and high concentration combinations of the three selected chemical parameters (oleic acid, total phenols and α-tocopherols). The selection was made to define the mixing proportions of the 6 "mother" samples suitable for the experimental aim (different shelf life conditions of temperature, exposure to light, dark, and time of storage). In order to evaluate the influence of the temperature in dark conditions on the shelf life of EVOOs, 5 EVOOs out of the 51 virtual blends were chosen according to the above D-optimal design. Finally, the selected 20 EVOO samples were obtained by blending the 6 "mother" EVOOs in the established proportions, and were analysed for their full characterisation. The legal quality parameters (free acidity content, PV, K_{232}, K_{270} and ΔK) and the compositions of fatty acids, diacylglycerides, phenolic compounds and α-tocopherol of the 20 EVOOs were determined as mentioned above; and their volatile profiles according to Servili et al., (2001). The 20 EVOOs were bottled in 500 mL-glass UVA grade bottles, sealed with screw cap and placed on shelves at different shelf life conditions regarding temperature, light exposure and time of storage. Light exposure was 500 lux for 12h/day. The 20 EVOOs selected were studied at 25ºC both in light and dark conditions during 12 months. The 5 EVOOs chosen to evaluate the influence of temperature in the dark were studied at 25ºC, 30°C and 35°C during 24 months. Thus, up to 50 EVOO samples of experimental blends of the 6 "mother" EVOOs were stored (20 samples exposed to light; and 30 samples in darkness at three different temperatures). One bottle of each EVOO was taken monthly for samples exposed to light and every two months for those stored in the dark to be analysed as described by Alonso-Salces et al. (2021).

For the external validation of the predictive model for the EVOO stored at 25ºC exposed to light, the identical protocol for the real-time shelf life study (see above) in terms of bottling, storage, management of samples and analytical evaluation was carried out using 18 industrial blends of EVOOs from different olive cultivars and geographical origin (EU and extra-EU countries), bottled at the beginning of their commercial life. At time zero, all the EVOOs studied were characterized by good quality indexes, including free fatty acid percentage and indicators of primary (PV and K_{232}) and secondary (K_{270}) oxidation products, all being below the upper legal limit values established in current EU Reg. 1830/2015. The set of EVOOs studied also presented a large variability in the composition of fatty acids, phenolic compounds, and α-tocopherol; as well as in the profiles of volatile compounds in terms of high contents of molecules responsible for the positive sensory attributes, and low ones of those considered to be related to off-flavours (Esposto et al., 2017). The total absence of the (*E,E*)-2,4-decadienal was also confirmed (Rosa María Alonso-Salces et al., 2021).

The ^{1}H-NMR experiments and the subsequent univariate (ANOVA, Fisher index, Box & Whisker plots and simple least squares regressions) and multivariate (principal component analysis (PCA), and supervised as partial least squares discriminant analysis (PLS-DA) and partial least squares regression (PLS-R)) data analysis were performed according to the methodology described by Alonso-Salces et al., (2021).

Freshness of Virgin Olive Oil

In most studies found in literature, the evaluation of VOO quality was performed by the analysis of the legal parameters (free acidity content, PV, K_{232}, K_{270} and ΔK) and its fatty acid profile. VOO health-promoting characteristics were also assessed by its phenolic profile and α-tocopherol content; and its sensory properties were defined by its volatile compounds. These measurements allowed to study the influence of light exposure, temperature, headspace oxygen and olive cultivar on both the quality and the stability of VOOs with a wide range of chemical compositions stored under simulated market shelf life conditions (Benito, Abenoza, Oria, & Sánchez-Gimeno, 2011; Conte, Milani, Calligaris, Rovellini, Lucci, & Nicoli, 2020; Esposto et al., 2017; Iqdiam, Welt, Goodrich-Schneider, Sims, Baker, & Marshall, 2020; Kotsiou & Tasioula-Margari, 2015; Stefanoudaki, Williams, & Harwood, 2010). The parameter K_{270}, phenolic compounds, α-tocopherol and volatile compounds were proved to be the most important factors for VOO quality control, since they were related to the resistance of VOO to oxidation and the rancid defect (Benito et al., 2011; Lobo-Prieto, Tena, Aparicio-Ruiz, Morales, & García-González, 2020; Morales et al., 2005). Besides, phenolics and volatiles were confirmed to be responsible for the bitter and pungent tastes of VOO (Servili et al., 2015). In the present work, these legal parameters and chemical measurements traditionally analysed in olive oil were used as a reference to establish whether the VOO samples were fresh or not according to the legal limits established by EU Reg. 1830/2015 (Rosa María Alonso-Salces et al., 2021). Then, the ^{1}H-NMR spectra of the VOOs exposed to light at 25ºC during 12 months or stored in darkness at different temperatures during 24 months were analysed by PLS-DA in order to build predictive models to determine VOO freshness. Satisfactory PLD-DA models for the determination of VOO freshness were achieved using the complete NMR dataset or subsets (light or dark stored samples), and the freshness information provided by the different traditional parameters determined in the samples (Table 2). These classification models were stable and robust since recognition and prediction abilities in crossvalidation were close to each other and the former was greater than the latter, for both categories. All models achieved similar results: 81–91% of fresh VOOs and 82–96% of non-fresh VOOs were correctly recognised; and 80–89% of fresh VOOs and 80–91% of non-fresh VOOs were correctly predicted in crossvalidation. All samples available for the external validation set were fresh VOOs. The predictions of all models in the external set were acceptable. All models presented better prediction abilities in the external validation than in crossvalidation, except for total phenol model for samples stored in the dark. Usually the opposite occurs, i.e., the predictions in crossvalidation are higher or equal than in the external validation. The reason for the overestimated results achieved in the external validation was the limited samples available in the external set (regarding both number and variability covered).

Since the chemical composition of fresh VOO evolves during storage, there may come a time when the oil can no longer be considered fresh according to legal, sensory and/or health-promoting aspects. Table 3 collects the most important variables on the PLS-DA models developed for the determination of VOO freshness (Table 2). All these variables were influential in the models built with all (light and dark) data or dark data. However, the NMR signals #1, #2, #3, #4, #5, #6, #9, #10, #11, #12, #17, #21, #22, #23, #25, #31, #33, #34 and #35 were not present or did not change significantly during the 12 months of storage in light conditions in order to differentiate fresh VOO from non-fresh VOO.

Table 2. PLS-DA models to discriminate between olive oils according to their freshness[1]

PLS-DA model	Data	PLS-comp	Boundary	Class[2]	Class code	n	p	%R	%P-CV	%P-EV
K_{270}	All	4	0.6663	non-fresh	0	137	0.22	82	80	
				fresh	1	483	0.78	81	80	100
Total phenol	All	4	0.5261	non-fresh	0	226	0.36	89	85	
				fresh	1	394	0.64	85	82	89
(*E,E*)-2,4-decadienal	All	5	0.6248	non-fresh	0	175	0.28	90	84	
				fresh	1	445	0.72	83	82	100
K_{270}	Light	5	0.5186	non-fresh	0	104	0.40	96	91	
				fresh	1	156	0.60	90	87	100
Total phenol	Light	4	0.5205	non-fresh	0	103	0.40	87	83	
				fresh	1	157	0.60	88	84	94
(*E*)-2-decenal	Light	3	0.5720	non-fresh	0	119	0.46	89	85	
				fresh	1	141	0.54	84	84	100
K_{270}	Dark	3	0.6205	non-fresh	0	91	0.24	90	86	
				fresh	1	289	0.76	88	84	100
Total phenol	Dark	2	0.5993	non-fresh	0	126	0.33	89	85	
				fresh	1	254	0.67	83	81	78
(*E,E*)-2,4-decadienal	Dark	5	0.5718	non-fresh	0	126	0.33	94	86	
				fresh	1	254	0.67	91	89	100

[1]Abbreviations: n, number of samples; PLS-comp, number of PLS components selected; p, prior probability; %R, % of recognition ability; %P-CV, % of prediction ability in 3-fold crossvalidation; %P-EV, % of prediction ability in the external validation.

[2]Freshness criteria: $K270<0.220$ in fresh VOO stored in the light or $K270 < 0.190$ in fresh VOO in the dark; total phenol content >250 mg/kg in fresh VOO; (E)-2-decenal content <410 μg/kg in fresh VOO (rancid defect threshold); (E,E)-2,4-decadienal content <354 μg/kg in fresh VOO (rancid defect threshold).

Table 3. Chemical shift assignments of the ^{1}H-NMR signals related to the most important variables (buckets of ^{1}H-NMR spectra) on the PLS-DA model for freshness (fresh VOO vs non-fresh VOO)

#	Bucket (ppm)	Multiplic.[a]	Functional group	Attribution
1	9.63	dd	-*CHO* (C3)	dialdehyde of secoiridoids (oleuropein, ligstroside) lacking a carboxymethyl group
		dd	-*CHO* (C1)	aldehydic form of secoiridoids (oleuropein, ligstroside)
2	9.51-9.49	d	-*CHO*	*E*-2-alkenals (*E*-2-hexenal)
3	9.23-9.21	d	-*CHO* (C1)	dialdehyde of secoiridoids (oleuropein, ligstroside) lacking a carboxymethyl group
4	8.11	bs	-OO*H* (hydroperoxide group)	hydroperoxide derivatives
5	7.57-7.55	s	=*CH*-O- (C3)	aldehydic form of secoiridoid (oleuropein, ligstroside)
6	7.05-7.03	dt	=*CH*- (C4', C8'; phenolic ring)	dialdehyde of ligstroside lacking a carboxymethyl group aldehydic form of ligstroside
7	6.79-6.73	d	=*CH*- (C5', C7'; phenolic ring)	dialdehyde of secoiridoids (oleuropein, ligstroside) lacking a carboxymethyl group aldehydic form of secoiridoid (oleuropein, ligstroside)
8	6.61-6.59	dd	=*CH*- (C8'; phenolic ring)	dialdehyde of oleuropein lacking a carboxymethyl group aldehydic form of oleuropein
9	6.55-6.53	t	-*CH*=*CH*-*CH*=*CH*- (*cis, trans* conjugated diene system)	hydroperoxide derivatives
10	5.99-5.95	t	-*CH*=*CH*-*CH*=*CH*- (*cis, trans* conjugated diene system)	hydroperoxide derivatives
11	5.57	m	-*CH*=*CH*-*CH*=*CH*- (*cis, trans* conjugated diene system)	hydroperoxide derivatives
12	5.49	m	-*CH*=*CH*- (^{13}C satellite of signal at 5.28-5.38 ppm, acyl group)	
13	5.25	m	>*CH*OCOR (glyceryl group)	triacylglycerides
14	5.15-5.05	m	>*CH*OCOR (glyceryl group)	*sn*-1,2-diacylglycerides

#	Bucket (ppm)	Multiplic.[a]	Functional group	Attribution
15	4.59-4.55			unsaturated alcohols
16	4.45-4.41	m	-CH_2OCOR (^{13}C satellite of signal at 4.24-4.32 ppm, glyceryl group)	triacylglycerides
17	4.35		>*CH*-OOH (methine proton of hydroperoxide group)	hydroperoxide derivatives
18	4.33-4.27	m	-CH_2OCOR (glyceryl group)	triacylglycerides
19	4.09-4.05	q	>*CH*-OH (glyceryl group)	*sn*-1,3-diacylglycerides
20	3.73-3.69	d	-CH_2OH (glyceryl group)	*sn*-1,2-diacylglycerides
21	3.61	m	-*CH*(OHC-C=)-CH_2-CHO (C5)	dialdehyde of secoiridoids (oleuropein, ligstroside) lacking a carboxymethyl group
22	2.81-2.79	m	-CO-CH_2-CH_2-CHO	4-oxo-alkanals
23	2.65-2.61	m	-CH_2-*CHOHC*-CH_2-	9,10-epoxystearate
24	2.55-2.49		-R'CHX-*CH*R-CHO	2,3-branched saturated aldehydes
25	2.43-2.41	m	-OCO-CH_2- (^{13}C satellite of signal at 2.26-2.32 ppm, acyl group)	
26	2.35-2.33		-R'*CH*X-CHR-CHO	2,3-branched saturated aldehydes
27	2.31-2.27	m	-OCO-CH_2- (acyl group)	
28	2.15	m	-OCO-CH_2- (^{13}C satellite of signal at 2.26-2.32 ppm, acyl group)	
29	1.97-1.95		-CH_2-CH=CH- (acyl group)	
30	1.67	s	-CH_3	squalene
31	1.65		-CH_2-CH_2-CHO	alkanals
32	1.61		-OCO-CH_2-CH_2- (acyl group)	
33	1.33		-$(CH_2)_n$- (acyl group)	
34	1.25		-$(CH_2)_n$- (acyl group)	oleic
35	0.97	t	-CH_3 (acyl group)	linolenic (or ω-3)

[a] Signal multiplicity: bs, broad signal; s, singlet; d, doublet; t, triplet; q, quadruplet; m, multiplet; dd, doublet-doublet; dt, doublet-triplet; R, R', H or alkyl chain; X, electron withdrawing group.

The appearance in the ^{1}H-NMR spectra of VOOs of some signals due to hydroperoxide protons (#4), protons of the conjugated diene systems of hydroperoxide derivatives (#9, #10 and #11), and the methine proton of the hydroperoxide group (#17) indicated that primary oxidative degradation of the oil took place in the dark at all temperatures studied. Besides, a decrease of the ^{1}H signals #9, #10, #11 and #17 with increasing temperature in dark conditions disclosed the degradation of hydroperoxide derivatives into secondary oxidation products, such as aldehydes (#22, #24, #26, #31), alcohols (#15) and epoxides (#23). Aldehydes signals #24 and #26 increased during storage with increasing temperature in darkness, as well as in light exposure. This evolution of primary and secondary oxidation products during storage of VOOs under the moderate temperature and light/dark conditions studied was in agreement with previous results reported in literature (Farhoosh et al., 2013; Korifi, Plard, Le Dréau, Rébufa, Rutledge, & Dupuy, 2016; Tena et al., 2017). It was proved that the oxidation of triacylglycerides or the free fatty acids resulted from the hydrolytic degradation of triacylglycerides lead to the formation of relative stable hydroperoxides at temperatures below 65ºC. This fact allows the detection of these primary oxidation products during shelf life storage conditions. Moreover, increasing temperature provides additional energy for activating the hydroperoxide decomposition into the secondary oxidation product, and extra free radicals that will accelerate the propagation reactions (Farhoosh et al., 2013). Furthermore, photo-oxidation of VOOs was confirmed to take place even at milder conditions (23ºC and 400 lux) than those of present study (Tena et al., 2017). Hydroperoxides are intermediate reaction products, thus during their oxidation, they could be formed and decomposed at the same time. Hence, the evolution of their contents that is observed depends on the formation and decomposition rates (Conte et al., 2020). In the present study, these rates were closed to each other during the one year storage of VOOs exposed to light at 25ºC, therefore the changes in the ^{1}H signals due to hydroperoxides were not significant in the classification model for light conditions. In contrast, the rise in the intensity of saturated aldehydes signals (#24 and #26) was relevant under these storage conditions. From the data at 25ºC (light *vs* dark), it was evidenced that the formation of hydroperoxides was enhanced by light; and that the formation and decomposition rates of hydroperoxides were similar in VOOs exposed to light, whereas the decomposition rate was higher than the formation one in darkness. Therefore the intensity of the hydroperoxides signals decreased during storage, meanwhile those of aldehydes increased, as previously reported (Esposto et al., 2017).

Changes in the intensities of the ^{1}H signals of unsaturated aldehydes responsible for the negative rancid off-flavours of VOOs (Esposto et al., 2017) were not detected in the VOOs stored during one year in the light or two years in the dark at different temperatures. This is explained by the fact that some protons in the alkenal structures overlap with those of major components of VOOs (Sacchi et al., 1996a). Moreover, some other aldehydic protons (9.3–9.9 ppm) from unsaturated aldehydes generated during the secondary oxidation reactions overlap with those signals of secoiridoids, present in higher concentrations, and the increase of those aldehydes during storage might have been not significant to be detected in contrast with the decrease of secoiridoids. These results related to the second oxidation products (aldehydes) measured in VOOs agreed with those achieved using other spectroscopic approach based on mesh cell-FTIR used to study the oxidation of VOO under accelerated mild storage conditions emulating transport and storage stages (Tena et al., 2017). Indeed, FTIR aldehyde band intensities of VOOs stored in similar conditions (dark at 23ºC and 35ºC, and light at 400 lux and 23ºC) to those of the present study were not significantly different among the studied conditions.

The ^{1}H signal intensities of VOO phenolic compounds (#1, #3, #5, #6, #7, #8 and #21) decreased during storage. This behaviour, which contributed to distinguish fresh and non-fresh VOOs, was in accordance with the role that these substances play as antioxidants during the oxidative degradation process of VOO (Esposto et al., 2017; Koidis & Boskou, 2014; Krichene et al., 2015; Stefanoudaki et al., 2010). Phenolic compounds act as primary antioxidants in VOOs, playing a critical role in VOO stability. Under dark conditions, the deterioration of VOO is mainly related to autoxidation of its antioxidants, preferentially its secoiridoid derivatives (Krichene et al., 2015; Stefanoudaki et al., 2010). In contrast, photo-oxidation is the main responsible of VOO deterioration in the presence of light (Esposto et al., 2017; Koidis et al., 2014). The involved ^{1}H signals were due to secoiridoids, which are dialdehydic forms of oleuropein and ligstroside lacking a carboxymethyl group (#1, #3, #6, #7, #8 and #21) and aldehydic forms of oleuropein and ligstroside (#1, #5, #6, #7 and #8). No significant differences were observed in the dialdehydic signals #1, #3, #7 and #21 between light and dark conditions during the first year of storage; whereas signals #3 and #21 decreased in darkness with increasing temperature (Krichene et al., 2015). The intensities of the aldehydic signal #5, #6 and #8 were slightly higher in VOOs in the dark than in those exposed to light, which means that less degradation of the VOOs happened in darkness (Esposto et al., 2017; Krichene et al., 2010; Krichene et al., 2015); no influence of temperature was noticed. The ^{1}H signals of both the aldehydic (#5) and dialdehydic forms of oleuropein (#8) and ligstroside (#6) derivatives, denoted that the degradation behaviour of both families of secoiridoids (hydroxytyrosol and tyrosol derivatives, respectively) was quite similar (Esposto et al., 2017; Krichene et al., 2015). Secoiridoids degradation rate was shown to be dependent of the initial concentrations in the VOO: higher contents of secoiridoids lead to higher resistance of the VOO to oxidation (Esposto et al., 2017; Krichene et al., 2015). The effects of light, temperature and initial phenolic contents on VOO stability are important issues to be considered to ensure the quality of VOO and its health-promoting properties. Hence, a VOO stored at reduced temperatures and protected from light increases its shelf life, and consequently will comply with the European Regulation Health Claim during a longer period of time.

As mentioned earlier, during the hydrolytic degradation of VOO, triacylglycerides hydrolyse resulting in a decrease of the intensity of the triacylglyceride signals. The glyceryl proton signals of triacylglycerides (#13 and #18), and the α-methylene (#27) and β-methylene (#32) protons of the acyl group of fatty acids disclosed the occurrence of hydrolytic degradation during storage both in the light and in the dark. The degradation of VOOs stored in darkness increased with increasing temperature. The decrease in the triacylglyceride contents of VOOs during storage was also reported by other authors (Aparicio et al., 2013; Farhoosh et al., 2013; Köseoğlu, Sevim, & Kadiroğlu, 2019; Méndez & Falqué, 2007). The intensities of the glyceryl proton signals of triacylglycerides (#16), the vinylic (#12), the α-methylene (#25 and #28), and the methylene (#33) proton of the acyl group of fatty acids were higher in VOOs kept in the dark than those in the light, suggesting less degradation in dark conditions during the first year of storage; little or no influence of temperature was observed on these signals during dark storage over two years. As a result of the hydrolysis reaction of triacylglycerides, fatty acids are released, increasing VOO acidity (Lukić, Ros, Guella, Camin, Masuero, Mulinacci, et al., 2020). The maximum acidity level of VOO is stablished by the EU Regulation 2015/1830. Several research studies showed that the free fatty acid content of VOOs increased throughout storage in both light and dark conditions (Baiano, Terracone, Viggiani, & Del Nobile, 2014; Ghanbari Shendi, Sivri Ozay, Ozkaya, & Ustunel, 2018; Köseoğlu et al., 2019; Méndez et al., 2007; Stefanoudaki et al., 2010). However, the hydrolytic process of triacylglycerides was confirmed to be sensitive to the action of light, which led to a greater increase in the acidity of VOOs exposed to light (Méndez et al., 2007). Besides, triacylglyceride hydrolysis occurred to a greater extent as the temperature and acidity of VOO increased; the former being the most significant factor (Pérez-Camino, Moreda, & Cert, 2001; Spyros, Philippidis, & Dais, 2004).

The signal intensities of the allylic (#29) protons of the acyl group, the methylene proton of oleic acyl group (#34) and the methyl proton of the linolenic acyl group (#35) were lower in non-fresh samples, revealing that fatty acids were degrading during storage in the light and in the dark, but no influence of temperature was discerned. The degradation of fatty acids is a consequence of their oxidation (Méndez et al., 2007).

During storage in both light and dark conditions and with increasing temperature, the intensity of the glyceryl proton signal of *sn*-1,3-diacylglycerides (#19) increased, corresponding to a decrease of *sn*-1,2-diacylglyceride signals (#14 and #20). This evidenced the loss of quality and freshness of the VOO, because fresh good quality olive oils contain mostly native *sn*-1,2-diacylglyceride and only small amounts of *sn*-1,3-diacylglyceride. Diacylglycerides result from either incomplete biosynthesis of triacylglycerides in olive fruits or their later acid-catalysed and/or enzymatic hydrolysis during processing and storage (Pérez-Camino et al., 2001; Spyros et al., 2004). 1,2-diacylglycerides are considered to come mainly from the incomplete biosynthesis of triacylglycerides, whilst 1,3-diacylglycerides from their enzymatic or chemical hydrolysis. During VOO storage, 1,2-isomers are transformed to the thermodynamically more stable 1,3-isomers by an intermolecular transposition (Circi, Ingallina, Vista, Capitani, Di Vecchia, Leonardi, et al., 2018; Pérez-Camino et al., 2001; Sacchi et al., 1996a; Spyros et al., 2004). Diacylglyceride isomerisation rate rises as temperature and free acidity increases (Pérez-Camino et al., 2001; Spyros et al., 2004). Besides, light exposure accelerates this isomerisation reaction (Spyros et al., 2004). All these trends and findings of former studies are confirmed by the results of present work. Several approaches regarding diacylglyceride ratios have been previously proposed to estimate VOO storage time and freshness (Spyros et al., 2004); (Caponio et al., 2013) (Castellani, Serrilli, Bonadies, & Bianco, 2008).

The signals due to the methyl proton of squalene (#30) and the aldehydic proton of (*E*)-2-hexenal (#2) decreased with the storage time in both light and dark, and little or no influence of temperature was noted, as previously observed (Frankel, 2010; Fregapane et al., 2013; Salvo, Rotondo, La Torre, Cicero, & Dugo, 2017; Stefanoudaki et al., 2010). The intensity of the squalene signal (#30) was higher in light exposed VOOs than in those stored in the dark. On the one hand, this observation contrasted with the results of a previous study, in which the squalene content decreased significantly during six month storage but was not affected by storage conditions regarding light/dark exposure (Silva, Anjos, Cavalcanti, & Celeghini, 2015). On the other hand, current observation agreed with the findings of another work, which reported that the presence of squalene seemed to protect purified olive oil triacylglyceride fraction from oxidation in the dark, justifying its lower contents in VOOs stored in dark conditions (Tsimidou, 2010). The main volatile compound in fresh VOO is the native (*E*)-2-hexenal, which is responsible of the positive green odour notes in VOOs (Fregapane et al., 2013; Stefanoudaki et al., 2010). (*E*)-2-hexenal, as well as other native volatiles compounds in VOOs, was also reported to decrease during storage (Frankel, 2010; Fregapane et al., 2013).

The validity and fundamentals of the classification models achieved for the determination of VOO freshness are supported by the justified behaviour of the most influential compounds in the models, which agrees with the findings and observations of previous studies. The conformity among the results provided by the predictive models obtained with the data of the different parameters related to the legal, health-promoting and sensory aspects considered for the evaluation of VOO quality (Servili et al., 2015), implies that all these aspects are included in the models. Therefore, the present approach based on ^{1}H-NMR fingerprinting and chemometrics can assess by itself the overall quality of VOOs.

Light Exposure of Virgin Olive Oil

Photo-oxidation is considered the most damaging factor for the oxidative sensibility of VOOs (Aparicio et al., 2013; Esposto et al., 2017; Tena et al., 2017). The hydrolytic degradation of acylglycerides in VOOs is sensitive to light, being more intense than in darkness (Méndez et al., 2007). However, the response of a VOO to these alteration processes depends on its chemical composition, particularly natural antioxidants and fatty acids, which can directly or indirectly influence those phenomena respectively (Choe et al., 2006; Choe & Min, 2009), as well as the presence of other pro-oxidant factors such as temperature, oxygen and activators such as chlorophylls and transition metals (Choe et al., 2006; Esposto et al., 2017; Psomiadou et al., 2002). Regarding all these premises, VOOs stored in different light exposure and temperature conditions are supposed to present characteristic chemical compositions, which may be reflected in their ^{1}H-NMR fingerprints. In this regard, ^{1}H-NMR spectral data of the VOOs stored in the light during 12 months and in the dark at different temperatures during 24 months were studied by pattern recognition. PLS-DA provided a robust classification model to distinguish the samples exposed to light from the samples stored in the dark at different temperatures (Table 4). The recognition and prediction abilities in crossvalidation of the PLS-DA model were 95% and 94% respectively for the VOO samples exposed to light, and 92% and 91% for those stored in darkness, respectively. Table 5 shows the most influential variables in the classification model. The signal intensities of the glyceryl protons of triacylglycerides (#16), and the vinylic (#12), α-methylene (#25), allylic (#36), β-methylene (#37) and methylene (#38) protons of the acyl groups of fatty acids were lower in oils exposed to light than in those stored in darkness, indicating that degradation arose to a greater extent under light exposure. No influence of temperature was observed on these signals during dark storage, except for signal #12 which slightly decreased at 35 °C. Spyros et al., (2004) suggested that the oxidation products of fatty acids may have a strong accelerating effect on triacylglyceride hydrolysis, in order to explain the increase of the triacylglyceride hydrolysis rate in VOOs exposed to light during their storage. In this sense, the decline in the contents of both triacylglycerides and fatty acids were larger in VOOs stored in the light than in the dark.

In contrast, the signals of α- and β-protons of 2,3-branched saturated aldehydes (#24 and #26), and β-methylene protons of alkanals (#31), which are secondary oxidation products, followed similar trends at 25ºC in both light and dark conditions, but their intensities increased with increasing temperature (25–35ºC) in darkness.

Table 4. PLS-DA models to discriminate between olive oils according to their exposure to light or dark conditions[1]

PLS-DA model	Data	PLS-comp	Boundary	Class	Class code	n	p	%R	%P
Exposure	All	3	0.5106	dark	0	360	0.58	92	91
				light	1	260	0.42	95	94

[1]Abbreviations: n, number of samples; PLS-comp, number of PLS components selected; p, prior probability; %R, % of recognition ability; %P, % of prediction ability; 3-fold crossvalidation.

Table 5. Weighted regression coefficient (Bw) of the most influent variables on the PLS-DA model built to discriminate between olive oils according to their exposure to light or dark conditions

#	Bucket (ppm)	Bw (3 PLS-comp)	Functional group	Attribution
12	5.49 5.47 5.43	-6.14E-03 -6.56E-03 -6.30E-03	-*CH*=*CH*- (^{13}C satellite of signal at 5.28-5.38 ppm, acyl group)	
13	5.25 5.21	6.83E-03 6.34E-03	>*CH*OCOR (glyceryl group)	triacylglycerides
14	5.05	9.15E-03	>*CH*OCOR (glyceryl group)	*sn*-1,2-diacylglycerides
16	4.39 4.37	-6.65E-03 -6.04E-03	-CH_2OCOR (^{13}C satellite of signal at 4.09-4.32 ppm, glyceryl group)	triacylglycerides
18	4.27	6.25E-03	-CH_2OCOR (glyceryl group)	triacylglycerides
20	3.75 3.73 3.71 3.69	7.13E-03 8.51E-03 8.08E-03 6.46E-03	-CH_2OH (glyceryl group)	*sn*-1,2-diacylglycerides
24	2.59 2.57 2.51 2.49	-7.58E-03 -7.07E-03 -6.56E-03 -8.06E-03	-R'CHX-*CH*R-CHO	2,3-branched saturated aldehydes
25	2.47 2.45 2.43 2.41 2.39	-7.61E-03 -7.74E-03 -7.36E-03 -7.24E-03 -8.07E-03	-OCO-CH_2- (^{13}C satellite of signal at 2.26-2.32 ppm, acyl group)	—
26	2.37 2.35 2.33	-8.37E-03 -7.64E-03 -6.22E-03	-R'*CH*X-CHR-CHO	2,3-branched saturated aldehydes
27	2.29	6.22E-03	-OCO-CH_2- (acyl group)	
36	2.09	-6.39E-03	-CH_2-CH=CH- (^{13}C satellite of signal at 1.96-2.07 ppm, acyl group)	
37	1.75 1.73	-6.60E-03 -6.98E-03	-OCO-CH_2-CH_2- (^{13}C satellite of signal at 1.51-1.65 ppm, acyl group)	
31	1.65	-7.19E-03	-CH_2-CH_2-CHO	alkanals
38	1.51 1.49 1.47 1.45 1.43 1.41 1.39	-6.96E-03 -6.48E-03 -6.65E-03 -8.17E-03 -8.49E-03 -7.99E-03 -7.44E-03	-$(CH_2)_n$- (^{13}C satellite of signal at 1.19-1.37 ppm, acyl group)	
07	0.89	6.26E-03	-CH_3 (acyl group)	saturated, oleic (or ω-9) and linoleic (or ω-6)

Indeed, temperature provides additional energy to form secondary oxidation products (Conte et al., 2020; Farhoosh et al., 2013). Secondary oxidation processes in VOOs were found not to be affected by light exposure conditions at ambient temperature, however the evolution of these processes were influenced by moderate temperatures (35–65ºC) (Tena et al., 2017).

The glyceryl proton signals of triacylglycerides (#13 and #18) and *sn*-1,2-diacylglycerides (#14 and #20), as well as the α-methylene (#27) and methyl (#07) protons of the acyl group of fatty acids, disclosed the occurrence of degradation of triacylglycerides and native diacylglycerides, which took place at similar rates at 25ºC in both light and dark conditions during the first year of storage; however degradation increased with increasing temperature in the dark. The changes in the contents of *sn*-1,2-diacylglycerides, fatty acids and triacylglycerides during storage are interrelated because the isomerisation of 1,2-diacylglycerides is acid-catalysed by the free fatty acids produced during the hydrolysis of triacylglycerides. Furthermore, diacylglyceride isomerisation was quantitatively proved to be strongly affected by temperature during dark storage (Spyros et al., 2004).

The ^{1}H-NMR buckets with negative Bw coefficients were due to signals of fatty acids, which were more degraded in VOO samples exposed to light than in those stored in the dark, and were slightly or not affected by the increase in temperatures. In contrast, the ^{1}H signals of triacylglycerides and *sn*-1,2-diacylglycerides presented positive Bw coefficients, because their degradation rate was similar in VOOs stored in the light to those in the dark, but their degradation increased with increasing temperature. Oppositely, the secondary oxidation products, even though being influenced by the storage temperature as well, presented negative Bw coefficients since their contents increased during VOO degradation. Thus, Bw coefficients of the classification model reflected the different behaviours of the compounds involved in oxidative and hydrolytic degradation of VOOs depending on the VOO storage conditions.

These results prove that VOOs exposed to light or stored in darkness experienced different chemical changes due to the hydrolytic processes of acylglycerides as well as the oxidative degradation of fatty acids, which have significant implications in VOO quality. Therefore, taking into account the importance of avoiding the exposure of VOO to light during its shelf life and storage until its consumption, the existence of analytical tools to check whether a VOO has been correctly stored since it was produced is of great interest in order to preserve the high quality standards of VOO. ^{1}H-NMR fingerprinting of VOO together with pattern recognition techniques provide a rapid high throughput approach to control the correct storage of VOO during its commercial life, and have an estimation of its genuine quality depending on its production date.

Storage Time of Virgin Olive Oil

Partial least squares regression (PLS-R) models were developed to predict how long a VOO had been stored, which was related to the degree of deterioration of its quality (Table 6). The most influential variables in the PLS-R models for each storage condition (light at 25°C or dark at 25, 30 and 35°C) are shown in Table 7. In most samples, the signals due to the glyceryl protons of triacylglycerides (#13 and #18) and *sn*-1,2-diacylglycerides (#14 and #20) and the α-methylene (#27), β-methylene (#32) and allylic (#36) protons of the acyl groups of fatty acids decreased with the time of storage (Caponio et al., 2013; Méndez et al., 2007; Pérez-Camino et al., 2001; Spyros et al., 2004), resulting in negative Bw coefficients. In contrast, the signals of those compounds that increased with degradation during the storage time presented positive Bw coefficients, such as the glyceryl protons of *sn*-1,3-diacylglycerides (#19) and the α-methylene protons of 2,3-branched saturated aldehydes (#24), as expected (Caponio et al., 2013; Pérez-Camino et al., 2001; Spyros et al., 2004; Tena et al., 2017). In VOOs stored at 35°C in the dark, other signals due to primary oxidation products such as hydroperoxide derivatives (#17), which degraded during storage to secondary oxidation products such as aldehydes (#24 and #26) (Conte et al., 2020; Choe et al., 2006; Farhoosh et al., 2013; Tena et al., 2017), were also among the most influential variables in the PLS-R model. Moreover, the degradation of secoiridoids (#21) was also important at 35°C (Krichene et al., 2015).

Table 6. PLS-R models to determine the storage time of a VOO kept at 25°C in the light or at different temperatures (25°C, 30°C and 35°C) in the dark[1]

PLS-R model2	Data3	n	PLS-comp	R-cal	R-val	RMSEP	Mean of |Yref-Ypred|	Confidence interval (95%)
NMR buckets (X) vs time (Y)	Light (0-12 months)	257	6	0.966	0.953	1.1	0.895	0.084
NMR buckets (X) vs time (Y)	Dark, 25°C (0-24 months)	260	5	0.946	0.926	2.8	2.27	0.21
NMR buckets (X) vs time (Y)	Dark, 30°C (0-24 months)	80	4	0.941	0.886	3.8	3.03	0.52
NMR buckets (X) vs time (Y)	Dark, 35°C (0-24 months)	80	5	0.958	0.925	3.1	2.61	0.39

[1]Abbreviations: n, number of samples; PLS-comp, number of PLS components; R-cal, correlation coefficient in calibration; R-val, correlation coefficient in validation; RMSEP, root mean square error in the prediction; 3-fold crossvalidation.

[2]X matrix = normalized intensities of the NRM buckets; Y matrix = storage time (month).

[3]Samples used to build the model.

Table 7. Weighted regression coefficient (Bw) of the most influent variables on the PLS-R model built to determinethe storage time of a VOO

Data	#	Bucket (ppm)	Bw	Functional group	Attribution
Light, 25°C (0-12 months)	14	5.07 5.05	-0.248 -0.328	>*CH*OCOR (glyceryl group)	*sn*-1,2-diacylglycerides
	18	4.33	-0.318	-CH_2OCOR (glyceryl group)	triacylglycerides
	19	4.07 4.05	0.176 0.179	>*CH*-OH (glyceryl group)	*sn*-1,3-diacylglycerides
	20	3.71 3.69	-0.419 -0.155	-CH_2OH (glyceryl group)	*sn*-1,2-diacylglycerides
	24	2.55 2.53	0.168 0.356	-R'CHX-*CH*R-CHO	2,3-branched saturated aldehydes
	27	2.31 2.29	-0.173 -0.165	-OCO-CH_2- (acyl group)	
	36	2.09	-0.177	-CH_2-CH=CH- (^{13}C satellite of signal at 1.96-2.07 ppm, acyl group)	
Dark, 25°C (0-24 months)	13	5.25	-0.304	>CHOCOR (glyceryl group)	triacylglycerides
	14	5.07 5.05	-0.430 -0.481	>CHOCOR (glyceryl group)	sn-1,2-diacylglycerides
	18	4.33 4.27	-0.550 -0.301	-CH2OCOR (glyceryl group)	triacylglycerides
	19	4.07 4.05	0.372 0.378	>CH-OH (glyceryl group)	sn-1,3-diacylglycerides
	20	3.71	-0.617	-CH2OH (glyceryl group)	sn-1,2-diacylglycerides
	24	2.53	0.306	-R'CHX-CHR-CHO	2,3-branched saturated aldehydes
	27	2.31 2.29	-0.339 -0.372	-OCO-CH2- (acyl group)	
	32	1.61	-0.332	-OCO-CH2-CH2- (acyl group)	
Dark, 30°C (0-24 months)	13	5.25	-0.299	>CHOCOR (glyceryl group)	triacylglycerides
	14	5.07	-0.327	>CHOCOR (glyceryl group)	sn-1,2-diacylglycerides
		5.05	-0.359		

Table 7. (Continued)

Data	#	Bucket (ppm)	Bw	Functional group	Attribution
Dark, 30°C (0-24 months)	18	4.33 4.27	-0.488 -0.309	-CH2OCOR (glyceryl group)	triacylglycerides
	19	4.09 4.07 4.05	0.313 0.351 0.343	>CH-OH (glyceryl group)	sn-1,3-diacylglycerides
	20	3.73 3.71	-0.270 -0.449	-CH2OH (glyceryl group)	sn-1,2-diacylglycerides
	24	2.53	0.371	-R'CHX-CHR-CHO	2,3-branched saturated aldehydes
	27	2.31 2.29	-0.346 -0.330	-OCO-CH2- (acyl group)	
	32	1.61	-0.294	-OCO-CH2-CH2- (acyl group)	
Dark, 35°C (0-24 months)	13	5.25 5.23	-0.229 -0.235	>CHOCOR (glyceryl group)	triacylglycerides
	16	4.47 4.45	-0.222 -0.225	-CH2OCOR (13C satellite of signal at 4.24-4.32 ppm, glyceryl group)	triacylglycerides
	17	4.37 4.35	-0.222 -0.224	>CH-OOH (methine proton of hydroperoxide group)	hydroperoxide derivatives
	18	4.29 4.27	-0.217 -0.267	-CH2OCOR (glyceryl group)	triacylglycerides
	19	4.07 4.05	0.254 0.228	>CH-OH (glyceryl group)	sn-1,3-diacylglycerides
Dark, 35°C (0-24 months)	21	3.61	-0.262	-CH(OHC-C=)-CH2-CHO (C5)	dialdehyde of secoiridoids (oleuropein, ligstroside) lacking a carboxymethyl group
	24	2.51	0.223	-R'CHX-CHR-CHO	2,3-branched saturated aldehydes
	26	2.35 2.33	0.216 0.234	-R'CHX-CHR-CHO	2,3-branched saturated aldehydes
	27	2.29 2.27	-0.237 -0.221	-OCO-CH2- (acyl group)	

Table 8. Intensity threshold values of 1H-NMR signals (buckets) related to the most important variables (buckets of 1H-NMR spectra) on the PLS-R model built to determine the storage time of a VOO

Data	Bucket (ppm)	Parameter	n	Freshness is lost when bucket intensity is … than the threshold	NMR signal intensity threshold (mean)	Confidence interval (95%)
Light, 25°C (0-12 months)	5.07	K270	17	<	7.66	0.42
	5.05	K270	17	<	1.273	0.041
	4.33	K270	17	<	5.16	0.13
	4.07	(E)-2-decenal	18	>	4.99	0.21
	4.05	(E)-2-decenal	18	>	3.05	0.12
	3.71	K270	17	<	9.19	0.53
	3.69	(E)-2-decenal	18	<	1.76	0.11
	2.55	(E)-2-decenal	18	>	1.20	0.12
	2.53	K270	17	>	1.88	0.12
	2.31	K270	17	<	888.7	1.7
	2.29	K270	17	<	1506.0	3.4
	2.09	K270	17	<	11.41	0.25
Dark, 25°C (0-24 months)	5.25	K270	17	<	285.59	0.67
	5.07	K270	17	<	7.56	0.44
	5.05	K270	17	<	1.270	0.032
	4.33	K270	17	<	5.503	0.096
	4.27	K270	17	<	516.7	1.3
	4.07	(E)-2-decenal	18	>	4.95	0.224
	4.05	(E)-2-decenal	18	>	3.15	0.154
	3.71	K270	17	<	9.69	0.41
	2.53	K270	17	>	1.404	0.083
	2.31	K270	17	<	891.4	1.8
	2.29	K270	17	<	1503.9	3.5
	1.61	K270	17	<	1043.9	2.9
	5.25	K270	4	<	285.0	2.7
	5.07	K270	4	<	7.3	1.2

Table 8. (Continued)

Data	Bucket (ppm)	Parameter	n	Freshness is lost when bucket intensity is … than the threshold	NMR signal intensity threshold (mean)	Confidence interval (95%)
Dark, 35°C (0-24 months)	5.05	K270	4	<	1.230	0.087
	4.33	K270	4	<	5.39	0.34
	4.27	K270	4	<	515.2	4.9
	4.09	(E)-2-decenal	5	>	7.92	0.20
	4.07	(E)-2-decenal	5	>	5.43	0.70
	4.05	(E)-2-decenal	5	>	3.41	0.48
	3.73	ΔK	4	<	1.691	0.095
	3.71	K270	4	<	9.1	1.4
	2.53	K270	4	<	1.54	0.30
	2.31	K270	4	<	889.5	8.9
	2.29	K270	4	<	1499.1	17.3
	1.61	K270	4	<	1041.4	18.5
	5.25	K270	4	<	284.3	2.5
	5.23	K270	4	<	81.8	2.5
	4.47	K270	4	<	0.493	0.039
	4.45	K270	4	<	1.646	0.063
	4.37	K270	4	<	1.49	0.16
	4.35	K270	4	<	2.28	0.32
	4.29	K270	4	<	330.6	3.0
	4.27	K270	4	<	514.3	4.8
	4.07	(E)-2-decenal	5	>	5.75	0.61
	4.05	(E)-2-decenal	5	>	3.69	0.55
	3.61	K270	4	<	0.660	0.081
	2.51	K270	4	<	1.00	0.17
	2.35	(E)-2-decenal	5	>	14.8	1.1
	2.33	(E)-2-decenal	5	>	50.8	2.8
	2.29	K270	4	<	1496	15
	2.27	K270	4	<	409.3	6.1

The signals of the compounds involved in the degradation of the VOO during storage are correlated to the "time of exceeding the threshold value" of the traditional parameters measured to establish when an oil is not "fresh" anymore according to the Commission Regulation (EEC) No 2568/91: 0.220 for the K_{270} for samples exposed to light or 0.190 for the K_{270} for samples stored in the dark (assumed since none of the samples has reached the legal limit of 0.220 stored in the dark); 0.01 for ΔK; 250 mg of total phenols/kg of oil (as required by the Health Claim according to the EU Reg 432/2012); and 410 μg of (*E*)-2-decenal/kg of oil or 354 μg of (*E,E*)-2,4-decadienal/kg of oil (representing the thresholds for the rancid defect) (Rosa María Alonso-Salces et al., 2021). Thus, the threshold of the intensity values of the most influential ^{1}H-NMR signals (buckets) on PLS-R models were estimated for the months at which the traditional parameters were exceeded. For this purpose, simple least squares regressions of the storage time versus the bucket intensity were built for each influential variable and VOO sample. This way, threshold values were determined for the principal ^{1}H signals to assess whether a VOO is fresh or not. Table 8 shows the intensity threshold value for each ^{1}H signal obtained with the data of the traditional parameter which afforded the best precision. These intensity threshold values were similar and presented satisfactory repeatability values (0.4–20% of relative standard deviation). The month of exceeding K_{270} threshold value (0.220) provided the best results to establish the intensity thresholds for the ^{1}H signals of triacylglycerides, *sn*-1,2-diacylglycerides, acyl groups of fatty acids and 2,3-branched saturated aldehydes; whereas the month of exceeding the (*E*)-2-decenal content threshold value (410 μg/kg), for the buckets of *sn*-1,3-diacylglycerides. Positive correlations had been already observed between light exposure duration and K_{270} and C_7-C_{11} aldehydes during VOO storage at ambient temperature (Esposto et al., 2017).

Estimation of the Best Before Date

The composition of VOO, once it has been extracted, influences and determines the evolution of VOO during its storage, which also depends on external factors such as light, temperature, oxygen availability and metal contamination from packing materials (Esposto et al., 2017; Velasco et al., 2002). Taking this into account, the ^{1}H-NMR spectral data of VOOs at time zero, i.e., at the moment of VOO production, would contain information related to the probable evolution of VOOs during their storage, and therefore, would allow to estimate the best before date for its consumption. In this sense, a PLS-R model was developed to determine the best before date of a VOO stored in the light at 25ºC (Table 9), using the data of the time exceeding the K_{270} parameter threshold for VOOs stored in this conditions. These preliminary results confirmed that ^{1}H-NMR spectral data contains the necessary information to achieve a regression model to estimate the best before date of a VOO by analysing the oil at time zero. However, this preliminary regression model should be put in perspective, and be aware that further research is required with a larger representative sample set of VOOs. These VOOs stored under different storage conditions (light exposure, temperature) have to be analysed at time zero, and then the evolution of their K_{270} parameter followed until the time this parameter is exceeded, in order to achieve robust PLS-R models with appropriate accuracies and precisions.

Table 9. PLS-R models to determine the best before date of a VOO stored in the light at 25ºC[1]

PLS-R model2	Data[3]	n	PLS-comp	R-cal	R-val	RMSEP	Mean of |Yactual-Ypred|	Confidence interval (95%)
NMR buckets (X) vs time (Y)	Light (time 0)	22	2	0.933	0.841	0.71	0.59	0.18

[1]Abbreviations: n, number of samples used to build the model; PLS-comp, number of PLS components; R-cal, correlation coefficient in calibration; R-val, correlation coefficient in validation; RMSEP, root mean square error in the prediction; leave-one-out crossvalidation.

[2]X matrix = normalized intensities of the NRM buckets; Y matrix = time (month) exceeding the threshold value of K_{270}.

[3]Samples used to build the model.

Conclusion

In the present chapter, the metabolic fingerprinting of VOO by ^{1}H-NMR spectroscopy combined with chemometrics is presented as a valuable analytical tool for the quality control and stability evaluation of VOOs. For this purpose, the most important factors that influence VOO quality and stability during storage, i.e., light exposure, temperature, time of storage and the initial composition of VOO were considered. ^{1}H-NMR has proved to provide data closely related with the evolution of the classical parameters, such as iodine value, peroxide value, conjugated dienes, etc., used to evaluate the oxidation stability of VOO. Principal component analysis of the ^{1}H-NMR fingerprints of VOO stored in the dark at room temperature allowed to visualize the evolution of VOO with time according to the changes in the ^{1}H-NMR signals of compounds involved in the hydrolytic and oxidative degradation of VOO. The potential of the classification and regression models achieved to determine whether a VOO is fresh, whether it has been stored in light or dark conditions, the storage time of VOO under different storage conditions and the best before date of a fresh VOO was proved. Besides, these predictive models disclosed the chemical compounds responsible for the compositional changes taking place in VOO due to its hydrolytic and oxidative degradation, which helps to understand the quality changes that VOO can experience during its shelf life. None of the signals present in the ^{1}H-NMR spectrum of VOOs at time zero disappeared during VOO storage over one year in the light and more than three and a half years in the dark. Over the time of storage, relatively small changes in the original ^{1}H signal intensities and the emergence of new low intensity signals indicated that some oxidative and hydrolytic degradation of VOO was occurring under normal market conditions. These results confirm the high oxidative stability of VOO at moderate temperatures, even exposed to light, and dismiss any significant changes which could render its consumption hazardous. However, those changes in the ^{1}H-NMR fingerprints of VOOs were evident enough to evaluate their quality according to legal, sensory and health-promoting aspects using pattern recognition techniques. Furthermore, this research corroborated that light and increasing temperature accelerate VOO degradation during its shelf life. Therefore, the use of packaging that protects VOO from light (e.g., dark glass bottles), minimal headspace and the control of temperature during transport, distribution and storage of VOO are highly recommended to guarantee VOO quality from production to consumption.

Acknowledgments

This work was developed in the framework of the project OLEUM "Advanced solutions for assuring authenticity and quality of olive oil at global scale" funded by the European Commission within the Horizon 2020 Programme (grant agreement No. 635690, 2014–2020). The information contained in this chapter reflects the authors' views; the European Commission is not liable for any use of the information contained herein. The authors would like to thank all producers that supplied the extra virgin olive oils for this study, and the technical and staff support provided by SGIker (UPV/EHU, MICINN, GV/EJ, ESF).

References

Alonso-Salces, R. M., Gallo, B., Collado, M. I., Sasía-Arriba, A., Viacava, G. E., García-González, D. L., Gallina Toschi, T., Servili, M., & Berrueta, L. Á. (2021). 1H–NMR fingerprinting and supervised pattern recognition to evaluate the stability of virgin olive oil during storage. *Food Control, 123*, 107831.

Alonso-Salces, R. M., Héberger, K., Holland, M. V., Moreno-Rojas, J. M., Mariani, C., Bellan, G., Reniero, F., & Guillou, C. (2010). Multivariate analysis of NMR fingerprint of the unsaponifiable fraction of virgin olive oils for authentication purposes. *Food Chemistry, 118*(4), 956-965.

Alonso-Salces, R. M., Holland, M. V., & Guillou, C. (2011). ^{1}H-NMR fingerprinting to evaluate the stability of olive oil. *Food Control, 22*(12), 2041-2046.

Alonso-Salces, R. M., Moreno-Rojas, J. M., Holland, M. V., Reniero, F., Guillou, C., & Heberger, K. (2010). Virgin olive oil authentication by multivariate analyses of ^{1}H NMR fingerprints and δ^{13}C and δ^2H data. *Journal of Agricultural and Food Chemistry, 58*(9), 5586-5596.

Alonso-Salces, R. M., Nicolas Segebarth,Sergio Garmón-Lobato, Margaret V. Holland, Jose M. Moreno-Rojas, Juan A. Fernández-Pierna, Vincent Baeten, Sandra R. Fuselli, Blanca Gallo, Luis Angel Berrueta, Fabiano Reniero, Claude Guillou, Károly Héberger. (2015). ^{1}H-NMR and isotopic fingerprinting of olive oil and its unsaponifiable fraction: Geographical origin of virgin olive oils by pattern recognition. *European Journal of Lipid Science and Technology, 117*(12), 1991-2006.

Aparicio, R., & Harwood, J. (2013). *Handbook of Olive Oil: Analysis and properties*. New York: Springer.

Baiano, A., Terracone, C., Viggiani, I., & Del Nobile, M. A. (2014). Changes produced in extra-virgin olive oils from cv. Coratina during a prolonged storage treatment. *Czech Journal of Food Sciences, 32*(1), 1-9.

Benito, M., Abenoza, M., Oria, R., & Sánchez-Gimeno, A. C. (2011). Influence of light on the stability and shelf life of olive oil. *Acta Horticulturae, 924*, 413-418.

Berrueta, L. A., Alonso-Salces, R. M., & Héberger, K. (2007). Supervised pattern recognition in food analysis. *Journal of Chromatography A, 1158*(1-2), 196-214.

Caponio, F., Paradiso, V. M., Bilancia, M. T., Summo, C., Pasqualone, A., & Gomes, T. (2013). Diacylglycerol isomers in extra virgin olive oil: Effect of different storage conditions. *Food Chemistry, 140*(4), 772-776.

Castellani, L., Serrilli, A. M., Bonadies, F., & Bianco, A. (2008). Natural phenols and diglycerides in virgin olive oil and their relation. *Natural Product Research, 22*(16), 1413-1417.

Circi, S., Ingallina, C., Vista, S., Capitani, D., Di Vecchia, A., Leonardi, G., D'Achille, G., Centauri, L., Camin, F., & Mannina, L. (2018). A multi-methodological protocol to characterize PDO Olive oils. *Metabolites, 8*(3), 43.

Conte, L., Milani, A., Calligaris, S., Rovellini, P., Lucci, P., & Nicoli, M. C. (2020). Temperature dependence of oxidation kinetics of extra virgin olive oil (EVOO) and shelf-life prediction. *Foods, 9*(3), 295.

Choe, E., & Min, D. B. (2006). Mechanisms and factors for edible oil oxidation. *Comprehensive Reviews in Food Science and Food Safety, 5*(4), 169-186.

Choe, E., & Min, D. B. (2009). Mechanisms of antioxidants in the oxidation of foods. *Comprehensive Reviews in Food Science and Food Safety, 8*(4), 345-358.

Christophoridou, S., & Dais, P. (2009). Detection and quantification of phenolic compounds in olive oil by high resolution ^{1}H nuclear magnetic resonance spectroscopy. *Analytica Chimica Acta, 633*(2), 283-292.

Dais, P., & Hatzakis, E. (2013). Quality assessment and authentication of virgin olive oil by NMR spectroscopy: A critical review. *Analytica Chimica Acta, 765*, 1-27.

Deiana, M., Rosa, A., Cao, C. F., Pirisi, F. M., Bandino, G., & Dessi, M. A. (2002). Novel approach to study oxidative stability of extra virgin olive oils: Importance of α-tocopherol concentration. *Journal of Agricultural and Food Chemistry, 50*(15), 4342-4346.

DG-AGRI. (2012). Action plan for the EU olive oil sector, DG Agriculture & Rural Development (Unit C-2).

D'imperio, M., Mannina, L., Capitani, D., Bidet, O., Rossi, E., Bucarelli, F., Quaglia, G., & Segre, A. (2007). NMR and statistical study of olive oils from Lazio: A geographical, ecological and agronomic characterization. *Food Chemistry, 105*(3), 1256-1267.

Esbensen, K. H., Guyot, D., Westad, F., & Houmøller, L. P. (2002). *Multivariate Data Analysis: In Practice: An Introduction to Multivariate Data Analysis and Experimental Design* (5th ed./Kim H. Esbensen/with contributions from Dominique Guyot, Frank Westad, Lars P. Houmøller. ed.). Oslo: Camo Process AS.

Esposto, S., Taticchi, A., Di Maio, I., Urbani, S., Veneziani, G., Selvaggini, R., Sordini, B., & Servili, M. (2015). Effect of an olive phenolic extract on the quality of vegetable oils during frying. *Food Chemistry, 176*, 184-192.

Esposto, S., Taticchi, A., Urbani, S., Selvaggini, R., Veneziani, G., Di Maio, I., Sordini, B., & Servili, M. (2017). Effect of light exposure on the quality of extra virgin olive oils according to their chemical composition. *Food Chemistry, 229*, 726-733.

Farhoosh, R., & Hoseini-Yazdi, S. Z. (2013). Shelf-life prediction of olive oils using empirical models developed at low and high temperatures. *Food Chemistry, 141*(1), 557-565.

Frankel, E. N. (2010). Chemistry of extra virgin olive oil: Adulteration, oxidative stability, and antioxidants. *Journal of Agricultural and Food Chemistry, 58*(10), 5991-6006.

Fregapane, G., Gómez-Rico, A., Inarejos, A. M., & Salvador, M. D. (2013). Relevance of minor components stability in commercial olive oil quality during the market period. *European Journal of Lipid Science and Technology, 115*(5), 541-548.

García-González, D. L., Vivancos, J., & Aparicio, R. (2011). Mapping Brain Activity Induced by Olfaction of Virgin Olive Oil Aroma. *Journal of Agricultural and Food Chemistry, 59*(18), 10200-10210.

Ghanbari Shendi, E., Sivri Ozay, D., Ozkaya, M. T., & Ustunel, N. F. (2018). Changes occurring in chemical composition and oxidative stability of virgin olive oil during storage. *OCL - Oilseeds and fats, Crops and Lipids, 25*(6).

Guillen, M. D., & Ruiz, A. (2001). High resolution H-1 nuclear magnetic resonance in the study of edible oils and fats. *Trends in Food Science & Technology, 12*(9), 328-338.

Guillen, M. D., & Ruiz, A. (2006). Study by means of H-1 nuclear magnetic resonance of the oxidation process undergone by edible oils of different natures submitted to microwave action. *Food Chemistry, 96*(4), 665-674.

Iqdiam, B. M., Welt, B. A., Goodrich-Schneider, R., Sims, C. A., Baker, G. L. I., & Marshall, M. R. (2020). Influence of headspace oxygen on quality and shelf life of extra virgin olive oil during storage. *Food Packaging and Shelf Life, 23*, 100433.

Kalua, C. M., Allen, M. S., Bedgood, J. D. R., Bishop, A. G., Prenzler, P. D., & Robards, K. (2007). Olive oil volatile compounds, flavour development and quality: A critical review. *Food Chemistry, 100*(1), 273-286.

Kjeldahl, K., & Bro, R. (2010). Some common misunderstandings in chemometrics. *Journal of Chemometrics, 24*(7-8), 558-564.

Koidis, A., & Boskou, D. (2014). Virgin olive oil: Losses of antioxidant polar phenolic compounds due to storage, packaging, and culinary uses. In *Processing and Impact on Active Components in Food*, (pp. 267-274).

Korifi, R., Plard, J., Le Dréau, Y., Rébufa, C., Rutledge, D. N., & Dupuy, N. (2016). Highlighting metabolic indicators of olive oil during storage by the AComDim method. *Food Chemistry, 203*, 104-116.

Köseoğlu, O., Sevim, D., & Kadiroğlu, P. (2019). Effects of filtration on the quality properties of extra virgin olive oils during storage. *JAOCS, Journal of the American Oil Chemists' Society, 96*(3), 291-301.

Kotsiou, K., & Tasioula-Margari, M. (2015). Changes occurring in the volatile composition of Greek virgin olive oils during storage: Oil variety influences stability. *European Journal of Lipid Science and Technology, 117*(4), 514-522.

Krichene, D., Allalout, A., Mancebo-Campos, V., Salvador, M. D., Zarrouk, M., & Fregapane, G. (2010). Stability of virgin olive oil and behaviour of its natural antioxidants under medium temperature accelerated storage conditions. *Food Chemistry, 121*(1), 171-177.

Krichene, D., Salvador, M. D., & Fregapane, G. (2015). Stability of virgin olive oil phenolic compounds during long-term storage (18 months) at temperatures of 5-50°C. *Journal of Agricultural and Food Chemistry, 63*(30), 6779-6786.

Lobo-Prieto, A., Tena, N., Aparicio-Ruiz, R., Morales, M. T., & García-González, D. L. (2020). Tracking sensory characteristics of virgin olive oils during storage: Interpretation of their changes from a multiparametric perspective. *Molecules, 25*(7), 1686.

López-Miranda, J., Pérez-Jiménez, F., Ros, E., De Caterina, R., Badimón, L., Covas, M. I., Escrich, E., Ordovás, J. M., Soriguer, F., Abiá, R., Alarcón de la Lastra, C., Battino, M., Corella, D., Chamorro-Quirós, J., Delgado-Lista, J., Giugliano, D., Esposito, K., Estruch, R., Fernandez-Real, J. M., Yiannakouris, N. (2010). Olive oil and health: Summary of the II international conference on olive oil and health consensus report,

Jaén and Córdoba (Spain) 2008. *Nutrition, Metabolism and Cardiovascular Diseases, 20*(4), 284-294.

Lukić, I., Da Ros, A., Guella, G., Camin, F., Masuero, D., Mulinacci, N., Vrhovsek, U., & Mattivi, F. (2020). Lipid profiling and stable isotopic data analysis for differentiation of extra virgin olive oils based on their origin. *Molecules, 25*(1), 4.

Mannina, L., & Segre, A. (2002). High resolution nuclear magnetic resonance: From chemical structure to food authenticity. *Grasas y Aceites, 53*(1), 22-33.

Mannina, L., Sobolev, A. P., & Segre, A. (2003). Olive oil as seen by NMR and chemometrics. *Spectroscopy Europe, 15*(3), 6-14.

Méndez, A. I., & Falqué, E. (2007). Effect of storage time and container type on the quality of extra-virgin olive oil. *Food Control, 18*(5), 521-529.

Morales, M. T., Aparicio-Ruiz, R., & Aparicio, R. (2013). Chromatographic Methodologies: Compounds for Olive Oil Odor Issues. In R. Aparicio & J. Harwood (Eds.), *Handbook of Olive Oil*, (pp. 261-309). Boston, MA: Springer.

Morales, M. T., Luna, G., & Aparicio, R. (2005). Comparative study of virgin olive oil sensory defects. *Food Chemistry, 91*(2), 293-301.

Owen, R. W., Haubner, R., Mier, W., Giacosa, A., Hull, W. E., Spiegelhalder, B., & Bartsch, H. (2003). Isolation, structure elucidation and antioxidant potential of the major phenolic and flavonoid compounds in brined olive drupes. *Food and Chemical Toxicology, 41*(5), 703-717.

Owen, R. W., Mier, W., Giacosa, A., Hull, W. E., Spiegelhalder, B., & Bartsch, H. (2000). Identification of lignans as major components in the phenolic fraction of olive oil. *Clinical Chemistry, 46*(7), 976-988.

Papadia, P., Del Coco, L., Muzzalupo, I., Rizzi, M., Perri, E., Cesari, G., Simeone, V., Mondelli, D., Schena, F. P., & Fanizzi, F. P. (2011). Multivariate analysis of ^{1}H-NMR spectra of genetically characterized extra virgin olive oils and growth soil correlations. *JAOCS, Journal of the American Oil Chemists' Society, 88*(10), 1463-1475.

Pérez-Camino, M. C., Moreda, W., & Cert, A. (2001). Effects of Olive Fruit Quality and Oil Storage Practices on the Diacylglycerol Content of Virgin Olive Oils. *Journal of Agricultural and Food Chemistry, 49*(2), 699-704.

Pérez-Trujillo, M., Gómez-Caravaca, A. M., Segura-Carretero, A., Fernández-Gutiérrez, A., & Parella, T. (2010). Separation and identification of phenolic compounds of extra virgin olive oil from *Olea europaea* L. by HPLC-DAD-SPE-NMR/MS. Identification of a new diastereoisomer of the aldehydic form of oleuropein aglycone. *Journal of Agricultural and Food Chemistry, 58*(16), 9129-9136.

Psomiadou, E., & Tsimidou, M. (1999). On the role of squalene in olive oil stability. *Journal of Agricultural and Food Chemistry, 47*(10), 4025-4032.

Psomiadou, E., & Tsimidou, M. (2002). Stability of virgin olive oil. 2. Photo-oxidation studies. *Journal of Agricultural and Food Chemistry, 50*(4), 722-727.

Rotondo, A., Salvo, A., Giuffrida, D., Dugo, G., & Rotondo, E. (2011). NMR analysis of aldehydes in sicilian extra-virgin olive oils by DPFGSE techniques. *AAPP Atti della Accademia Peloritana dei Pericolanti, Classe di Scienze Fisiche, Matematiche e Naturali, 89*(1), C1A8901002.

Sacchi, R., Patumi, M., Fontanazza, G., Barone, P., Fiordiponti, P., Mannina, L., Rossi, E., & Segre, A. L. (1996a). A high-field ^{1}H nuclear magnetic resonance study of the minor components in virgin olive oils. *JAOCS, Journal of the American Oil Chemists' Society, 73*(6), 747-758.

Sacco, A., Brescia, M. A., Liuzzi, V., Reniero, F., Guillou, G., Ghelli, S., & van der Meer, P. (2000). Characterization of Italian olive oils based on analytical and nuclear magnetic resonance determinations. *Journal of the American Oil Chemists Society, 77*(6), 619-625.

Salvo, A., Rotondo, A., La Torre, G. L., Cicero, N., & Dugo, G. (2017). Determination of 1,2/1,3-diglycerides in Sicilian extra-virgin olive oils by ^{1}H-NMR over a one-year storage period. *Natural Product Research, 31*(7), 822-828.

Servili, M., Esposto, S., Fabiani, R., Urbani, S., Taticchi, A., Mariucci, F., Selvaggini, R., & Montedoro, G. F. (2009). Phenolic compounds in olive oil: antioxidant, health and organoleptic activities according to their chemical structure. *Inflammopharmacology, 17*(2), 76-84.

Servili, M., Esposto, S., Taticchi, A., Urbani, S., Di Maio, I., Sordini, B., & Selvaggini, R. (2014). The effect of diverse agricultural and technological factors on olive oil quality and yield. *Acta Horticulturae, 1057*, 603-618.

Servili, M., Esposto, S., Taticchi, A., Urbani, S., Di Maio, I., Veneziani, G., & Selvaggini, R. (2015). New approaches to virgin olive oil quality, technology, and by-products valorization. *European Journal of Lipid Science and Technology, 117*(11), 1882-1892.

Shi, T., Zhu, M., Zhou, X., Huo, X., Long, Y., Zeng, X., & Chen, Y. (2019). ^{1}H NMR combined with PLS for the rapid determination of squalene and sterols in vegetable oils. *Food Chemistry, 287*, 46-54.

Silva, S. F., Anjos, C. A. R., Cavalcanti, R. N., & Celeghini, R. M. D. S. (2015). Evaluation of extra virgin olive oil stability by artificial neural network. *Food Chemistry, 179*, 35-43.

Spyros, A., Philippidis, A., & Dais, P. (2004). Kinetics of diglyceride formation and isomerization in virgin olive oils by employing ^{31}P NMR spectroscopy. Formulation of a quantitative measure to assess olive oil storage history. *Journal of Agricultural and Food Chemistry, 52*(2), 157-164.

Stefanoudaki, E., Williams, M., & Harwood, J. (2010). Changes in virgin olive oil characteristics during different storage conditions. *European Journal of Lipid Science and Technology, 112*(8), 906-914.

Tena, N., Aparicio, R., & García-González, D. L. (2017). Virgin olive oil stability study by mesh cell-FTIR spectroscopy. *Talanta, 167*, 453-461.

Tsimidou, M. Z. (2010). Squalene and tocopherols in olive oil: Importance and methods of analysis. In V. Preedy & R. Watson (Eds.), *Olives and olive oil in health and disease prevention*, (pp. 561-567). London: Academic Press.

Ün, İ., & Ok, S. (2018). Analysis of olive oil for authentication and shelf life determination. *Journal of Food Science and Technology, 55*(7), 2476-2487.

Velasco, J., & Dobarganes, C. (2002). Oxidative stability of virgin olive oil. *European Journal of Lipid Science and Technology, 104*(9-10), 661-676.

Chapter 4

A Chemometric Study of Chemical Research of Element Accumulation in Mushrooms

Marija V. Dimitrijevć and Dragoljub L. Miladinović*

Department of Pharmacy, Faculty of Medicine,
University of Niš, Niš, Serbia

Abstract

Chemometrics is a discipline of chemistry that finds correlation between specific data using mathematical and statistical methods. During any thorough research, the scientists are handling vast amounts of data related to the samples which are being researched. In this type of research, finding the correlation (similarities or differences) between analyzed samples and data is of great importance. The contribution of chemometrics is in its way to objectively analyze the results and groups them based on their common characteristics and not per assumptions of scientists. This prevents subjective analysis of given results and shows correlation between the results which cannot be always identified easily in matrix that can often consist of a lot of data. For this type of analysis, many different methods can be used, which are divided in unsupervised and supervised.

In this chapter, we will be talking about unsupervised methods where principal component analysis (PCA) and hierarchical cluster analysis (HCA) are part of. Using these methods, we can spot the correlation between the samples and underlying data structures without the potential bias of scientists about the previous knowledge of data of the samples. Theoretical basics and practical contribution of these methods can be better understood in a specific example. The PCA and HCA methods

* Corresponding Author's Email: dragoljubm@gmail.com.

In: Chemometrics
Editor: Larry D. Crenshaw
ISBN: 979-8-88697-274-0
© 2022 Nova Science Publishers, Inc.

were applied to a set of data on element accumulation in a great number of mushrooms. Both review and our own data were used regarding the wild edible mushrooms gathered in the forests of European countries (Serbia, Poland, and Turkey) and China. The essential elements were used as variables, and depending on their content in mushrooms, the groups of mushrooms were formed. The graphical representation of HCA method gives the ability of determining the number of clusters, but not the reason for clustering of mushrooms. Unlike HCA, the graphical representation of PCA gives more information and explanation about grouping of certain mushroom species. The PCA method shows a certain clustering of samples which can be explained by minimal or maximum content of certain elements which are characteristic for that cluster.

Keywords: chemometrics, PCA, HCA, element accumulation, mushrooms

Introduction

Any research leads to a set of data which are characteristic for a specific sample. In order to better understand the data and eliminate the results which are the consequence of the serious error, statistical analysis is used. A statistic is a numerical measurement describing some characteristic of a sample (Berenson et al., 2011). Development of computers and computer software has attributed to the greater usage of statistics. Statistical data analysis, by itself, does not represent a replacement for chemical, biochemical, physical, biological and other analysis (Pastor, 2018).

Keller (2014) divided statistics in two groups: descriptive, which has the task of making the data easily understandable and inferential statistics which is about making conclusions about the group based on the given data samples. The characteristics of samples which are studied are called variables (Babbie, 2009), and they represent variable characteristics of the studied sample – variability. In the beginning of statistical data analysis, only two variables were used, so the statistics itself was nothing more than a descriptive analysis of some data. With the development of technology and the appearance of analysis where the end result was a set of data with many variables, the new multivariate statistical analysis (MSA) started emerging with the task of determining correlation of multiple variables. Multivariable analysis can be defined as a skill of highlighting chemically relevant information from a lot of data which are gathered from the chemical experiments, using statistical and mathematical means (Pastor, 2018).

A chemical discipline that uses statistical and mathematical methods, to design or select optimum procedures and experiments, and to provide maximum chemical information by analyzing chemical data is chemometrics (Varmuza and Filzmose, 2009). The most important task of chemometric is the application of multivariate analysis to the results obtained by chemical research.

Unlike the traditional statistical methods, multivariate analysis makes multidisciplinary approach possible, which includes detection of the hidden correlation among the variables and dividing useful from useless information. All of the MSA methods can be split into two groups: supervised methods, which seek the correlation between independent and dependent variables, and non-supervised. The supervised methods have a specific correlation among the variables, where the variation of dependent and independent variables is tracked.

Unlike supervised, non-supervised methods determine the structure of the data set based on measurements, and they determine whether there is a relationship between variables that are of equal importance and are considered as independent. These interdependent methods are important because they enable the discovery of, at first glance, hidden connections among complex data. They reveal the pattern of grouping among the analyzed data and indicate the variables that have the most significant influence on the formation of these patterns (Aparicio and Aparicio-Ruiz, 2002). This group of methods includes: principal component analysis, cluster, factor and correspondence analysis, structural equations and multidimensional scaling. Using MSA, the volume of data is decreasing while retaining only the important and useful information that make interpretation of different chemical and physical analytical data easier.

Cluster Analysis

Cluster analysis (CA) is a statistical technique within multivariatie analysis that is applied in order to reduce the volume of analyzed data (e.g., experimental results), which primary purpose is to group data into clusters based on the characteristics they possess (Apostolov, 2014). However, clustering will not always provide the desired solution. For this reason, many clustering algorithms have been proposed in the literature that do not only perform differently, but that even work on different principles (Varmuza and

Filzmose, 2009). Data grouping depends on the chosen method, so it is possible to classify the data based on their simmilarities or dissimilarities.

The choice of method depends solely on the researcher and what he wants to show with cluster analysis. Regardless of the chosen method, the end goal is always clustering and the construction of a hierarchy of a treelike structure – dendrogram (Figure 1).

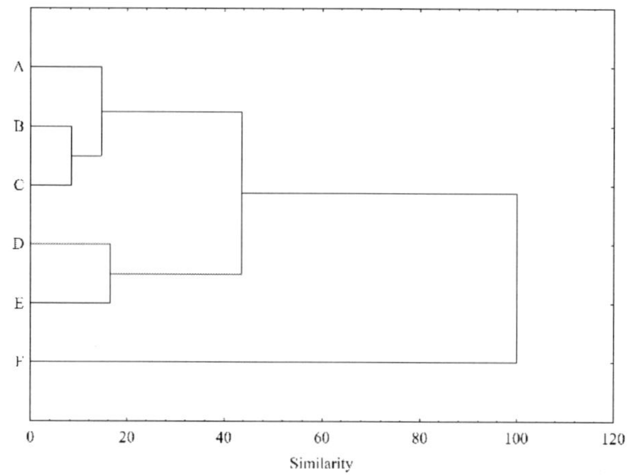

Figure 1. Example of clustering treelike structure – dendrogram.

On the dendrogram, the distance for each cluster is read from the horizontal axis. When connecting objects is needed, it is important to determine the distance between them. This can be calculated using different mathematical calculations, but the most common is Euclid's distance.

According to the applied algorithms on the basis of which the grouping is performed, cluster analysis can be divided into hierarchical and non-hierarchical analysis. The difference between these two methods is in the way the variables are connected. Hierarchical method is based on calculating the distance between the objects, after which the technic of merging object into groups is applied, and finally, merging close groups in one larger group which is then presented as a dendrogram (Mrmošanin, 2019).

Hierarchical methods of groupping can be agglomerative and divisive. The most commonly used grouping method is agglomerative, in which each unit is considered as a separate group that, based on similarity, is combined into a larger cluster, until a single cluster is obtained. Divisional technique is a separation technique where a group of smaller clusters is obtained from one cluster in which all data are contained.

Linkage criterion can be different and they depend on the analyzed data and the purpose of its application. The most commonly used linkage criterion are:

Single linkage clustering-nearest neighbour – based on the highest simmilartiy among the analyzed objects, e.g., their shortest distance (Figure 2). In this method, the cluster is formed and then compared with the next analyzed object and in this way the result are the so called clusters of "long links/chains."

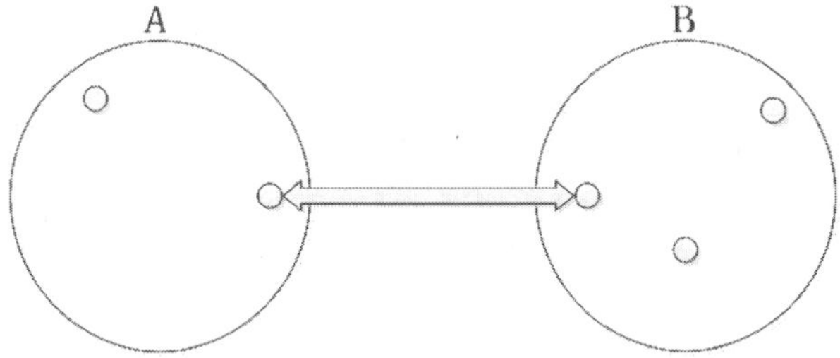

Figure 2. Conceptual diagram of the single linkage.

Complete linkage clustering is based on the least simmilarity, e.g., the objects with the greatest distance are being connected (Figure 3).

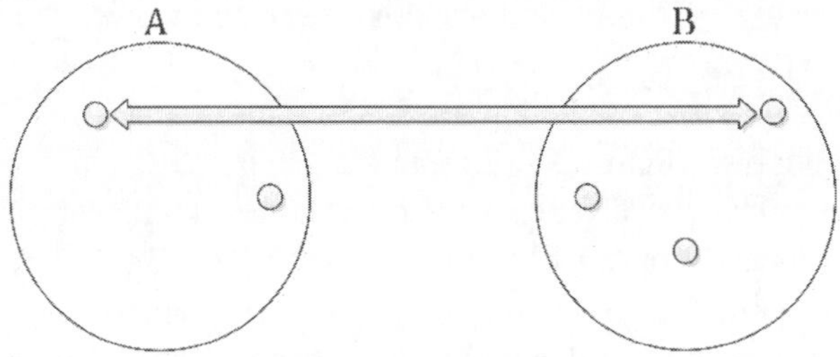

Figure 3. Conceptual diagram of the complete linkage.

Ward's method is based on analysis of variance where the distance between two clusters is determined as the sum of the squares between all the variables that make up the cluster (Figure 4). Two clusters are merged if their association resulted in a minimal increase in the sum of squares within the cluster in relation to the increase in the sum of squares that would occur if any other two clusters merged.

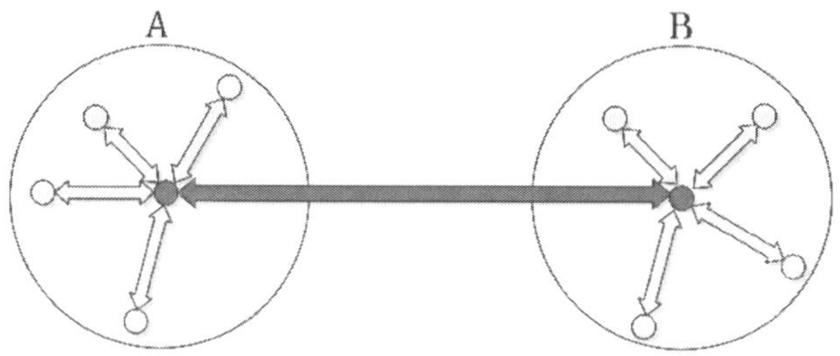

Figure 4. Conceptual diagram of the Ward's methods.

Figure 5 (p. 186) shows differences between results of applying single linkage (a), complete linkage (b) and Ward's method (c) on the same results, explained by Landau and Chis Ster, 2010.

The results of clustering with Ward's method were most consistent with the original samples. This is because the method required the distance between samples in Euclidean distance, and the distance between two clusters was significantly affected by the number of samples in the two clusters (Bu et al., 2020). A Its dendrogram is structured, direct, accurate, and clearly represents the classification results. In the event of little differentiation in samples, this method enlarges the differences between clusters and captures the essential attributes of clusters, thereby providing accurate and reliable classification results (Bu et al., 2016). In addition, Ward's method is interesting because it looks for clusters in multivariate Euclidean space which it is also the reference space in multivariate ordination methods, and in particular in principal component analysis (PCA) (Murtagh and Legendre, 2014).

Non-hierarchical analysis is less common in natural science and is used when the number of analyzed data is over 200. The principal of grouping is opposite of hierarchical. First, the central cluster is determined, in which all observation units are located, and then the second cluster is determined, and if the objects of observation are closer to it, they are arranged in that other cluste (Nikolić, 2018).

During the analysis, variables of samples that have different orders of magnitude and measurement units are often obtained. Clustering in cluster analysis based on the similarity between samples leads to variables with large values being given more weight than small ones. Standardization of variables would be necessary in these cases. The purpose of standardization is to equalize the variability of the variables which will be further used for cluster analysis.

Standardization of variables involves dividing the value of each variable for every data point by a measure of the variance for that variable, usually the standard deviation (Adamson and Bawden, 1981). There is no change in scale in the variables during standardization. With unstandardized values, the clustering is most common completely featureless and valueless. With standardized data, all the clustering techniques, gave well-structured classifications, based upon substituent type although all differ in detail (Adamson and Bawden, 1981).

Principal Component Analysis

Principal component analysis (PCA) can be considered as "the mother of all methods in multivariate data analysis" (Varmuza and Filzmose, 2009). It is a statistical method used in comprehensive research where the result is a set of complex data that are mathematically transformed to a smaller number of variables, while retaining the original information. This not only reduces the number of variables in the analysis, but also makes progress in understanding the structure of the studied phenomenon, so it can be said that the PCA method is a research tool used to generate hypotheses about the studied phenomenon (Delić, 2003).

The main goal of the PCA method is to identify the factors, which are difficult to notice, in the original data set for easier interpretation. The second goal is to determine several linear combinations of original variables, which, in addition to having maximum variance, will be uncorrelated with each other and will lose as little as possible information contained in the set of original variables. The end result of the analysis is transformed variables called principal components (PCs) and representing a linear combination of the original variables. The total number of newly formed PCs is equal to the number of the initial variables. However, not all of the PCs are of equal importance, only the first few of them contain the important data.

The first principal component, PC1, stands out with the largest variance of the original variables, while all subsequent main components are formed by including the maximum share of the remaining variables.

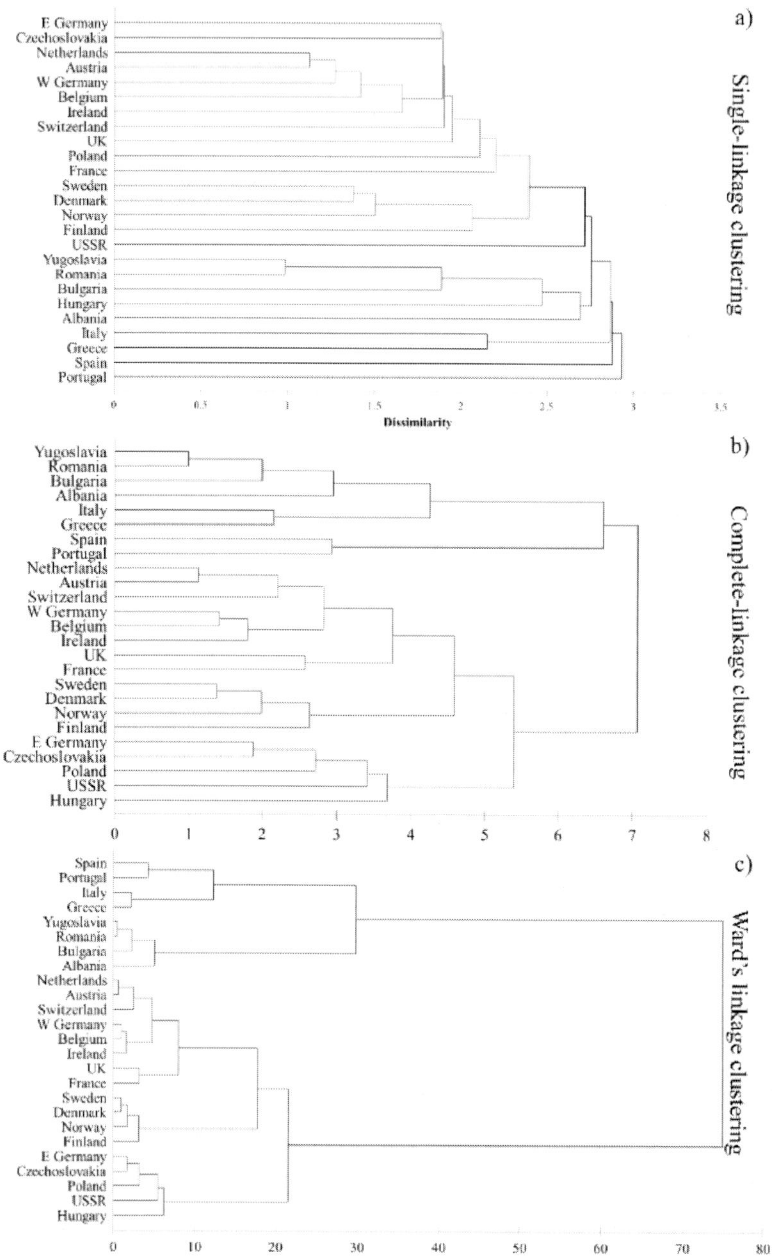

Figura 5. Dendograms showing partitions resulting from (a) single linkage, (b) complete linkage, and (c) Ward's method applied to protein consumption proximity matrix (Landau and Chis Ster, 2010).

Every principal component is defined by eigenvector which is determined by eigenvalue. Eigenvalue determines the degree of variation described using factorial coordinates, where the biggest impact has the first component (Arfken, 1985). The first principal component always has the highest eigenvalue. It is necessary to determine the number of eigenvalues during the analysis. The number of significant components can be determined using the criteria established by Kaiser (Kaiser, 1960) and, according to him, the important principal components are those whose value is higher than one. Principal components with low eigenvalue usually explain the mistakes during the measuring process. Since one of the goals of the PCA method is better interpretation of data, the possibility of the graphical presentation of the data is of great importance. The PCA diagram consists of higher or lower number of points which represent the analyzed variables which, if close to each other, show the similarities of the analyzed sample, or differences, if they are far apart. Vectors that describe a variable with their orientation in space indicate an increasing value of that variable, while the angle of the vector between the variables shows the degree of correlation that is greater the smaller the angle is.

The squared cosine shows the importance of a component for a given observation and indicates the contribution of a component to the squared distance of the observation to the origin (Abdi and Williams, 2010). Also, based on the square of the cosines of the angle between two vectors of variables, the mutual correlation can be determined. When this value is > 0,8 it is considered that the proporties are presented in the best possible way, between 0,8 and 0,6 – the proporties are well presented, between 0,6 and 0,4 the proporties are more or less adequatly presented and if the value is < 0,4 the proporties are presented poorly.

Results and Discussion

Element Accumulation in Mushrooms via Chemometrics

In this chapter, we will be talking about the practical appliance of chemometrics, and discussion of the results obtained in software. For better understanding, the PCA and HCA methods were applied to a set of data on element accumulation in a great number of mushrooms (35 mushroom species). The goal of multivariate unsupervised methods is to evaluate whether clustering exists in a dataset without using class membership information in the calculations (Beebe, Pell and Seasholtz, 1998).

Table 1. Accumulation of essential elements in wild edible mushrooms gathered in the forests of European countries (Serbia, Poland, and Turkey) and China (mg/kg)

Species	Cu	Fe	K	Mn	Zn	Reference
B. speciosus	44	990	7561	25.5	102	Wang et al. 2015
B. griseus	32	1180	16309	55.5	165	Wang et al. 2015
B. luridus	42.5	1358.5	9400	76	91	Wang et al. 2015
B. umbriniporus	38.5	552	9550	23.5	61.5	Wang et al. 2015
B. tomentipes	38	563.5	11131.5	27.5	102.5	Wang et al. 2015
B. edulis	41.5	2565.5	7139	48	64.5	Wang et al. 2015
B. pallidus	58.5	1649	17146.5	37.5	41	Wang et al. 2015
B. brunneissimus	24.5	721.5	11038.5	33.5	127.5	Wang et al. 2015
B. rubellus	45.5	2462	11169	56	177	Wang et al. 2015
B. aereus	49	1488.5	9635	51.5	120	Wang et al. 2015
B. edulis	109.1	148.3	5471.9	33.3	275.8	Turfan et al. 2018
C. cornucopioides	108.5	598.3	9319.2	147.6	378.5	Turfan et al. 2018
L. deliciosus	31.2	298.7	4848.4	66.1	393.9	Turfan et al. 2018
C. cibarius	54	330	49000	30	82	Brzezicha-Cirocka et al. 2016
B. edulis	37	200	27000	21	160	Brzezicha-Cirocka et al. 2016
B. edulis	36	47	32000	8.6	210	Brzezicha-Cirocka et al. 2016
L. auranticum	41	270	35000	24	112	Brzezicha-Cirocka et al. 2016
C. cibarius	58.8	165.1	41823.3	12.74	86.9	Cvetković et al. 2015
B. edulis	12.2	45.9	11266.9	9	38.7	Cvetković et al. 2015
L. piperatus	42.6	78.8	32827.1	14.5	45.2	Cvetković et al. 2015
L. aurantiacum	12.7	158.5	23187.1	23.3	72	Mleczek et al. 2013
X. badius	11.6	152	22552.5	21.8	31.9	Mleczek et al. 2013
L. deliciosus	14	132.6	24987.5	30.2	88.7	Mleczek et al. 2013
B. edulis	10.7	94.1	21308.5	22.7	51.6	Mleczek et al. 2013
C. cibarius	8.2	98.1	21017.9	35.7	26.5	Mleczek et al. 2013
B. edulis	15.5	55.21	11781.2	5.05	69.9	Dimitrijević et al. 2015
B. impolitus	11.9	46.2	15871.6	20	173.4	Dimitrijević et al. 2015
B. regius	10.6	56.9	13676.1	8.2	78.7	Dimitrijević et al. 2015
L. crocipodium	17.3	26.3	18856.9	11.5	68.5	Dimitrijević et al. 2015
L. pseudoscaber	27.2	514.8	21192.3	14.7	130.3	Dimitrijević et al. 2015
X. badius	4.7	45.7	16588.8	23.8	53.7	Dimitrijević et al. 2015
L. sangufluus	0.3	307	17278	115	43	Dimitrijević 2021
L. semisanguifluus	5.7	202	17426	9.4	32	Dimitrijević 2021
L. deliciosus	5	27	14431	7.5	150	Dimitrijević 2021
L. piperatus	47	114	16845	9.8	91	Dimitrijević 2021

Both sets of data used, the review and our own, were regarding the wild edible mushrooms gathered in the forests of European countries (Serbia, Poland, and Turkey) and China (Table 1).

For chemometrics analysis, the essential elements of choice were copper, iron, potassium, manganese, and zinc, which were used as variables. The reason why only these elements were chosen is that only these five elements are present in all of the chosen manuscripts related to accumulation of elements in wild edible mushrooms. Using HCA and PCA, we can spot the correlation between the samples and underlying data structures and eliminate the potential bias of scientists about the previous knowledge of data of the samples. Contribution of these methods can be better understood in an example.

HCA Analysis

To classify mushroom species in the space of studied elements, the HCA was applied. Ward's method was used to classify mushroom species based on the content of selected elements, using Euclidean distances as a measure of similarity. Euclidean distances were presented as a ratio $(D_{link}/D_{max}) \times 100$, where D_{link} is the distance between variables that are grouped, and D_{max} is the maximum distance between variables (Nikolić, 2018).

Since the Euclidean distance measured is considerably affected by scaling factors, the classifications obtained using raw (i.e., unstandardized) and standard data are usually different when this coefficient is used (Adamson and Bawden, 1981). Therefore, cluster analysis was done after the standardization of variables.

The results obtained following HCA are shown as a dendrogram (Figure 6) and analyzed mushroom species are grouped in four statistically significant clusters at $(D_{link}/D_{max})100 < 50$. Cluster A was constituted by *C. cibarius, B. edulis, L. deliciosus, X. badius, L. aurantiacum,* (mushrooms harvested in Poland), *L. deliciosus, B. impolitus, L. semisanguifluus, X. badius, L. crocipodium, B. regius, B. edulis and B. edulis* (mushrooms harvested in Serbia). Within this cluster, there are the greatest similarities between mushrooms. The disadvantage of cluster analysis is that it does not give us a reason to associate the samples but indicates which samples show the greatest similarity.

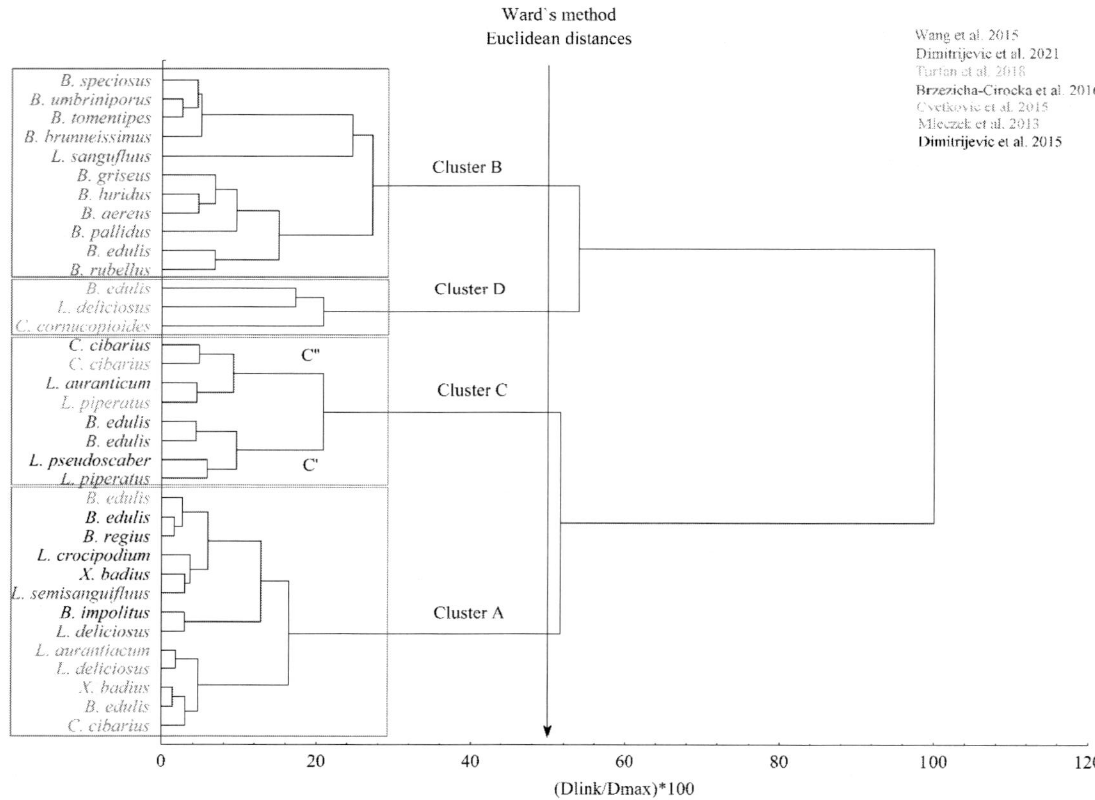

Figure 6. Dendrogram of hierarchical cluster analysis of selected mushroom species.

Based on the dendrogram and the Euclidean distance (Miladinović et al., 2012), it can be concluded that within cluster A, the greatest similarity is shown by *X. badius* and *B.edulis,* mushrooms from Poland (Euclide distance is 0.28). Then follow mushrooms collected in Serbia, *B. edulis* and *B. regius* (Euclide distance is 0.30) and *L. aurantiacum* and *L. deliciosus* (mushrooms from Poland) (Euclide distance is 0.35).

Cluster B was constituted by mushrooms from China (*B. speciosus, B. umbriniporus, B. tomentipes, B. brunneissimus, B. griseus, B. luridus, B. aereus, B. pallidus, B. edulis* and *B. rubellus*), except one species from Serbia (*L. sanguifluus*). In this cluster, the species *B. umbriniporus* and *B. tomentipes* show the greatest similarity with 0.51 Euclide distance.

Cluster C contains two almost identical subclusters, C' and C''. Two samples of *B. edulis* from Poland show the greatest similarity in the subcluster C' (Euclide distance is 0.88), while two samples of *C. cibarius*, one from Poland and another from Serbia show similarity in the subcluster C'' (Euclide distance is 0.96).

Mushroom samples from Turkey, *B. edulis, L. deliciosus* and *C. cornucopioides* were grouped into a separate cluster D. There is the greatest variation in the content of elements within this cluster. This fact is confirmed by the values of the Euclidean distance, which are the largest for the species of mushrooms that form this cluster.

As it can be seen from the given example, HCA method enables analysis of large volume of mushroom samples and visually systemize easily. Since the mushroom samples are collected from different locations, the accumulation of elements is also different, however, some similarities were observed using the HCA method. As can be concluded from the discussion, cluster analysis connects the variables used into groups of objects (i.e., clusters), based on their similarities but it does not provide any information about which specific similarities are used to group in cluster.

PCA Analysis

Whereas research in any field of chemistry provides large data sets, it is necessary to reduce the amount of data and still preserve the information that indicates that research. The PCA technique is most commonly used for these purposes. Its idea is simple - reduce the dimensionality of a dataset, while preserving as much ‘variability’ (i.e., statistical information) as possible (Jolliffe and Cadima, 2016).

Results of element accumulation in different mushroom species were subjected to the extraction of a principal component. Based on the concentrations of elements found in the different species, PCA was applied to visualize the arrangement of samples in an n-dimensional space and grouped samples according to similarities. Analysis was applied to the data matrix having five variables (copper, iron, potassium manganese, and zinc) and 35 cases (review data). For visual analysis, usually two-dimensional or three-dimensional projection of samples is constructed having the axes (principal components, PC) as the factors (Granato et al., 2018).

The first step in interpreting the results is to determine how many PCs need to be examined, i.e., how much should be reduced from the total number of variables because the point of PCA is to reduce dimensionality. Calculating the percent of total variance explained by each PC it is noted that the contribution is not equal for all PCs (Table 2).

Table 2. Eigenvalues of correlation matrix

Principal Component	Eigenvalue	Total variance %	Cumulative Eigenvalue	Cumulative %
PC1	2.28	45.58	2.28	45.58
PC2	1.10	22.08	3.38	67.66
PC3	0.80	16.09	4.19	83.75
PC4	0.55	10.95	4.74	94.71
PC5	0.26	5.29	5.00	100.00

The percentage of each principal component shows its contribution along a given axis to the variability of the sample (Miladinović et al., 2013). Based on the given values, it can be noticed that the first principal components have a significant contribution in the modeling of variables.

The elements in the diagonal of the matrix are the eigenvalues, that is, the variance explained by each principal component[1] The number of eigenvalues is determined by the number of variables used in the analysis and, it value decreases from the first principal component to the last. Eigenvalues are commonly plotted on a scree plot to show the decreasing rate at which variance is explained by additional principal components (Figure 7)[2]

The first thing to consider in the scree plot is a marked drop in the percentage of explained variation. The number of eigenvalues that will be considered in the PCA analysis and their contribution in total variance was

[1] http://strata.uga.edu/8370/lecturenotes/principalComponents.html.
[2] http://strata.uga.edu/8370/lecturenotes/principalComponents.html.

determined. One possible way to determine the number of PC is based on the Kaiser criterion (Kaiser, 1960): eigenvalues higher than 1 are considered as "significant" in the PCA analysis.

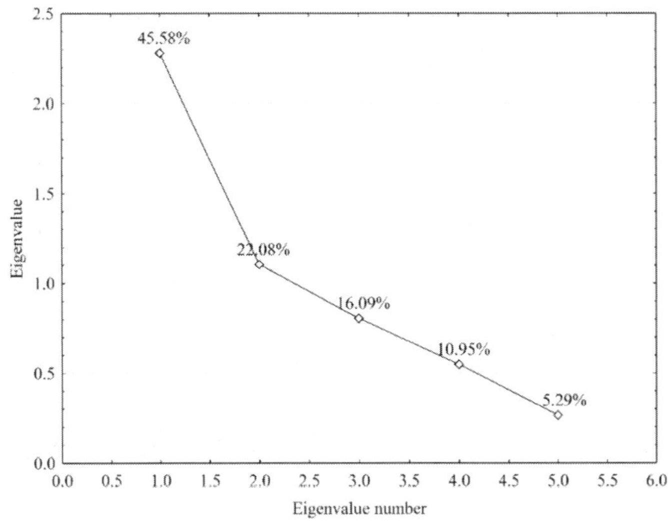

Figure 7. Eigenvalues of correlation matrix.

This statistical analysis resulted in a two-component model that explained 67.66% of information regarding the element compositional variability in mushrooms. The eigenvalue of the first principal component is the largest and indicates its most significant contribution in the total variance. Science two principal components with eigenvalues exceeding one (2.28 and 1.10) were extracted, according to Kaiser Criterion and explained 67.66%, only they will be used for further consideration because the information contained in the remaining variables is not of great importance for analysis. The first principal component (PC1) explained 45.6% of the total variance, and the second 22.1%. PC1 is generally better correlated with the variables than PC2. This is expected because PCs are extracted successively, each one accounting for as much of the remaining variance as possible (Hossain et al., 2011).

Manganese had the most significant contribution on PC1 (26.8%), while the highest contribution on PC2 had variables K (30.3%), and the Fe (30.5%) (Table 3).

Since the number of the PCs has been determined, loading analysis is started in order to find the underlying relationships in the original data structure and to be explained how our variables contribute to each of the PCs, and this is revealed by the loadings. Positive loadings indicate a variable, and a principal component are positively correlated: an increase in one result in an increase in the other. Negative loadings indicate a negative correlation. Large (either positive or negative) loadings indicate that a variable has a strong effect on that principal component[3].

In this step, loading could be visualized as a regression vector (a vector of correlation coefficients between the original variables with each PC-score) (Granato et al., 2018).

Table 3. Loadings of the variables for the first two principal components

Variable	Factor loading		Variable contributions (%)	
	Factor 1	Factor 2	Factor 1	Factor 2
Cu	-0.713	0.447	22.3	18.1
Fe	-0.582	-0.580	14.9	30.5
K	0.541	0.579	12.8	30.3
Mn	-0.782	-0.023	26.8	0.0
Zn	-0.726	0.482	23.1	21.0

Table 3 presents loadings for principal components that were used, and large loadings are highlighted in boldface to emphasize the variables that contribute to each principal component.

It can be observed that PC1, with high loadings >0.7 (except for P -0.582), explained 45.58% of data variation and has a strong negative loading for copper, iron, manganese, and zinc. Similarly, PC 2 has a positive loading for potassium (0.579) and explained 22.08% of the total variance. The positive factor loadings indicate that the factor will be higher in the positive axis of that PC (Granato et al., 2018). Loadings values higher than 0.75 are considered strong, between 0.75–0.5 moderate and 0.5–0.3 are weak based on their absolute values (Nair et al., 2010).

In the end, it remains to visualize the distribution of the samples. A projection of data sets on the factors plane allowed the visualization of the relationship between the element's concentrations and the mushroom species (Drewnowska and Falandysz, 2015). In the Figure 8, the location of samples and vectors that correspond to the loadings can be seen. These vectors illustrate how an increase in each variable influences where a sample plots in this space (Drewnowska and Falandysz, 2015).

[3] http://strata.uga.edu/8370/lecturenotes/principalComponents.html.

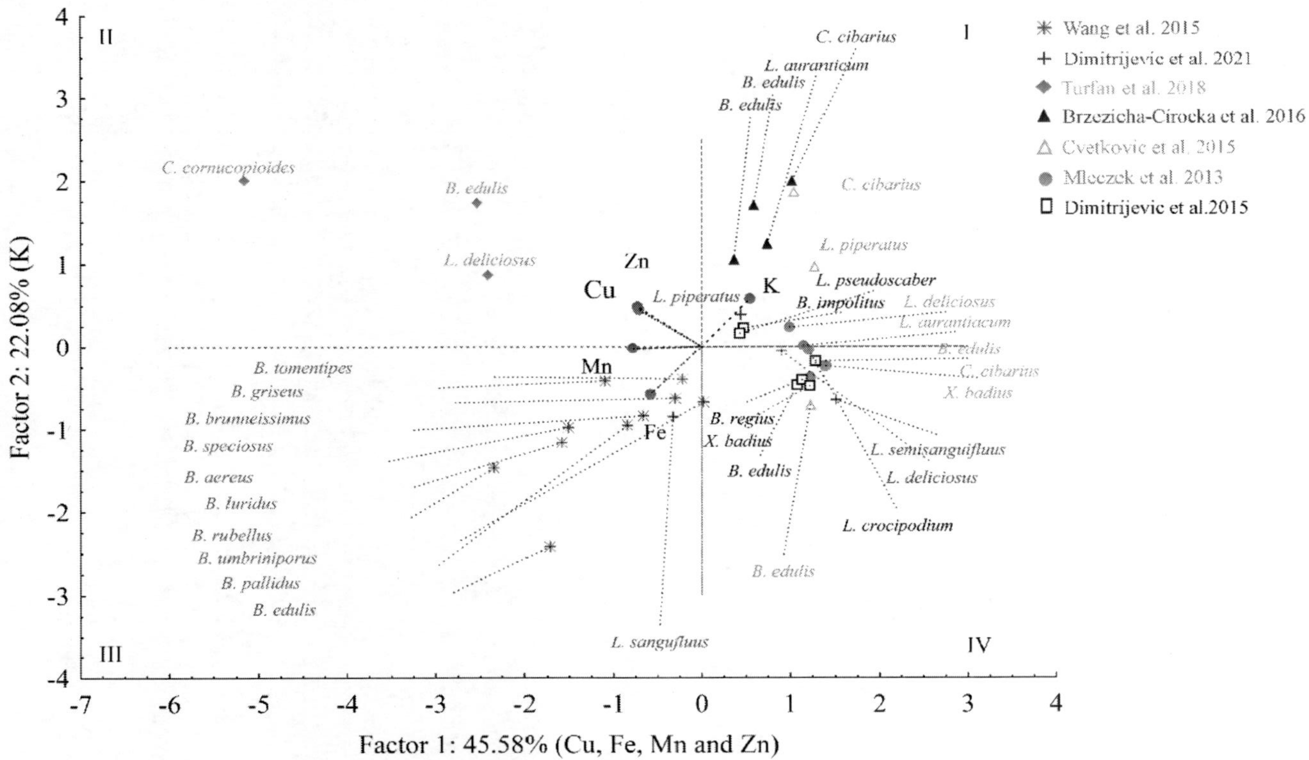

Figure 8. PCA diagram of mushrooms species based on elements accumulation.

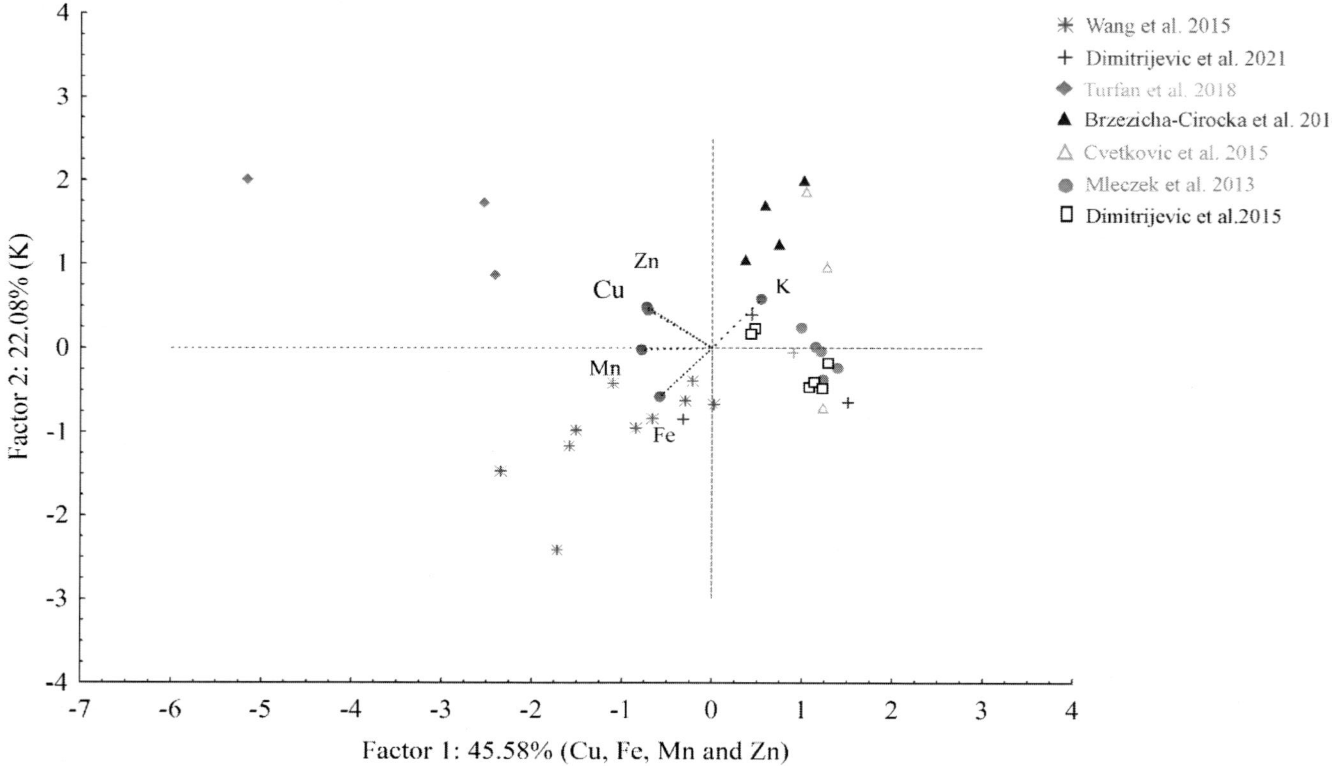

Figure 9. PCA diagram of mushrooms species based on elements accumulation.

Figure 8 shows there is a slight tendency to group the samples into four groups. Each group is determined by the species and the element's content in them. As can be deduced from the figure, some mushrooms tend to cluster together.

If the PCA diagram is overloaded with sample names, the biplot can be displayed as in the Figure 9. Making a biplot using this approach has a great advantage - it makes it possible to color-code samples by external variables to check if those samples are sorted according to those variables.

Projection of data on the PC1 and PC2 diagram split mushrooms species collected in Chine and Turkey and species collected in Serbia and Poland. The first mentioned samples showed negative PC1 scores, while the rest showed positive PC1 scores (Table 4). This separation was highly influenced by the content of copper, iron, manganese, and zinc because these elements have high loading on PC1.

The group of mushrooms located on the positive side of PC1 (with positive scores) was very influenced by the content of potassium.

Potassium, with a factor loading of 0.579 obtained for PC1, separated the samples located on the right-hand side from the sample located on the left-hand side because they have higher concentrations of potassium. On the contrary, these mushroom species have a low iron content, so they are located diametrically opposite to the iron.

Regarding the samples with positive PC1, a separation between these mushrooms can be observed. Mushroom species (Quadrant I) with highest positive scores on PC2, confirming that their characteristic was the highest content in potassium and lowest content in manganese. The concentration of potassium in species located on the negative side of PC2 (Quadrant IV) decreased concerning the mentioned species but compared to other species, it's still higher. This location suggests that these mushrooms have a lower content of zinc and copper, which are located diametrically opposite to group Quadrant IV.

Figure 8 shows the score plot for PC1 versus PC2 in which the tendency to separate the group of samples from Chine and samples from Turkey is noted. Based on variables with negative loading mushroom samples were differentiated according to the contents of copper, iron, and zinc. Mushroom species from Turkey (Quadrant II) had the highest negative scores on PC1, and the content of copper and zinc highly influenced them. This group of mushrooms has higher concentrations of copper and zinc but lower iron content than Chines mushrooms.

Table 4. Factor coordinates of cases, based on correlations

Cases	Country	Factor loading	
		PC 1	PC 2
B. speciosus	China	-0.661	-0.841
B. griseus	China	-1.099	-0.419
B. luridus	China	-1.584	-1.164
B. umbriniporus	China	0.019	-0.673
B. tomentipes	China	-0.216	-0.395
B. edulis	China	-1.718	-2.419
B. pallidus	China	-0.845	-0.954
B. brunneissimus	China	-0.300	-0.629
B. rubellus	China	-2.343	-1.469
B. aereus	China	-1.511	-0.982
B. edulis	Turkey	-2.537	1.727
C. cornucopioides	Turkey	-5.163	2.007
L. deliciosus	Turkey	-2.410	0.865
C. cibarius	Poland	1.019	1.995
B. edulis	Poland	0.365	1.048
B. edulis	Poland	0.585	1.696
L. auranticum	Poland	0.742	1.235
C. cibarius	Serbia	1.042	1.862
B. edulis	Serbia	1.230	-0.709
L. piperatus	Serbia	1.274	0.961
L. aurantiacum	Poland	1.147	0.011
X. badius	Poland	1.393	-0.242
L. deliciosus	Poland	0.991	0.234
B. edulis	Poland	1.285	-0.183
C. cibarius	Poland	1.226	-0.376
B. edulis	Serbia	1.078	-0.469
B. impolitus	Serbia	0.472	0.228
B. regius	Serbia	1.132	-0.406
L. crocipodium	Serbia	1.206	-0.045
L. pseudoscaber	Serbia	0.430	0.162
X. badius	Serbia	1.221	-0.480
L. sangufluus	Serbia	-0.323	-0.852
L. semisanguifluus	Serbia	1.505	-0.649
L. deliciosus	Serbia	0.901	-0.061
L. piperatus	Serbia	0.443	0.387

The group formed by Chines mushrooms presents higher iron concentrations, which is expected because they are co-located in the same quadrant as an iron (Quadrant II). Since this group is located diametrically opposite to potassium, it indicates a low potassium content.

PCA technique enabled visualization of complex experimental data and relationships responsible for grouping observed (Miladinović et al., 2021). A part from observed groups of mushroom species, PCA method also explains that content of potassium is the main reason for thier groupping.

Conclusion

Based on results shown in this manuscript, it can be concluded that PCA and HCA are very useful for studying the data structures and finding similarities among samples, especially when there is a large volume of them. Using these chemometric methods, obtained results are objectively analyzed, which eliminates any subjective conclusions which might come from the researchers. The main difference between these two methods is that HCA gives quantitative approach, e.g., based on Euclid's distance, the similarities/dissimilarities between the samples can be determined, while PCA gives qualitative approach, and explains the measure of association. This means that, if it is necessary to group the similar samples, the HCA method is sufficient. However, for determining what clusters them, the PCA method is used. Based on this, it can be concluded that spatial (e.g., PCA) and clustering (e.g., Ward's) methods use different yet complementary models which use the same mathematical functions. Therefore, in practice, the results of Ward's clustering are likely to reproduce the clusters that visually the regions of high point density in the PCA diagram. The results obtained via PCA analysis confirm those obtained via HCA analysis assuring the correctness of the conclusion.

In this particular example, using chemometrics analysis, it can be simply and easily concluded that the main difference is accumulation of potassium between the samples collected in China and Turkey, and samples collected in Serbia and Poland.

Results obtained in research give the contribution to the science, but chemometric analysis brings the results one step higher.

Acknowledgements

The authors would like to thank the Ministry of Education, Science and Technological Development of Republic of Serbia (Grant No: 451-03-68/2022-14/200113).

References

Abdi, Hervé, and Lynne J. Williams. 2010. "Principal Component Analysis." *Wiley Interdisciplinary Reviews: Computational Statistics* 2: 433–459. doi:10.1002/wics.101.

Adamson, George W., and David Bawden. 1981. "Comparison of Hierarchical Cluster Analysis Techniques for Automatic Classification of Chemical Structures." *Journal of Chemical Information and Computer Sciences* 21: 204–209. doi:10.1021/ci00032a005.

Aparicio, Ramon and Ramon Aparicio-Ruiz. 2002. "Chemometrics as an aid in authentication." In: *Oils and Fats Authentication*, edited by Michael Jee, 156-181. Oxford, UK, Blackwell publishing.

Apostolov, Suzana. 2014. "The study of retention and solvatochromic properties of the newly synthesized phenylacetamide derivatives." PhD dis., University of Novi Sad.

Arfken, George. 1985. "Eigenvectors, Eigenvalues." In: *Mathematical Methods for Physicists*, edited by George Arfken G 229-237, Florida, Academic Press.

Babbie, Earl. 2009. *The Practice of Social Research*. California, Wadsworth Publishing.

Beebe, Kenneth R., Randy J. Pell, Mary Beth Seasholtz. 1998. *Chemometrics: A Practical Guide*. New York: Wiley & Sons.

Berenson, Mark L., David M. Levine, Timothy C. Krehbiel. 2011. *Basic Business Statistics: Concepts and Applications*. New Jersey; Prentice Hall.

Brzezicha-Cirocka, Justyna, Małgorzata Mędyk, Jerzy Falandysz, and Piotr Szefer. 2016. "Bio- and Toxic Elements in Edible Wild Mushrooms from Two Regions of Potentially Different Environmental Conditions in Eastern Poland." *Environmental Science and Pollution Research* 23: 21517–21522. doi:10.1007/s11356-016-7371-0.

Bu, Jianwei, Wei Liu, Zhao Pan, and Kang Ling. 2020. "Comparative Study of Hydrochemical Classification Based on Different Hierarchical Cluster Analysis Methods." *International Journal of Environmental Research and Public Health* 17: 9515. doi:10.3390/ijerph17249515.

Bu, Jianwei, Ziyong Sun, Aiguo Zhou, Youning Xu, Rui Ma, Wenhao Wei, and Meng Liu. 2016. "Heavy Metals in Surface Soils in the Upper Reaches of the Heihe River, Northeastern Tibetan Plateau, China." *International Journal of Environmental Research and Public Health* 13: 247. doi:10.3390/ijerph13030247.

Cvetkovic, Jelena S., Violeta D. Mitic, Vesna P. Stankov-Jovanovic, Marija V. Dimitrijevic, and Snezana D. Nikolic-Mandic. 2015. "Elemental Composition of Wild Edible Mushrooms from Serbia." *Analytical Letters* 48: 2107–2121. doi:10.1080/00032719.2015.1010118.

Delić, Milica V. 2003. "Geodemografski model tržišnog prostora Srbije [*Geodemographic model of market space of Serbia*]." *Management - časopis za teoriju i praksu menadžmenta* 8: 35-43.

Dimitrijević, Marija. 2021. "Comparative research of the content of elements and antioxidant activity of selected mushroom species: a chemometric approach", PhD dis., University of Niš, PhD thesis.

Dimitrijevic, Marija V., Violeta D. Mitic, Jelena S. Cvetkovic, Vesna P. Stankov Jovanovic, Jelena J. Mutic, and Snezana D. Nikolic Mandic. 2015. "Update on

Element Content Profiles in Eleven Wild Edible Mushrooms from Family Boletaceae." *European Food Research and Technology* 242: 1–10. doi:10.1007/s00217-015-2512-0.

Drewnowska, Małgorzata, and Jerzy Falandysz. 2015. "Investigation on Mineral Composition and Accumulation by Popular Edible Mushroom Common Chanterelle (Cantharellus Cibarius)." *Ecotoxicology and Environmental Safety* 113: 9–17. doi:10.1016/j.ecoenv.2014.11.028.

Granato, Daniel, Jânio S. Santos, Graziela B. Escher, Bruno L. Ferreira, and Rubén M. Maggio. 2018. "Use of Principal Component Analysis (PCA) and Hierarchical Cluster Analysis (HCA) for Multivariate Association Between Bioactive Compounds and Functional Properties in Foods: A Critical Perspective." *Trends in Food Science & Technology* 72: 83–90. doi:10.1016/j.tifs.2017.12.006.

Hossain, Mohammad B., Ankit Patras, Catherine Barry-Ryan, Ana B. Martin-Diana, and Nigel P. Brunton. 2011. "Application of Principal Component and Hierarchical Cluster Analysis to Classify Different Spices Based on in Vitro Antioxidant Activity and Individual Polyphenolic Antioxidant Compounds." *Journal of Functional Foods* 3: 179–189. doi:10.1016/j.jff.2011.03.010.

Jolliffe, Ian T., and Jorge Cadima. 2016. "Principal Component Analysis: a Review and Recent Developments." *Philosophical Transactions of the Royal Society A: Mathematical, Physical and Engineering Sciences* 374: 20150202. doi:10.1098/rsta.2015.0202.

Kaiser, Henry F. 1960. "The Application of Electronic Computers to Factor Analysis." *Educational and Psychological Measurement* 20: 141–151. doi:10.1177/001316446002000116.

Keller, Gerald. 2014. *Statistics for Management and Economics*. USA, Cengage Learning.

Landau, S., and I. Chis Ster. 2010. "Cluster Analysis: Overview." *International Encyclopedia of Education* 72–83. doi:10.1016/b978-0-08-044894-7.01315-4.

Miladinović D, Ilić B, Miladinović Lj, Kocić B, Ćirić V, Stankov-Jovanović V, Cvetković O. 2013. "Antibacterial activity of *Thymus pulegioides* essential oil and its synergistic potential with antibiotics: a chemometric approach." In: *Recent Progress in Medicinal Plants vol. 38: Essential Oils III and Phytopharmacology*, edited by Narenda Govil and Sanjib Bhattacharya, Houston, Studium Press LLC.

Miladinović, Dragoljub L., Budimir S. Ilić, Tatjana M. Mihajilov-Krstev, Nikola D. Nikolić, Ljiljana C. Miladinović, and Olga G. Cvetković. 2012. "Investigation of the Chemical Composition–antibacterial Activity Relationship of Essential Oils by Chemometric Methods." *Analytical and Bioanalytical Chemistry* 403: 1007–1018. doi:10.1007/s00216-012-5866-1.

Miladinović, Dragoljub L., Marija V. Dimitrijević, Tatjana M. Mihajilov-Krstev, Marija S. Marković, and Vojislav M. Ćirić. 2021. "The Significance of Minor Components on the Antibacterial Activity of Essential Oil via Chemometrics." *LWT - Food Science and Technology* 136: 110305. doi:10.1016/j.lwt.2020.110305.

Mleczek, Mirosław, Marek Siwulski, Kinga Stuper-Szablewska, Krzysztof Sobieralski, Zuzanna Magdziak, and Piotr Goliński. 2013. "Accumulation of Elements by Edible Mushroom Species II. A Comparison of Aluminium, Barium and Nutritional Element

Contents." *Journal of Environmental Science and Health, Part B* 48: 308–317. doi:10.1080/03601234.2013.743799.

Mrmošanin, J. 2019. Analysis of catechins, procyanidins, macro, and microelements in dark, milk and white chocolate and cocoa powder and their antioxidant potential, PhD diss., University of Niš, PhD thesis.

Murtagh, Fionn, and Pierre Legendre. 2014. "Ward's Hierarchical Agglomerative Clustering Method: Which Algorithms Implement Ward's Criterion?" *Journal of Classification* 31: 274–295. doi:10.1007/s00357-014-9161-z.

Nair, Indu V, Singh Kailash, Arumugam M, Kocherla Gangadhar, David Clarson. 2010. "Trace metal quality of Meenachil River at Kottayam, Kerala (India) by principal component analysis." *World Applied Sciences Journal* 9:100–1107.

Nikolić, J. 2018. "Optimization of the soil samples preparations for the determination of polycyclic aromatic hydrocarbons in soil by gas chromatography-mass spectrometry." PhD diss., University of Niš.

Pastor, Kristian. 2018. "Authenticity of cereals and pseudocereals -development of new methods for the analysis of flour and final bakery products." PhD dis., University of Novi Sad.

Turfan, Nezahat, Aysun Pekşen, Beyhan Kibar, and Sabri Ünal. 2018. "Determination of nutritional and bioactive properties in some selected wild growing and cultivated mushrooms from turkey." *Acta Scientiarum Polonorum Hortorum Cultus* 17: 57–72. doi:10.24326/asphc.2018.3.6.

Varmuza, Kurt, Peter Filzmoser. 2009. *Introduction to Multivariate Statistical Analysis in Chemometrics*. USA, Taylor & Francis Group.

Wang, Xue-Mei, Ji Zhang, Tao Li, Yuan-Zhong Wang, and Hong-Gao Liu. 2015. "Content and Bioaccumulation of Nine Mineral Elements in Ten Mushroom Species of the GenusBoletus." *Journal of Analytical Methods in Chemistry* 2015: 1–7. doi:10.11 55/2015/165412.

Biographical Sketches

Dragoljub L. Miladinović

Affiliation: Department of Pharmacy, Faculty of Medicine, University of Niš

Education: Department of Chemistry, Faculty of Science and Mathematics, University of Novi Sad

Contact Information: dragoljubm@gmail.com

Research and Professional Experience: Natural Products Chemistry

Professional Appointments: Full Professor, Dr.

Publications from the Last 3 Years:

1. Miladinović DL, Dimitrijević MV, Mihajilov-Krstev TM, Marković MS, Cirić VM. The significance of minor components on the antibacterial activity of essential oil via chemometrics. *LWT - Food Science and Technology* (2021) 136:110305.
2. Marković MS, Pljevljakusić DS, Nikolić BM, Miladinović DL, Djokić MM, Rakonjac LjB, Stankov Jovanović VP. Ethnoveterinary knowledge in Pirot County (Serbia). *South African Journal of Botany* (2021) 137:278-289.
3. Kocić BD. Stanković Dordević DM, Dimitrijević MV, Marković MS, Miladinović DL. Potentially Effective and Safe Anti-*Helicobacter pylori* Natural Products: Chemometric Study. *Revista de Chimie* (2020) 71:267-273.
4. Mihajilov-Krstev T, Jovanović B, Zlatković B, Matejić J, Vitorović J, Cvetković V, Ilić B, Đorđević Lj, Joković N, Miladinović D, Jakšić T, Stanković N, Stankov Jovanović V. Nirit Bernstein. Phytochemistry, toxicology and therapeutic value of *Petasites hybridus* subsp. Ochroleucus (Common Butterbur) from the Balkans. *Plants* (2020) 9: 700.
5. Kocić BD, Dimitrijević MV, Miladinović LjC, Marković MS, Ranković GŽ, Miladinović DL. *In Vitro* Anti-*Helicobacter pylori* Activity of Berberine and Barberry Extracts: A Preliminary Report. *Natural Product Communications* (2019) 1–3.
6. Dimitrijević MV, Mitić VD, Ranković GŽ, Miladinović DL. Survey of Antioxidant Properties of Barberry: A Chemical and Chemometric Approach. *Analytical Letters* (2019) 53:1-12.

Marija V. Dimitrijević

Affiliation: Department of Pharmacy, Faculty of Medicine, University of Niš

Education: Department of Chemistry, Faculty of Science and Mathematics, University of Niš

Contact Information: marija.dimitrijevic@medfak.ni.ac.rs

Research and Professional Experience: General and Inorganic Chemistry

Professional Appointments: Teaching and Research Assistant, Dr.

Publications from the Last 3 Years:

1. Stanković M, Mitić V, Stankov Jovanović V, Dimitrijević M, Nikolić J, Stojanović G. Selected fungi of the genus Lactarius - screening of antioxidant capacity, antimicrobial activity, and genotoxicity. *Journal of Toxicology and Environmental Health, Part A* (2022) https://doi.org/10.1080/15287394.2022.2075502
2. Miladinović DL, Dimitrijević MV, Mihajilov-Krstev TM, Marković MS, Cirić VM. The significance of minor components on the antibacterial activity of essential oil via chemometrics. *LWT - Food Science and Technology* (2021) 136:110305.
3. Dimitrijević M, Mitić V, Đorđević D, Popović G, Krstić N, Nikolić J, Stankov Jovanović V. Macroelements versus toxic elements in selected wild edible mushrooms of the Russulacea family from Serbia. *Journal of the Serbian Chemical Society* (2021) 86: 927–940.
4. Kocić BD. Stankovic Dordevic DM, Dimitrijevic M, Markovic M, Miladinovic D. Potentially Effective and Safe Anti-*Helicobacter pylori* Natural Products: Chemometric Study. *Revista de Chimie* (2020) 71:267-273.
5. Kocić BD, Dimitrijević MV, Miladinović LjC, Marković MS, Ranković GŽ, Miladinović DL. *In Vitro* Anti-*Helicobacter pylori* Activity of Berberine and Barberry Extracts: A Preliminary Report. *Natural Product Communications* (2019) 1–3.
6. Dimitrijević MV, Mitić VD, Ranković GŽ, Miladinović DL. Survey of Antioxidant Properties of Barberry: A Chemical and Chemometric Approach. *Analytical Letters* (2019) 53:1-12.

Bibliography

A comprehensive guide to natural products

LCCN	2020032611
Type of material	Book
Main title	A comprehensive guide to natural products / Silje A. Dahl and Adam M. Frandsen, editors.
Published/Produced	New York: Nova Science Publishers, [2020]
Description	xi, 194 pages: illustrations; 23 cm.
ISBN	9781536184181 (paperback) (adobe pdf)
LC classification	QK865 .C66 2020
Related names	Dahl, Silje A., editor.
Summary	"In this compilation, the role of liquid chromatography, mass spectrometry and chemometrics for the analysis and characterization of plant natural products is addressed. The authors provide a comprehensive review of the pharmacological activity of cetrarioid lichens and their major secondary metabolites as antioxidants to prevent and treat oxidative stress-related diseases. Following this, the way in which the detection of various secondary metabolites and bioactive compounds in some plants can reduce sickle cells in vitro is studied. In addition, the efficiency of green and conventional solvent systems concerning the three classes of phytochemicals (phenols, alkaloids, and flavonoids) is described. Lastly, a brief history of antibiotics and the spread of resistance is provided, and future strategies to combat drug-resistant microbes are discussed"-- Provided by publisher.
LC Subjects	Botanical chemistry. Phytochemicals. Natural products--Biotechnology.
Notes	Includes bibliographical references and index.

Additional formats	Online version: A comprehensive guide to natural products Hauppauge: Nova Science Publishers, 2020. 9781536184327 (DLC) 2020032612
Series	Chemistry research and applications

A comprehensive guide to natural products

LCCN	2020032612
Type of material	Book
Main title	A comprehensive guide to natural products / Silje A. Dahl and Adam M. Frandsen, (editors).
Published/Produced	New York: Nova Science Publishers, [2020]
Description	1 online resource
ISBN	9781536184327 (adobe pdf) (paperback)
LC classification	QK865
Related names	Dahl, Silje A., editor.
Summary	"In this compilation, the role of liquid chromatography, mass spectrometry and chemometrics for the analysis and characterization of plant natural products is addressed. The authors provide a comprehensive review of the pharmacological activity of cetrarioid lichens and their major secondary metabolites as antioxidants to prevent and treat oxidative stress-related diseases. Following this, the way in which the detection of various secondary metabolites and bioactive compounds in some plants can reduce sickle cells in vitro is studied. In addition, the efficiency of green and conventional solvent systems concerning the three classes of phytochemicals (phenols, alkaloids, and flavonoids) is described. Lastly, a brief history of antibiotics and the spread of resistance is provided, and future strategies to combat drug-resistant microbes are discussed"-- Provided by publisher.
LC Subjects	Botanical chemistry. Phytochemicals. Natural products--Biotechnology.
Notes	Includes bibliographical references and index.

Additional formats	Print version: A comprehensive guide to natural products New York: Nova Science Publishers, [2020] 9781536184181 (DLC) 2020032611
Series	Chemistry research and applications

Acoustic Cavitation and Bubble Dynamics

LCCN	2019760315
Type of material	Book
Personal name	Yasui, Kyuichi, author.
Main title	Acoustic Cavitation and Bubble Dynamics / by Kyuichi Yasui.
Edition	1st ed. 2018.
Published/Produced	Cham: Springer International Publishing: Imprint: Springer, 2018.
Description	1 online resource (VIII, 124 pages 84 illustrations) text file PDF
ISBN	9783319682372
Summary	This brief explains in detail fundamental concepts in acoustic cavitation and bubble dynamics, and describes derivations of the fundamental equations of bubble dynamics in order to support those readers just beginning research in this field. Further, it provides an in-depth understanding of the physical basis of the phenomena. With regard to sonochemistry, the brief presents the results of numerical simulations of chemical reactions inside a bubble under ultrasound, especially for a single-bubble system and including unsolved problems. Written so as to be accessible both with and without prior knowledge of fundamental fluid dynamics, the brief offers a valuable resource for students and researchers alike, especially those who are unfamiliar with this field. A grasp of fundamental undergraduate mathematics such as partial derivative and fundamental integration is advantageous; however, even without any background in mathematics, readers can skip the equations and still understand the fundamental physics of the phenomena using the book's wealth

	of illustrations and figures. As such, it is also suitable as an introduction to the field. .
Contents	Acoustic Cavitation - Bubble Dynamics - Unsolved Problems.
LC Subjects	Acoustics. Chemistry, Physical and theoretical. Chemometrics. Fluid mechanics.
Other Subjects	Math. Applications in Chemistry. Acoustics. Engineering Fluid Dynamics. Theoretical and Computational Chemistry.
Additional formats	Print version: Acoustic cavitation and bubble dynamics. 9783319682365 (DLC) 2017952916 Printed edition: 9783319682365 Printed edition: 9783319682389
Series	Ultrasound and Sonochemistry, 2511-123X Ultrasound and Sonochemistry, 2511-123X

Analytical Chemistry Applied to Emerging Pollutants

LCCN	2019768860
Type of material	Book
Personal name	Vaz Jr., Sílvio, author.
Main title	Analytical Chemistry Applied to Emerging Pollutants / by Sílvio Vaz Jr.
Edition	1st ed. 2018.
Published/Produced	Cham: Springer International Publishing: Imprint: Springer, 2018.
Description	1 online resource (XI, 122 pages 66 illustrations, 48 illustrations in color.) text file PDF
ISBN	9783319744032
Summary	This book addresses the highly relevant subject of emerging pollutants, which are especially alarming since most of the available treatment technologies are unable to degrade them. It discusses the sources of these pollutants and their fate in the environment, and the main tools available for their analysis. It also describes the representative environmental

	matrices (air, soil and water) and appropriate analytical methods for each matrix. Furthermore, it examines aspects of toxicology, chemometrics, sample preparation and green analytical chemistry. As such, it provides a broad overview of the potential analytical approaches for monitoring and controlling emerging pollutants. This book fills a gap in the literature, and is a valuable resource for all professionals concerned with emerging pollutant control in real-world situations.
Contents	Chapter 1: What are Emerging Pollutants (EPs) and their Fate on the Environment - Chapter 2: Fundamental Aspects of Analytical Chemistry for Environmental Analysis - Chapter 3: Analytical Techniques - Chapter 4: The Main Environmental Matrices: Air, Soil and Water - Chapter 5: Degradation Processes of EPs - Chapter 6: General Remarks and Conclusions - Glossary.
LC Subjects	Analytical chemistry. Environmental chemistry. Pharmacology.
Other Subjects	Analytical Chemistry. Environmental Chemistry. Pharmacology/Toxicology.
Additional formats	Print version: Analytical chemistry applied to emerging pollutants 9783319744025 (DLC) 2018939302 Printed edition: 9783030089825 Printed edition: 9783319744025 Printed edition: 9783319744049

Applications of Advanced Oxidation Processes (AOPs) in Drinking Water Treatment

LCCN	2019747154
Type of material	Book
Main title	Applications of Advanced Oxidation Processes (AOPs) in Drinking Water Treatment / edited by Antonio Gil, Luis Alejandro Galeano, Miguel Ángel Vicente.

Edition	1st ed. 2019.
Published/Produced	Cham: Springer International Publishing: Imprint: Springer, 2019.
Description	1 online resource (XVIII, 429 pages 70 illustrations, 61 illustrations in color.) text file PDF
ISBN	9783319768823
Related names	Galeano, Luis Alejandro. editor. Gil, Antonio. editor. Vicente, Miguel Ángel. editor.
Summary	This volume reviews the drinking water treatments in which AOPs display a high application potential. Firstly it reveals the typical supply sources and limitations of conventional technologies and critically reviews natural organic matter characterization and removal techniques, focusing mainly on AOP treatments. It then explores using AOPs for simultaneous inactivation/disinfection of several types of microorganisms, including highly resistant Cryptosporidium protozoa. Lastly, it discusses relevant miscellaneous topics, like the most promising AOP solid catalysts, the regime change of Fenton-like processes toward continuous reactors, the application of chemometrics for process optimization, the impact on disinfection byproducts and the tracing of toxicity during AOP treatments. This work is a useful reference for researchers and students involved in water technologies, including analytical and environmental chemistry, chemical and environmental engineering, toxicology, biotechnology, and related fields. It is intended to encourage industrial and public-health scientists and decision-makers to accelerate the application of AOPs as technological alternatives for the improvement of drinking water treatment plants.

Contents Surface and groundwater sources for drinking water - Limitations of conventional drinking water technologies in pollutants removal - Natural Organic Matter: characterization and removal by AOPs to assist drinking water facilities - Natural organic matter removal by heterogeneous catalytic wet peroxide oxidation (CWPO) - Separation and characterization of NOM intermediates along AOPs oxidation - Photo (Catalytic) Oxidation Processes for the Removal of Natural Organic Matter and Contaminants of Emerging Concern from Water - Homogeneous Fenton and photo-Fenton disinfection of surface and ground water - AOPs methods for the removal of taste and odor compounds - Wastewater Treatment by Heterogeneous Fenton-like Processes in Continuous Reactors - Disinfection by chemical oxidation methods - Inactivation of Cryptosporidium by Advanced Oxidation Processes - Cost-effective catalytic materials for AOP-treatment units - Impact on Disinfection Byproducts Using Advanced Oxidation Processes for Drinking Water Treatment - Evolution of toxicity and estrogenic activity throughout AOP's surface and drinking water treatment - Chemometric methods for the optimization of the advanced oxidation processes for the treatment of drinking and wastewater.

LC Subjects Catalysis.
Chemical engineering.
Water pollution.
Water quality.

Other Subjects Waste Water Technology / Water Pollution Control / Water Management / Aquatic Pollution.
Catalysis.
Industrial Chemistry/Chemical Engineering.
Water Quality/Water Pollution.

Additional formats	Print version: Applications of advanced oxidation processes (AOPS) in drinking water treatment 9783319768816 (DLC) 2018946685 Printed edition: 9783030083175 Printed edition: 9783319768816 Printed edition: 9783319768830
Series	The Handbook of Environmental Chemistry, 1867-979X; 67 The Handbook of Environmental Chemistry, 1867-979X; 67

Applications of Quantum Dynamics in Chemistry

LCCN	2019744628
Type of material	Book
Personal name	Gatti, Fabien. author.
Main title	Applications of Quantum Dynamics in Chemistry / by Fabien Gatti, Benjamin Lasorne, Hans-Dieter Meyer, André Nauts.
Edition	1st ed. 2017.
Published/Produced	Cham: Springer International Publishing: Imprint: Springer, 2017.
Description	1 online resource (XVI, 429 pages 110 illustrations) text file PDF
ISBN	9783319539232
Related names	Lasorne, Benjamin. author. Meyer, Hans-Dieter. author. Nauts, André. author.
Summary	This book explains the usage and application of Molecular Quantum Dynamics, the methodology where both the electrons and the nuclei in a molecule are treated with quantum mechanical calculations. This volume of Lecture Notes in Chemistry addresses graduate students and postdocs in the field of theoretical chemistry, as well as postgraduate students, researchers and teachers from neighboring fields, such as quantum physics, biochemistry, biophysics, or anyone else who is interested in this rising method in theoretical chemistry, and who wants to gain experience in the

opportunities it can offer. It can also be useful for teachers interested in illustrative examples of time-dependent quantum mechanics as animations of realistic wave packets have been designed to assist in visualization. Assuming a basic knowledge about quantum mechanics, the authors link their explanations to recent experimental investigations where Molecular Quantum Dynamics proved successful and necessary for the understanding of the experimental results. Examples including reactive scattering, photochemistry, tunneling, femto- and attosecond chemistry and spectroscopy, cold chemistry or crossed-beam experiments illustrate the power of the method. The book restricts complicated formalism to the necessary and in a self-contained and clearly explained way, offering the reader an introduction to, and instructions for, practical exercises. Continuative explanation and math are optionally supplemented for the interested reader. The reader learns how to apply example simulations with the MCTDH program package (Multi Configuration Time Dependent Hartree calculations). Readers can thus obtain the tools to run their own simulations and apply them to their problems. Selected scripts and program code from the examples are made available as supplementary material. This book bridges the gap between the existing textbooks on fundamental theoretical chemistry and research monographs focusing on sophisticated applications. It is a must-read for everyone who wants to gain a sound understanding of Molecular Quantum Dynamics simulations and to obtain basic experience in running their own simulations.

Contents	Part I Introduction - Part II Concepts and Methods: Quantum effects - Electronic states and potential energy operators - The Choice of coordinates - Kinetic energy operators - Introduction to molecular symmetry - Introduction to numerical methods and to MCTDH - Part III Applications: Infrared spectroscopy - Quantum control with laser pulses in the electronic ground state - Photodissociation spectra - Cross sections for reactive scattering - Quantum control with laser pulses for electronically excited states - Non-adiabatic photochemistry - Optimal control and quantum computers - Part IV Conclusion.
LC Subjects	Chemistry, Physical and theoretical. Chemoinformatics. Chemometrics. Quantum physics.
Other Subjects	Theoretical and Computational Chemistry. Computer Applications in Chemistry. Math. Applications in Chemistry. Quantum Physics.
Additional formats	Print version: Applications of quantum dynamics in chemistry. 9783319539218 (DLC) 2017932629 Printed edition: 9783319539218 Printed edition: 9783319539225 Printed edition: 9783319852775
Series	Lecture Notes in Chemistry, 0342-4901; 98 Lecture Notes in Chemistry, 0342-4901; 98

Applied Compositional Data Analysis: With Worked Examples in R

LCCN	2019743635
Type of material	Book
Personal name	Filzmoser, Peter. author.
Main title	Applied Compositional Data Analysis: With Worked Examples in R / by Peter Filzmoser, Karel Hron, Matthias Templ.
Edition	1st ed. 2018.
Published/Produced	Cham: Springer International Publishing: Imprint: Springer, 2018.

Description	1 online resource (XVII, 280 pages 74 illustrations, 57 illustrations in color.) text file PDF
ISBN	9783319964225
Related names	Hron, Karel. author. Templ, Matthias. author.
Summary	This book presents the statistical analysis of compositional data using the log-ratio approach. It includes a wide range of classical and robust statistical methods adapted for compositional data analysis, such as supervised and unsupervised methods like PCA, correlation analysis, classification and regression. In addition, it considers special data structures like high-dimensional compositions and compositional tables. The methodology introduced is also frequently compared to methods which ignore the specific nature of compositional data. It focuses on practical aspects of compositional data analysis rather than on detailed theoretical derivations, thus issues like graphical visualization and preprocessing (treatment of missing values, zeros, outliers and similar artifacts) form an important part of the book. Since it is primarily intended for researchers and students from applied fields like geochemistry, chemometrics, biology and natural sciences, economics, and social sciences, all the proposed methods are accompanied by worked-out examples in R using the package robCompositions.
Contents	Preface - Acknowledgements - Compositional data as a methodological concept - Analyzing compositional data using R - Geometrical properties of compositional data - Exploratory data analysis and visualization - First steps for a statistical analysis - Cluster analysis - Principal component analysis - Correlation analysis - Discriminant analysis - Regression analysis - Methods for high-dimensional compositional data - Compositional tables - Preprocessing issues - Index.-.

LC Subjects	Geochemistry. Statistics.
Other Subjects	Statistics for Engineering, Physics, Computer Science, Chemistry and Earth Sciences. Geochemistry. Statistical Theory and Methods. Statistics and Computing/Statistics Programs. Statistics for Life Sciences, Medicine, Health Sciences. Statistics for Social Sciences, Humanities, Law.
Additional formats	Print version: Applied compositional data analysis: with worked examples in R 9783319964201 (DLC) 2018952636 Printed edition: 9783319964201 Printed edition: 9783319964218
Series	Springer Series in Statistics, 0172-7397 Springer Series in Statistics, 0172-7397

Authentication and detection of adulteration of olive oil

LCCN	2020685834
Type of material	Book
Main title	Authentication and detection of adulteration of olive oil / Michael Kontominas, editor.
Published/Produced	New York: Nova Science Publishers, [2019]
Description	1 online resource.
ISBN	9781536145977 ebook
LC classification	TP683
Related names	Kontominas, Michael, editor.
Contents	A review of olive oil authentication / Apostolos Kiritsakis, Evangelos Evangelou and Amani Taamalli - Fingerprinting of virgin olive oils for authenticity and quality assessment / Rosa María Alonso-Salces, Margaret Veronica Holland, Sandra Rosa Fuselli and Claude Guillou - LC fingerprinting approach for authenticating olive oil / Ana M. Jiménez-Carvelo, Antonio González-Casado, Estefanía Pérez-Castaño and Luis Cuadros-Rodríguez - Authentication of olive oil with mid-infrared spectroscopy / Oguz Uncu and Banu Ozen

- Characterization and differentiation of Greek virgin olive oils from five cultivars based on selected physicochemical parameters combined with chemometrics / Ioanna Kosma, Anastasia Badeka, Stavros Kontakos and Michael Kontominas - Authentication and adulteration of olive oil / Lanfranco Conte, Paolo Lucci, Erica Moret and Sabrina Moret - Application of vibrational spectroscopy techniques / Astrid Maléchaux, Nathalie Dupuy and Jacques Artaud - Application of thermal and physical techniques for olive oil adulteration and authenticity / Maria Paciulli, Augusta Caligiani, Maria Teresa Rodriguez Estrada, Lorenzo Cerretani and Emma Chiavaro - Olive oil adulteration / Dimitrios Boskou - Advances in DNA extraction for detecting extra-virgin olive oil adulteration / Cinzia Montemurro, Wilma Sabetta, Valentina di Rienzo and Antonella Pasqualone - Sesquiterpene hydrocarbons, a promising tool for virgin olive oil geographical authetication / Stefania Vichi, Alba Tres, Beatriz Quintanilla-Casas, Julen Bustamante and Francesc Guardiola - The use of pyropheophytins and diacylglycerol testing to detect old, deodorised, and/or poorly stored extra virgin olive oils / Claudia M. Guillaume and Leandro M. Ravetti - Application of phycicochemical parameters combined with chemometrics for the differentiation of geographical origin of extra virgin olive oils of the Koroneiki cultivar / Ioanna Kosma, Anatasia Badeka, Stavros Kontakos and Michael Kontominas.

LC Subjects	Olive oil. Food adulteration and inspection.
Notes	Includes bibliographical references and index.
Additional formats	Print version: Authentication and detection of adulteration of olive oil New York: Nova Science Publishers, [2019] 9781536145960 (DLC) 2018960055

Authentication and detection of the adulteration of olive oil

LCCN	2018960055
Type of material	Book
Main title	Authentication and detection of the adulteration of olive oil / Michael G. Kontominas, editor.
Published/Produced	New York: Nova Science Publishers, [2019]
Description	xii, 397 pages: illustrations; 26 cm.
ISBN	9781536145960 (hbk.)
LC classification	TP683 .A93 2019
Variant title	Authentication and detection of adulteration of olive oil
Related names	Kontominas, Michael, editor.
Contents	A review of olive oil authentication / Apostolos Kiritsakis, Evangelos Evangelou and Amani Taamalli - Fingerprinting of virgin olive oils for authenticity and quality assessment / Rosa María Alonso-Salces, Margaret Veronica Holland, Sandra Rosa Fuselli and Claude Guillou - LC fingerprinting approach for authenticating olive oil / Ana M. Jiménez-Carvelo, Antonio González-Casado, Estefanía Pérez-Castaño and Luis Cuadros-Rodríguez - Authentication of olive oil with mid-infrared spectroscopy / Oguz Uncu and Banu Ozen - Characterization and differentiation of Greek virgin olive oils from five cultivars based on selected physicochemical parameters combined with chemometrics / Ioanna Kosma, Anastasia Badeka, Stavros Kontakos and Michael Kontominas - Authentication and adulteration of olive oil / Lanfranco Conte, Paolo Lucci, Erica Moret and Sabrina Moret - Application of vibrational spectroscopy techniques / Astrid Maléchaux, Nathalie Dupuy and Jacques Artaud - Application of thermal and physical techniques for olive oil adulteration and authenticity / Maria Paciulli, Augusta Caligiani, Maria Teresa Rodriguez Estrada, Lorenzo Cerretani and Emma Chiavaro - Olive oil adulteration / Dimitrios Boskou - Advances in DNA extraction for detecting extra-

	virgin olive oil adulteration / Cinzia Montemurro, Wilma Sabetta, Valentina di Rienzo and Antonella Pasqualone - Sesquiterpene hydrocarbons, a promising tool for virgin olive oil geographical authetication / Stefania Vichi, Alba Tres, Beatriz Quintanilla-Casas, Julen Bustamante and Francesc Guardiola - The use of pyropheophytins and diacylglycerol testing to detect old, deodorised, and/or poorly stored extra virgin olive oils / Claudia M. Guillaume and Leandro M. Ravetti - Application of phycicochemical parameters combined with chemometrics for the differentiation of geographical origin of extra virgin olive oils of the Koroneiki cultivar / Ioanna Kosma, Anatasia Badeka, Stavros Kontakos and Michael Kontominas.
LC Subjects	Olive oil. Food adulteration and inspection.
Notes	Includes bibliographical references and index.
Series	Food science and technology

Chemometric monitoring: product quality assessment, process fault detection, and applications

LCCN	2017015762
Type of material	Book
Personal name	Kundu, Madhusree, author.
Main title	Chemometric monitoring: product quality assessment, process fault detection, and applications / Madhusree Kundu, Palash Kumar Kundu, Seshu K. Damarla.
Published/Produced	Boca Raton: CRC Press, 2018.
ISBN	9781498780070 (hardback: acid-free paper)
LC classification	QD75.4.C45 K86 2018
Related names	Kundu, Palash Kumar, author. Damarla, Seshu K., author.
LC Subjects	Chemometrics.
Notes	Includes bibliographical references and index.

Chemometrics: data driven extraction for science

LCCN	2017059486
Type of material	Book
Personal name	Brereton, Richard G., author.
Main title	Chemometrics: data driven extraction for science / Richard G. Brereton.
Edition	Second edition.
Published/Produced	Hoboken, NJ: John Wiley & Sons, 2018.
Description	1 online resource.
ISBN	9781118904688 (epub) 9781118904671 (pdf)
LC classification	QD75.4.C45
LC Subjects	Chemometrics--Data processing. Chemical processes--Statistical methods--Data processing.
Notes	Originally published in 2003 as: Chemometrics: data analysis for the laboratory and chemical plant.
Additional formats	Print version: Brereton, Richard G., author. Chemometrics Second edition. Hoboken, NJ: John Wiley & Sons, 2018 9781118904664 (DLC) 2017054468

Chemometrics: data driven extraction for science

LCCN	2017054468
Type of material	Book
Personal name	Brereton, Richard G., author.
Main title	Chemometrics: data driven extraction for science / Richard G. Brereton.
Edition	Second edition.
Published/Produced	Hoboken, NJ: John Wiley & Sons, 2018.
ISBN	9781118904664 (pbk.)
LC classification	QD75.4.C45 B74 2018
LC Subjects	Chemometrics--Data processing. Chemical processes--Statistical methods--Data processing.
Notes	Originally published in 2003 as: Chemometrics: data analysis for the laboratory and chemical plant.

Additional formats Online version: Brereton, Richard G., author. Chemometrics Second edition. Hoboken, NJ: John Wiley & Sons, 2018 9781118904688 (DLC) 2017059486

Chemometrics: methods, applications, and new research

LCCN	2016050466
Type of material	Book
Main title	Chemometrics: methods, applications, and new research / Aderval S. Luna, editor.
Published/Produced	New York: Nova Science Publishers, Inc., [2017]
Description	1 online resource.
ISBN	9781536105278
LC classification	QD75.4.C45
Related names	Luna, Aderval S. (Aderval Severino), editor.
LC Subjects	Chemometrics. Chemistry--Mathematics. Chemistry, Analytic.
Notes	Includes index.
Additional formats	Print version: Chemometrics Hauppauge, New York: Nova Science Publishers, Inc., [2017] 9781536105186 (DLC) 2016050153
Series	Analytical chemistry and microchemistry

Chemometrics: methods, applications, and new research

LCCN	2016050153
Type of material	Book
Main title	Chemometrics: methods, applications, and new research / Aderval S. Luna, editor.
Published/Produced	New York: Nova Science Publishers, Inc., [2017] ©2017
Description	vi, 237 pages: illustrations; 26 cm.
ISBN	9781536105186 (hardcover) 153610518X (hardcover)
LC classification	QD75.4.C45 C488 2017
Related names	Luna, Aderval S. (Aderval Severino), editor.
LC Subjects	Chemometrics. Chemistry--Mathematics. Chemistry, Analytic.

Notes	Includes bibliographical references and index.
Additional formats	Online version: Chemometrics Hauppauge, New York: Nova Science Publishers, Inc., [2017] 9781536105278 (DLC) 2016050466
Series	Analytical chemistry and microchemistry

Chemometrics and cheminformatics in aquatic toxicology

LCCN	2021010980
Type of material	Book
Main title	Chemometrics and cheminformatics in aquatic toxicology / [edited by] Kunal Roy.
Edition	1st edition.
Published/Produced	Hoboken, NJ: Wiley, 2021.
Description	1 online resource
ISBN	9781119681601 (epub) 9781119681649 (adobe pdf) (hardback)
LC classification	QH90.8.T68
Related names	Roy, Kunal, 1971- author.
Summary	"This book introduces readers to the existing and emerging problems of contamination of the aquatic environment due to various metal and organic pollutants (including industrial chemicals, pharmaceuticals, cosmetics, biocides, nanomaterials, pesticides, surfactants, dyes, etc.) and resultant effects on water quality, chemical threat to the aquatic organisms and consequent effects on human health. The book discusses different chemometric (classification and pattern recognition tools, clustering techniques, principal component analysis, multivariate regression analysis, etc.) and cheminformatic (toxicophore, QSAR, data mining, etc.) tools for the non-experts and their application in analyzing and modeling toxicity data of chemicals (including metal and organic contaminants) to various aquatic organisms. Split into four sections the book covers chemometric and cheminformatic tools and protocols, case studies and literature reports, and

tools and databases. Multispecies aquatic toxicity modeling will demonstrate the data gap filling in absence of toxicity data for a particular aquatic species. Various aquatic toxicity databases and chemometric software tools and webservers are covered and practical examples of model development with illustrations are provided"-- Provided by publisher.

Contents Water quality and contaminants of emerging concern (CECs) / Antonio Juan García-Fernández, Silvia Espín, Pilar Gómez-Ramírez, Pablo Sánchez-Virosta, Isabel Navas - The effects of contaminants of emerging concern on water quality / Heiko L. Schoenfuss - Chemometrics: multivariate statistical analysis of analytical chemical and biomolecular data / Richard Brereton - Chemometrics and cheminformatics--an introduction / Chanin Nantasenamat - An introduction to some basic chemometric tools / Lennart Eriksson, Erik Johansson, Johan Trygg - From data to models: mining experimental values with machine learning tools / G. Gini, E. Benfenati - Machine learning approaches in computational toxicology studies / Pravin Ambure, Stephen Jones Barigye, Rafael Gozalbes - Counter-propagation neural networks for modelling and read across in aquatic (fish) toxicity / Marjan Vracko, Viktor Drgan - Aiming high versus aiming all: aquatic toxicology and QSAR multitarget models / Ana S. Moura and M. Natalia D.S. Cordeiro - Chemometric approaches to evaluate interspecies relationships and extrapolation in aquatic toxicity / S. Raimondo, C.M. Lavelle, M.G. Barron - The QSAR paradigm to explore and predict aquatic toxicity / Fotios Tsopelas, Anna Tsantili-Kakoulidou - Application of cheminformatics to model fish toxicity / Sorin Avram, Simona Funar-Timofei, Gheorghe Ilia - Chemometric modeling of algal and daphnia toxicity / Luminita Crisan, Ana Borota, Alina Bora,

Simona Funar-Timofei, Gheorghe Ilia - Chemometric modeling of algal toxicity / Gülçin Tuğcu, Serli Önlü, M.T. Sacan - Chemometric modelling of daphnia toxicity / Amit Kumar Halder and M. Natalia D.S. Cordeiro - Chemometric modeling of Daphnia toxicity: quantum-mechanical insights / Reenu and Vikas - Chemometric modeling of toxicity of chemicals to tadpoles / Kabiruddin Khan and Kunal Roy - Chemometric modeling of chemical toxicity to marine bacteria / Kabiruddin Khan and Kunal Roy - Chemometric modeling of pesticide aquatic toxicity / Alina Bora, Simona Funar-Timofei - Contribution of chemometric modeling to chemical risks assessment for aquatic plants: state of the art / Mabrouk Hamadache, Abdeltif Amrane, Othmane Benkortbi, Salah Hanini - Application of 3D QSAR approaches to classification and prediction of aquatic toxicity / S. Lee, M.G. Barron - Aquatic ecotoxicity QSAR methods for cationic polymers / Hans Sanderson, Pathan Mohsin Khan, Supratik Kar, Kunal Roy, Anna Magdalene Brun Hansen, Kristin Connors, Scott Belanger - In silico platforms for predictive ecotoxicology: from machine learning to deep learning / Yong Oh Lee, Baeckkyoung Sung - The use and evolution of web tools for aquatic toxicology studies / Renata Priscila Costa Barros, Luana de Morais e Silva, Natalia Ferreira de Sousa, Luciana Scotti, Marcus Tullius Scotti - The tools for aquatic toxicology within the VEGAHUB system / A. Lombarodo, D. Baderna, E. Benfenati - Aquatic toxicology databases / Supratik Kar, Kerzy Leszczynski - Computational tools for the assessment and substitution of biocidal active substances of ecotoxicological concern: the COMBASE project / María Blázquez, Oscar Andreu-Sánchez, Arantxa Ballesteros, María Luisa Fernández-Cruz, Carlos Fito, Sergi Gómez-Ganau, Rafael Gozalbes, David Hernández-Moreno, Anna

	Lombardo, Marco Marzo, Irati Ranero, Emilio Benfenati - Image analysis and deep learning web services for nanoinformatics / Anastasios G. Papadiamantis, Antreas Afantitis, Andreas Tsoumanis, Pantelis Karatzas, Philip Doganis, Dimitra-Danai Varsou, Haralambos Sarimveis, Laura-Jayne A. Ellis, Eugenia Valsami-Jones, Iseult Lynch, Georgia Melagraki.
LC Subjects	Water--Pollution--Toxicology--Mathematical models. Water--Pollution--Toxicology--Data processing. Water quality--Mathematical models. Water quality--Measurement--Data processing. Chemometrics. Cheminformatics.
Notes	Includes bibliographical references and index.
Additional formats	Print version: Chemometrics and cheminformatics in aquatic toxicology 1st edition. Hoboken, NJ: Wiley, 2021 9781119681595 (DLC) 2021010979

Chemometrics and cheminformatics in aquatic toxicology

LCCN	2021010979
Type of material	Book
Main title	Chemometrics and cheminformatics in aquatic toxicology / [edited by] Kunal Roy.
Edition	1st edition.
Published/Produced	Hoboken, NJ: Wiley, 2021.
ISBN	9781119681595 (hardback) (adobe pdf) (epub)
LC classification	QH90.8.T68 C44 2021
Related names	Roy, Kunal, 1971- author.
Summary	"This book introduces readers to the existing and emerging problems of contamination of the aquatic environment due to various metal and organic pollutants (including industrial chemicals, pharmaceuticals, cosmetics, biocides, nanomaterials, pesticides, surfactants, dyes, etc.) and resultant effects on water quality, chemical

threat to the aquatic organisms and consequent effects on human health. The book discusses different chemometric (classification and pattern recognition tools, clustering techniques, principal component analysis, multivariate regression analysis, etc.) and cheminformatic (toxicophore, QSAR, data mining, etc.) tools for the non-experts and their application in analyzing and modeling toxicity data of chemicals (including metal and organic contaminants) to various aquatic organisms. Split into four sections the book covers chemometric and cheminformatic tools and protocols, case studies and literature reports, and tools and databases. Multispecies aquatic toxicity modeling will demonstrate the data gap filling in absence of toxicity data for a particular aquatic species. Various aquatic toxicity databases and chemometric software tools and webservers are covered and practical examples of model development with illustrations are provided"--Provided by publisher.

Contents Water quality and contaminants of emerging concern (CECs) / Antonio Juan García-Fernández, Silvia Espín, Pilar Gómez-Ramírez, Pablo Sánchez-Virosta, Isabel Navas - The effects of contaminants of emerging concern on water quality / Heiko L. Schoenfuss - Chemometrics: multivariate statistical analysis of analytical chemical and biomolecular data / Richard Brereton - Chemometrics and cheminformatics--an introduction / Chanin Nantasenamat - An introduction to some basic chemometric tools / Lennart Eriksson, Erik Johansson, Johan Trygg - From data to models: mining experimental values with machine learning tools / G. Gini, E. Benfenati - Machine learning approaches in computational toxicology studies / Pravin Ambure, Stephen Jones Barigye, Rafael Gozalbes - Counter-propagation neural networks for modelling and read across in aquatic (fish) toxicity / Marjan Vracko, Viktor Drgan - Aiming high versus aiming all: aquatic toxicology and QSAR multitarget models / Ana S. Moura and M. Natalia D.S. Cordeiro - Chemometric approaches to evaluate interspecies relationships and extrapolation in aquatic toxicity / S. Raimondo, C.M. Lavelle, M.G. Barron - The QSAR paradigm to explore and predict aquatic toxicity / Fotios Tsopelas, Anna Tsantili-Kakoulidou - Application of cheminformatics to model fish toxicity / Sorin Avram, Simona Funar-Timofei, Gheorghe Ilia - Chemometric modeling of algal and daphnia toxicity / Luminita Crisan, Ana Borota, Alina Bora, Simona Funar-Timofei, Gheorghe Ilia - Chemometric modeling of algal toxicity / Gülçin Tuğcu, Serli Önlü, M.T. Sacan - Chemometric modelling of daphnia toxicity / Amit Kumar Halder and M. Natalia D.S. Cordeiro - Chemometric modeling of Daphnia toxicity: quantum-mechanical insights / Reenu and Vikas - Chemometric modeling of toxicity of chemicals to tadpoles / Kabiruddin Khan and Kunal Roy - Chemometric modeling of chemical toxicity to marine bacteria / Kabiruddin Khan and Kunal Roy - Chemometric modeling of pesticide aquatic toxicity / Alina Bora, Simona Funar-Timofei - Contribution of chemometric modeling to chemical risks assessment for aquatic plants: state of the art / Mabrouk Hamadache, Abdeltif Amrane, Othmane Benkortbi, Salah Hanini - Application of 3D QSAR approaches to classification and prediction of aquatic toxicity / S. Lee, M.G. Barron - Aquatic ecotoxicity QSAR methods for cationic polymers / Hans Sanderson, Pathan Mohsin Khan, Supratik Kar, Kunal Roy, Anna Magdalene Brun Hansen, Kristin Connors, Scott Belanger - In silico platforms for predictive ecotoxicology: from machine learning to deep learning / Yong Oh Lee, Baeckkyoung Sung - The use and evolution of web

	tools for aquatic toxicology studies / Renata Priscila Costa Barros, Luana de Morais e Silva, Natalia Ferreira de Sousa, Luciana Scotti, Marcus Tullius Scotti - The tools for aquatic toxicology within the VEGAHUB system / A. Lombarodo, D. Baderna, E. Benfenati - Aquatic toxicology databases / Supratik Kar, Kerzy Leszczynski - Computational tools for the assessment and substitution of biocidal active substances of ecotoxicological concern: the COMBASE project / María Blázquez, Oscar Andreu-Sánchez, Arantxa Ballesteros, María Luisa Fernández-Cruz, Carlos Fito, Sergi Gómez-Ganau, Rafael Gozalbes, David Hernández-Moreno, Anna Lombardo, Marco Marzo, Irati Ranero, Emilio Benfenati - Image analysis and deep learning web services for nanoinformatics / Anastasios G. Papadiamantis, Antreas Afantitis, Andreas Tsoumanis, Pantelis Karatzas, Philip Doganis, Dimitra-Danai Varsou, Haralambos Sarimveis, Laura-Jayne A. Ellis, Eugenia Valsami-Jones, Iseult Lynch, Georgia Melagraki.
LC Subjects	Water--Pollution--Toxicology--Mathematical models. Water--Pollution--Toxicology--Data processing. Water quality--Mathematical models. Water quality--Measurement--Data processing. Chemometrics. Cheminformatics.
Notes	Includes bibliographical references and index.
Additional formats	Online version: Roy, Kunal, 1971- Chemometrics and cheminformatics in aquatic toxicology. 1st edition Hoboken, NJ: Wiley, 2021 9781119681649 (DLC) 2021010980

Chemometrics and numerical methods in LIBS

LCCN	2022030870
Type of material	Book
Main title	Chemometrics and numerical methods in LIBS / edited by Vincenzo Palleschi.

Published/Produced	Hoboken, NJ: Wiley, 2022.
ISBN	9781119759584 (cloth) (adobe pdf) (epub)
LC classification	QD96.A8 C44 2022
Related names	Palleschi, V., editor.
Summary	"The LIBS community is constantly growing and it now widely recognized that LIBS techniques provide large sets of data statistical methods for simplification and analysis are essential; however, there is still some confusion on the proper application of these methods to LIBS analysis and, most of all, on the procedures that must be applied for the validation of the results obtained. The book will be organized in three parts, dealing with the main applications of chemometrics in LIBS: simplification of the spectral information, classification of spectra, quantitative analysis. Each part will be divided in sections, describing the different techniques with practical examples. For example, in Part 2, the techniques described will be discussed in the perspective of waste sorting by LIBS and LIBS micro-imaging of geological material, to mention just two of the main applications which exploits at the best the chemometric techniques of classification."--Provided by publisher.
LC Subjects	Laser-induced breakdown spectroscopy. Chemometrics.
Notes	Includes bibliographical references and index.
Additional formats	Online version: Chemometrics and numerical methods in LIBS Hoboken, NJ: Wiley, 2022 9781119759560 (DLC) 2022030871

Chemometrics in chromatography

LCCN	2017037096
Type of material	Book
Main title	Chemometrics in chromatography / edited by Łukasz Komsta, Yvan Vander Heyden, Joseph Sherma.
Published/Produced	Boca Raton: CRC Press, Taylor & Francis Group, an informa business, 2018.
Description	xix, 505 pages; 26 cm.
ISBN	9781498772532 (hardback: alk. paper)
LC classification	QD75.4.C45 C48448 2018
Related names	Komsta, Lukasz, editor. Vander Heyden, Yvan, editor. Sherma, Joseph, editor.
LC Subjects	Chemometrics. Chromatographic analysis.
Notes	Includes bibliographical references.
Series	Chromatographic science series

Chemometrics in spectroscopy

LCCN	2018930967
Type of material	Book
Personal name	Mark, Howard, author.
Main title	Chemometrics in spectroscopy / Howard Mark, President, Mark Electronics, Suffern, NY, United States, Jerry Workman, Jr., Executive Vice President, Research & Development Unity Scientific and Process Sensors Corp., Milford, MA, United States, Adjunct Professor, Health and Human Services, National University, San Diego, CA, United States.
Edition	Second edition.
Published/Produced	London, United Kingdom; San Diego, CA, United States: Elsevier: Academic Press, [2018]
Description	xxxi, 1057 pages: illustrations; 24 cm
ISBN	9780128053096
LC classification	QD95.5.S75 M35 2018
Related names	Workman, Jerry, author.

LC Subjects	Spectrum analysis--Statistical methods. Chemometrics.
Notes	Includes bibliographical references and index.

Contributions to Partial Differential Equations and Applications

LCCN	2019768502
Type of material	Book
Main title	Contributions to Partial Differential Equations and Applications / edited by B. N. Chetverushkin, W. Fitzgibbon, Y.A. Kuznetsov, P. Neittaanmäki, J. Periaux, O. Pironneau.
Edition	1st ed. 2019.
Published/Produced	Cham: Springer International Publishing: Imprint: Springer, 2019.
Description	1 online resource (XI, 452 pages 142 illustrations, 75 illustrations in color.) text file PDF
ISBN	9783319783253
Related names	Chetverushkin, B. N, editor. Fitzgibbon, W, editor. Kuznetsov, Y.A, editor. Neittaanmäki, P, editor. Periaux, J, editor. Pironneau, O, editor.
Summary	This book treats Modelling of CFD problems, Numerical tools for PDE, and Scientific Computing and Systems of ODE for Epidemiology, topics that are closely related to the scientific activities and interests of Prof. William Fitzgibbon, Prof. Yuri Kuznetsov, and Prof. O. Pironneau, whose outstanding achievements are recognised in this volume. It contains 20 contributions from leading scientists in applied mathematics dealing with partial differential equations and their applications to engineering, ab-initio chemistry and life sciences. It includes the mathematical and numerical contributions to PDE for applications presented at the ECCOMAS thematic conference "Contributions to PDE for Applications" held at Laboratoire

Jacques Louis Lions in Paris, France, August 31-September 1, 2015, and at the Department of Mathematics, University of Houston, Texas, USA, February 26-27, 2016. This event brought together specialists from universities and research institutions who are developing or applying numerical PDE or ODE methods with an emphasis on industrial and societal applications. This volume is of interest to researchers and practitioners as well as advanced students or engineers in applied and computational mathematics. All contributions are written at an advanced scientific level with no effort made by the editors to make this volume self-contained. It is assumed that the reader is a specialist already who knows the basis of this field of research and has the capability of understanding and appreciating the latest developments in this field.

Contents

Preface - Career papers - The career of Prof. W. Fitzgibbon, by Jeff Morgan, Jacques Periaux - The career of Prof. Yuri Kuznetsov, by Boris Chetverushkin, William Fitzgibbon, Jacques Periaux, and Olivier Pironneau - The career of Prof. Olivier Pironneau, by William Fitzgibbon and Jacques Periaux - Scientific contributions - Mean field games for modeling crowd motion, by Yves Achdou and Jean-Michel Lasry - Remarks about spatially structured SI model systems with cross diffusion, by Vanaya Anaya, Mostafa Bendahmane, Michel Langlais, and Mauricio Sepúlveda - Automatic clustering in large sets of time series, by Robert Azencott, Viktoria Muravina, Rasoul Hekmati, Wei Zhang, and Michael Paldino - Zero viscosity boundary effect limit and turbulence, by Claude Bardos - Parabolic equations with quadratic growth in Rn, by Alain Bensoussan, Jens Frehse, Shige Peng, and Sheung Chi Phillip Yam - On the sensitivity to the filtering radius in Leray models of incompressible flow, by Luca Bertagna, Annalisa

Quaini, Leo G. Rebholz, and Alessandro Veneziani - Model order reduction for problems with large convection effects, by Nicolas Cagniart, Yvon Maday, and Benjamin Stamm - Parametric optimization of pulsating jets in unsteady flow by Multiple-Gradient Descent Algorithm (MGDA), by Jean-Antoine Désidéri and Regis Duvigneau - Mixed formulation of a linearized lubrication fracture model in a poro-elastic medium, by Vivette Girault, Mary Wheeler, Kundan Kumar, and Gurpreet Singh - Two decades of wave-like equation for the numerical simulation of incompressible viscous flow: A review, by Roland Glowinski and Tsorng-Whay Pan - An arbitrary Lagrangian-Eulerian Finite Element method preserving convex invariants of hyperbolic systems, by Jean-Luc Guermond, Bojan Popov, Laura Saavedra, and Yong Yang - Dual-primal isogeometric tearing and interconnecting methods, by Christoph Hofer and Ulrich Langer - C0 Interior Penalty Discontinuous Galerkin approximation of a sixth order Cahn-Hilliard equation modeling microemulsification processes, by Ronald Hoppe and Christopher Linsenmann - On existence "in the large" of a solution to modified Navier-Stokes equations, by George Kobelkov - An algebraic solver for the Oseen problem with application to hemodynamics, by Igor Konshin, Maxim Olshanskii, and Yuri Vassilevski - Martin's problem for volume-surface reaction-diffusion systems, by Jeff Morgan and Vandana Sharma - A posteriori error estimates for the electric field integral equation on polyhedral, by Ricardo Nochetto and Benjamin Stamm - On some weighted Stokes problems: applications on Smagorinsky models, by Jacques Rappaz and Jonathan Rochat - Poincaré type inequalities for vector functions with zero mean normal traces on the boundary and applications to interpolation methods, by Sergey

	Repin - The ensemble interpretation of quantum mechanics and the two-slit experiment, by Glenn Webb.
LC Subjects	Biology--Technique. Chemometrics. Computer mathematics. Epidemiology. Fluid mechanics. Partial differential equations.
Other Subjects	Engineering Fluid Dynamics. Biological Techniques. Computational Science and Engineering. Epidemiology. Math. Applications in Chemistry. Partial Differential Equations.
Additional formats	Print version: Contributions to partial differential equations and applications. 9783319783246 (DLC) 2018935846 Printed edition: 9783030086756 Printed edition: 9783319783246 Printed edition: 9783319783260
Series	Computational Methods in Applied Sciences, 1871-3033; 47 Computational Methods in Applied Sciences, 1871-3033; 47

Data science in chemistry: artificial intelligence, big data, chemometrics, and quantum computing with Jupyter

LCCN	2020942815
Type of material	Book
Personal name	Gressling, Thorsten, author.
Main title	Data science in chemistry: artificial intelligence, big data, chemometrics, and quantum computing with Jupyter / Thorsten Gressling.
Published/Produced	Berlin; Boston: Walter de Gruyter GmbH, [2021]
Description	xviii, 522 pages: illustrations; 24 cm
ISBN	9783110629392 paperback 3110629399 paperback
LC classification	QD39.3.E46 G74 2021

LC Subjects	Chemistry--Data processing.
Other Subjects	Chemistry--Data processing.
Notes	Includes bibliographical references and index.
Series	De Gruyter graduate De Gruyter graduate.

Electronic Nose: Algorithmic Challenges

LCCN	2019743574
Type of material	Book
Personal name	Zhang, Lei. author.
Main title	Electronic Nose: Algorithmic Challenges / by Lei Zhang, Fengchun Tian, David Zhang.
Edition	1st ed. 2018.
Published/Produced	Singapore: Springer Singapore: Imprint: Springer, 2018.
Description	1 online resource (XV, 339 pages 89 illustrations, 70 illustrations in color.) text file PDF
ISBN	9789811321672
Related names	Tian, Fengchun. author. Zhang, David. author.
Summary	This book presents the key technology of electronic noses, and systematically describes how e-noses can be used to automatically analyse odours. Appealing to readers from the fields of artificial intelligence, computer science, electrical engineering, electronics, and instrumentation science, it addresses three main areas: First, readers will learn how to apply machine learning, pattern recognition and signal processing algorithms to real perception tasks. Second, they will be shown how to make their algorithms match their systems once the algorithms don't work because of the limitation of hardware resources. Third, readers will learn how to make schemes and solutions when the acquired data from their systems is not stable due to the fundamental issues affecting perceptron devices (e.g., sensors). In brief, the book presents and discusses the key technologies and new algorithmic

challenges in electronic noses and artificial olfaction. The goal is to promote the industrial application of electronic nose technology in environmental detection, medical diagnosis, food quality control, explosive detection, et cetera and to highlight the scientific advances in artificial olfaction and artificial intelligence. The book offers a good reference guide for newcomers to the topic of electronic noses, because it refers to the basic principles and algorithms. At the same time, it clearly presents the key challenges - such as long-term drift, signal uniqueness, and disturbance - and effective and efficient solutions, making it equally valuable for researchers engaged in the science and engineering of sensors, instruments, chemometrics, et cetera.

Contents	Part 1: Overview - Chapter 1. Introduction - Chapter 2. Literature Review - Part 2: E-nose Odor Recognition and Prediction: Challenge I - Chapter 3. Heuristic and Bio-inspired Neural Network Model - Chpater 4. Chaos based Neural Network Optimization Approach - Chapter 5. Multilayer Perceptrons based Concentration Estimation, et cetera.
LC Subjects	Bioinformatics. Biometrics (Biology). Health informatics. Pattern recognition.
Other Subjects	Pattern Recognition. Biometrics. Computational Biology/Bioinformatics. Health Informatics.
Additional formats	Print version: Electronic nose. 9789811321665 (DLC) 2018954598 Printed edition: 9789811321665 Printed edition: 9789811321689 Printed edition: 9789811347412

Emerging techniques for food processing, quality, and safety assurance

LCCN	2021015576
Type of material	Book
Main title	Emerging techniques for food processing, quality, and safety assurance / edited by Monika Sharma, Preeti Birwal, Megh R. Goyal.
Edition	First edition.
Published/Produced	Palm Bay, FL, USA: Apple Academic Press, [2021]
Description	1 online resource
ISBN	9781003184720 (ebook) (hardcover) (hardcover)
LC classification	TP370.5
Related names	Sharma, Monika (Food scientist), editor. Birwal, Preeti, editor. Goyal, Megh R., editor.
Summary	"The Handbook of Research on Food Processing and Preservation Technologies covers a vast abundance of information on various design, development, and applications of novel and innovative strategies for food processing and preservation. The roles and applications of minimal processing techniques (such as ozone treatment, vacuum drying, osmotic dehydration, dense phase carbon dioxide treatment, pulsed electric field, and high-pressure assisted freezing) are discussed, along with a wide range of applications. The handbook also explores some exciting computer-aided techniques emerging in the food processing sector, such as robotics, radio frequency identification (RFID), three-dimensional food printing, artificial intelligence, etc. Some emphasis has also been given on nondestructive quality evaluation techniques (such as image processing, terahertz spectroscopy imaging technique, near infrared, Fourier transform infrared spectroscopy technique, etc.) for food quality and safety evaluation. The significant roles of food properties in the design of specific foods and edible films have

been elucidated as well. Volume 5: Emerging Techniques for Food Processing, Quality, and Safety Assurance discusses various emerging techniques for food preservation, formulation, and nondestructive quality evaluation techniques. Each chapter covers major aspects pertaining to principles, design, and applications of various food processing methods, such as low temperature-based-ultrasonic drying of foods, hypobaric processing of foods, viability of high-pressure technology, application of pulsed electric fields in food preservation, green nanotechnology for food processing and preservation, advanced methods of encapsulation, basics and methods of food authentication, imaging techniques for quality inspection of spices and nuts, FTIR coupled with chemometrics for food quality and safety, and the use of robotic engineering for quality and safety"--Provided by publisher.

Contents Low Temperature Based Ultrasonic Drying of Foods / Pavan M. Gundu, Preeti Birwal, Chaitradeepa G. Mestri, and Abila Krishna - Hypobaric Processing of Food / Sharanabasava Kumbar and Menon Rekha Ravindra - Viability of High-Pressure Technology in the Food Industry / Mohona Munshi, Madhusudan Sharma, and Saptashish Deb - Potential of Pulsed Electric Fields in Food Preservation / Pranali Nikam, Suvartan Ranvir, John David, Thejus Jacob, and Raman Seth - Pulsed Light Technology in Food Processing and Preservation / Mahendra Gunjal, Humeera Tazeen, Pavan M. Gundu, Preeti Birwal, and Abila Krishna - Potential of Green Nanotechnology in Food Processing and Preservation / Shikha Pandhi, Arvind Kumar, Sadhna Mishra, and Dinesh Chandra Rai - Food Encapsulation: Principles, Novel Methods, and Applications / Nikunj Sharma, Syed Mansha Rafiq, and Syed Insha Rafiq - Food Authentication: Basics and Detection Methods /

	Sangita Bansal, Kanika Sharma, and Era V. Malhotra - Imaging Techniques for Quality Assessment of Spices and Nuts / Leena Kumari, Monika Sharma, and Gajanan Deshmukh - Fourier Transform Infrared (FTIR) Spectroscopy with Chemometrics: Evaluation of Food Quality and Safety / Neelam Upadhyay, C. G. Harshitha, Nilesh Kumar Pathak, and Rajan Sharma - Robotic Engineering: A Tool for Quality and Safety of Foods / Ravi Prakash and Menon Rekha Ravindra.
LC Subjects	Food industry and trade--Technological innovations. Food--Preservation.
Notes	Includes bibliographical references and index.
Additional formats	Issued also in electronic format. Print version: Emerging techniques for food processing, quality, and safety assurance First edition Palm Bay, FL, USA: Apple Academic Press, [2021] 9781774630358 (DLC) 2021015575
Series	Handbook of research on food processing and preservation technologies; volume 5 Innovations in agricultural and biological engineering Handbook of research on food processing and preservation technologies; volume 5. Innovations in agricultural and biological engineering.

Emerging techniques for food processing, quality, and safety assurance

LCCN	2021015575
Type of material	Book
Main title	Emerging techniques for food processing, quality, and safety assurance / edited by Monika Sharma, Preeti Birwal, Megh R. Goyal.
Edition	First edition.
Published/Produced	Palm Bay, FL, USA: Apple Academic Press, [2021]
ISBN	9781774630358 (hardcover) 9781774638552 (hardcover) (ebook)

LC classification	TP370.5 .E455 2021
Related names	Sharma, Monika (Food scientist), editor. Birwal, Preeti, editor. Goyal, Megh R., editor.
Summary	"The Handbook of Research on Food Processing and Preservation Technologies covers a vast abundance of information on various design, development, and applications of novel and innovative strategies for food processing and preservation. The roles and applications of minimal processing techniques (such as ozone treatment, vacuum drying, osmotic dehydration, dense phase carbon dioxide treatment, pulsed electric field, and high-pressure assisted freezing) are discussed, along with a wide range of applications. The handbook also explores some exciting computer-aided techniques emerging in the food processing sector, such as robotics, radio frequency identification (RFID), three-dimensional food printing, artificial intelligence, etc. Some emphasis has also been given on nondestructive quality evaluation techniques (such as image processing, terahertz spectroscopy imaging technique, near infrared, Fourier transform infrared spectroscopy technique, etc.) for food quality and safety evaluation. The significant roles of food properties in the design of specific foods and edible films have been elucidated as well. Volume 5: Emerging Techniques for Food Processing, Quality, and Safety Assurance discusses various emerging techniques for food preservation, formulation, and nondestructive quality evaluation techniques. Each chapter covers major aspects pertaining to principles, design, and applications of various food processing methods, such as low temperature-based-ultrasonic drying of foods, hypobaric processing of foods, viability of high-pressure technology, application of pulsed electric fields in food preservation, green nanotechnology for food

processing and preservation, advanced methods of encapsulation, basics and methods of food authentication, imaging techniques for quality inspection of spices and nuts, FTIR coupled with chemometrics for food quality and safety, and the use of robotic engineering for quality and safety"--Provided by publisher.

Contents Low Temperature Based Ultrasonic Drying of Foods / Pavan M. Gundu, Preeti Birwal, Chaitradeepa G. Mestri, and Abila Krishna - Hypobaric Processing of Food / Sharanabasava Kumbar and Menon Rekha Ravindra - Viability of High-Pressure Technology in the Food Industry / Mohona Munshi, Madhusudan Sharma, and Saptashish Deb - Potential of Pulsed Electric Fields in Food Preservation / Pranali Nikam, Suvartan Ranvir, John David, Thejus Jacob, and Raman Seth - Pulsed Light Technology in Food Processing and Preservation / Mahendra Gunjal, Humeera Tazeen, Pavan M. Gundu, Preeti Birwal, and Abila Krishna - Potential of Green Nanotechnology in Food Processing and Preservation / Shikha Pandhi, Arvind Kumar, Sadhna Mishra, and Dinesh Chandra Rai - Food Encapsulation: Principles, Novel Methods, and Applications / Nikunj Sharma, Syed Mansha Rafiq, and Syed Insha Rafiq - Food Authentication: Basics and Detection Methods / Sangita Bansal, Kanika Sharma, and Era V. Malhotra - Imaging Techniques for Quality Assessment of Spices and Nuts / Leena Kumari, Monika Sharma, and Gajanan Deshmukh - Fourier Transform Infrared (FTIR) Spectroscopy with Chemometrics: Evaluation of Food Quality and Safety / Neelam Upadhyay, C. G. Harshitha, Nilesh Kumar Pathak, and Rajan Sharma - Robotic Engineering: A Tool for Quality and Safety of Foods / Ravi Prakash and Menon Rekha Ravindra.

LC Subjects	Food industry and trade--Technological innovations. Food--Preservation.
Notes	Includes bibliographical references and index.
Additional formats	Issued also in electronic format. Online version: Emerging techniques for food processing, quality, and safety assurance First edition Palm Bay, FL, USA: Apple Academic Press, [2021] 9781003184720 (DLC) 2021015576
Series	Handbook of research on food processing and preservation technologies; volume 5 Innovations in agricultural and biological engineering Handbook of research on food processing and preservation technologies; volume 5. Innovations in agricultural and biological engineering.

Food chemistry chemometrics

LCCN	2016951121
Type of material	Book
Main title	Food chemistry chemometrics / [edited by] Gonzalo Delgado.
Published/Produced	New York, NY: Arcler Press LLC, 2017.
ISBN	9781680944242 (hardcover)

Innovations in coffee quality

LCCN	2020029143
Type of material	Book
Main title	Innovations in coffee quality / Daneysa Lahis Kalschne, Marinês Paula Corso and Rafael Carlos Eloy Dias, editors.
Published/Produced	New York: Nova Science Publishers, [2020]
Description	xx, 203 pages: illustrations; 23 cm.
ISBN	9781536182231 (hardcover) (adobe pdf)
LC classification	TP645 .I56 2020
Related names	Kalschne, Daneysa, editor.

Summary "Innovation in Coffee Quality presents recent contributions regarding the quality of the raw coffee material, the commercial coffees, and the sensory attributes of the coffee beverages. Modifications to industrial processes are proposed to use defective coffee beans. Traditional methods of sensory evaluation of coffee are compared with new methods. Advanced sensory analysis such as Flash Profile and Temporal Dominance of Sensations have been applied to assess sensory properties of coffee, proving effective in identifying sensory attributes in cold coffee beverage, for example. Other sensory tests, such as SCA, are efficient to evaluate the sensorial quality of coffee brews in the preparation of special coffees and blends of Arabica and Robusta species. From an industrial point of view, computer vision system is a remarkably interesting technique for future applications in coffee sector. Thus, methods employing image analysis obtained from smartphone with application of projection pursuit analysis to determine the quality of raw material, replacing traditional sensory testing are also described. This could also be an alternative method to evaluate the raw material online in coffee processes. A fast method based on the Vis-NIR spectrum and chemometrics for quality control of commercial roasted coffee is presented, in a less wasteful and more ecofriendly direction. Concerning to the health of coffee consumers, the new demands for conscious coffee consumption are discussed, and the beneficial effect of high-quality coffee due to the presence of antioxidant compounds in instant coffee is evaluated. In this way, Innovations in coffee quality brings on innovative methods applied to monitoring the quality of coffee, being fundamental for new researchers and students who seek expressive advances in coffee science"-- Provided by publisher.

LC Subjects	Coffee. Coffee--Processing. Coffee industry--Quality control. Coffee industry--Technological innovation.
Notes	Includes bibliographical references and index.
Additional formats	Online version: Innovations in coffee quality Hauppauge: Nova Science Publishers, 2020. 9781536183191 (DLC) 2020029144
Series	Food science and technology

Innovations in coffee quality

LCCN	2020029144
Type of material	Book
Main title	Innovations in coffee quality / editors, Daneysa Kalschne, Universidade Tecnológica Federal do Paraná, Campus Medianeira, Department of Foods, Avenida Brasil, Parque, Brazil, Marinês Corso, Universidade Tecnológica Federal do Paraná, Campus Medianeira, Department of Foods, Avenida Brasil, Parque, Brazil, Rafael Dias, Instituto Federal Catarinense, Campus Araquari, Department of Chemistry, Araquari, Santa Catarina, Brazil.
Published/Produced	New York: Nova Science Publishers, [2020]
Description	1 online resource
ISBN	9781536183191 (adobe pdf) (hardcover)
LC classification	TP645
Related names	Kalschne, Daneysa, editor.
Summary	"Innovation in Coffee Quality presents recent contributions regarding the quality of the raw coffee material, the commercial coffees, and the sensory attributes of the coffee beverages. Modifications to industrial processes are proposed to use defective coffee beans. Traditional methods of sensory evaluation of coffee are compared with new methods. Advanced sensory analysis such as Flash Profile and Temporal Dominance of Sensations have been applied to assess sensory properties of

coffee, proving effective in identifying sensory attributes in cold coffee beverage, for example. Other sensory tests, such as SCA, are efficient to evaluate the sensorial quality of coffee brews in the preparation of special coffees and blends of Arabica and Robusta species. From an industrial point of view, computer vision system is a remarkably interesting technique for future applications in coffee sector. Thus, methods employing image analysis obtained from smartphone with application of projection pursuit analysis to determine the quality of raw material, replacing traditional sensory testing are also described. This could also be an alternative method to evaluate the raw material online in coffee processes. A fast method based on the Vis-NIR spectrum and chemometrics for quality control of commercial roasted coffee is presented, in a less wasteful and more ecofriendly direction. Concerning to the health of coffee consumers, the new demands for conscious coffee consumption are discussed, and the beneficial effect of high-quality coffee due to the presence of antioxidant compounds in instant coffee is evaluated. In this way, Innovations in coffee quality brings on innovative methods applied to monitoring the quality of coffee, being fundamental for new researchers and students who seek expressive advances in coffee science"-- Provided by publisher.

LC Subjects	Coffee. Coffee--Processing. Coffee industry--Quality control. Coffee industry--Technological innovation.
Notes	Includes bibliographical references and index.
Additional formats	Print version: Innovations in coffee quality New York: Nova Science Publishers, [2020] 9781536182231 (DLC) 2020029143
Series	Food science and technology

Introduction to computational physical chemistry

LCCN	2017010486
Type of material	Book
Personal name	Schrier, Joshua, 1979-
Main title	Introduction to computational physical chemistry / Joshua Schrier, Haverford College.
Published/Produced	Mill Valley, California: University Science Books, [2017] ©2017
Description	xxvii, 474 pages: illustrations; 27 cm
ISBN	9781938787904 (alk. paper) 1938787900
LC classification	QD39.3.M3 S327 2017
LC Subjects	Chemistry--Mathematics--Textbooks. Chemistry, Physical and theoretical--Textbooks. Chemometrics--Textbooks.
Notes	Includes bibliographical references and indexes.
Additional formats	Online version: Schrier, Joshua, 1979- Introduction to computational physical chemistry Mill Valley, California: University Science Books, [2017] 9781938787911 (DLC) 2017010818

Introduction to Multivariate Calibration: A Practical Approach

LCCN	2019742450
Type of material	Book
Personal name	Olivieri, Alejandro C. author.
Main title	Introduction to Multivariate Calibration: A Practical Approach / by Alejandro C. Olivieri.
Edition	1st ed. 2018.
Published/Produced	Cham: Springer International Publishing: Imprint: Springer, 2018.
Description	1 online resource (XVII, 243 pages 158 illustrations, 156 illustrations in color.) text file PDF
ISBN	9783319970974

Summary	This book offers an introductory-level guide to the complex field of multivariate analytical calibration, with particular emphasis on real applications such as near infrared spectroscopy. It presents intuitive descriptions of mathematical and statistical concepts, illustrated with a wealth of figures and diagrams, and consistently highlights physicochemical interpretation rather than mathematical issues. In addition, it describes an easy-to-use and freely available graphical interface, together with a variety of appropriate examples and exercises. Lastly, it discusses recent advances in the field (figures of merit, detection limit, non-linear calibration, method comparison), together with modern literature references.
Contents	Chapter1: Chemometrics and multivariate calibration - Chapter2: The classical least-squares model - Chapter3: The inverse least-squares model - Chapter4: Principal component analysis - Chapter5: Principal component regression - Chapter6: The optimum number of latent variables - Chapter7: The partial least-squares model - Chapter8: Comparison of multivariate models - Chapter9: Data pre-processing. Part 1: samples and sensors - Chapter10: Data pre-processing. Part 2: mathematical filters.-Chapter11: Analytical figures of merit - Chapter12: MVC1: a software for multivariate calibration - Chapter13: Non-linearity and artificial neural networks - Chapter14: Solutions to exercises.
LC Subjects	Analytical chemistry. Statistics.
Other Subjects	Analytical Chemistry. Statistics for Engineering, Physics, Computer Science, Chemistry and Earth Sciences.

Additional formats Print version: Introduction to multivariate calibration: a practical approach 9783319970967 (DLC) 2018950544
Printed edition: 9783030073022
Printed edition: 9783319970967
Printed edition: 9783319970981

Many-Body Methods for Atoms, Molecules and Clusters

LCCN 2019759268
Type of material Book
Personal name Schirmer, Jochen, author.
Main title Many-Body Methods for Atoms, Molecules and Clusters / by Jochen Schirmer.
Edition 1st ed. 2018.
Published/Produced Cham: Springer International Publishing: Imprint: Springer, 2018.
Description 1 online resource (XII, 332 pages 53 illustrations, 2 illustrations in color.)
text file PDF
ISBN 9783319936024
Summary This book provides an introduction to many-body methods for applications in quantum chemistry. These methods, originating in field-theory, offer an alternative to conventional quantum-chemical approaches to the treatment of the many-electron problem in molecules. Starting with a general introduction to the atomic and molecular many-electron problem, the book then develops a stringent formalism of field-theoretical many-body theory, culminating in the diagrammatic perturbation expansions of many-body Green's functions or propagators in terms of Feynman diagrams. It also introduces and analyzes practical computational methods, such as the field-tested algebraic-diagrammatic construction (ADC) schemes. The ADC concept can also be established via a wave-function based procedure, referred to as intermediate state representation (ISR), which bridges the gap between propagator and wave-

function formulations. Based on the current rapid increase in computer power and the development of efficient computational methods, quantum chemistry has emerged as a potent theoretical tool for treating ever-larger molecules and problems of chemical and physical interest. Offering an introduction to many-body methods, this book appeals to advanced students interested in an alternative approach to the many-electron problem in molecules, and is suitable for any courses dealing with computational methods in quantum chemistry.

Contents

Part I Many-Electron Systems and the Electron Propagator - Systems of identical particles - Second quantization - One-particle Green's function - Part II Formalism of Diagrammatic Perturbation Theory - Perturbation theory for the electron propagator - Introducing diagrams - Feynman diagrams - Time-ordered or Goldstone diagrams - Part III Approximation and Computational Schemes - Self-energy and the Dyson equation - Algebraic-diagrammatic construction (ADC) - Direct ADC procedure for the electron propagator - Intermediate-state representation (ISR) - Order relations and separability - Part IV N-Electronic excitations - Polarization propagator - ADC and ISR approaches to the polarization propagator - Random-phase approximation (RPA) - Part V. A look at related methods - Algebraic propagator methods - Coupled-cluster methods for generalized excitations - Appendix.

LC Subjects

Atomic structure.
Chemistry, Physical and theoretical.
Chemometrics.
Mathematical physics.
Molecular structure.
Physical chemistry.

Other Subjects	Theoretical and Computational Chemistry. Atomic/Molecular Structure and Spectra. Math. Applications in Chemistry. Mathematical Applications in the Physical Sciences. Physical Chemistry. Theoretical, Mathematical and Computational Physics.
Additional formats	Print version: Many-body methods for atoms, molecules and clusters. 9783319936017 (DLC) 2018945078 Printed edition: 9783030066918 Printed edition: 9783319936017 Printed edition: 9783319936031
Series	Lecture Notes in Chemistry, 0342-4901; 94 Lecture Notes in Chemistry, 0342-4901; 94

Metabolomics: From Fundamentals to Clinical Applications

LCCN	2019761923
Type of material	Book
Main title	Metabolomics: From Fundamentals to Clinical Applications / edited by Alessandra Sussulini.
Edition	1st ed. 2017.
Published/Produced	Cham: Springer International Publishing: Imprint: Springer, 2017.
Description	1 online resource (X, 350 pages 56 illustrations, 54 illustrations in color.) text file PDF
ISBN	9783319476568
Related names	Sussulini, Alessandra. editor.
Summary	This book provides a comprehensive view of metabolomics, from the basic concepts, through sample preparation and analytical methodologies, to data interpretation and applications in medicine. It is the first volume to cover metabolomics clinical applications while also emphasizing analytical and statistical features. Moreover, future trends and perspectives in clinical metabolomics are also presented. For researches already experienced in

metabolomics, the book will be useful as an updated definitive reference. For beginners in the field and graduate students, the book will provide detailed information about concepts and experimental aspects in metabolomics, as well as examples and perspectives of applications of this strategy to clinical questions.

Contents Part I: Fundamentals and analytical methodologies in metabolomics - 1. Metabolomics: definitions and significance in systems biology - 2. Collection and preparation of clinical samples for metabolomics - 3. NMR strategies in metabolomics - 5. Strategies involving MS combined with chromatography in metabolomics - 6.MS combined with capillary electrophoresis in metabolomics - Part II: Statistical analysis and data interpretation - 7. Data pretreatment: How to prepare a metabolomics dataset for statistical analysis? - 8. Chemometrics in metabolomics - Understanding metabolomics data: Metabolite identification and biological relevance - Part III: Applications and future trends of metabolomics in clinical cases - 9. Applications of metabolomics in cancer studies - 10. Chronic diseases and lifestyle biomarkers identification by metabolomics - 11. Metabolomics in neurological and psychiatric diseases - 12. Metabolite profiling by MS imaging - 13. Single cell metabolomics .

LC Subjects Gene therapy.
Medical genetics.
Metabolism.

Other Subjects Metabolomics.
Gene Function.
Gene Therapy.

Additional formats Print version: Metabolomics: from fundamentals to clinical applications 9783319476551 (DLC) 2016963672
Printed edition: 9783319476551
Printed edition: 9783319476575
Printed edition: 9783319837925

Series Proteomics, Metabolomics, Interactomics and Systems Biology, 2730-6216; 965
Proteomics, Metabolomics, Interactomics and Systems Biology, 2730-6216; 965

Modelling with the Master Equation: Solution Methods and Applications in Social and Natural Sciences
LCCN 2019759202
Type of material Book
Personal name Haag, Günter, author.
Main title Modelling with the Master Equation: Solution Methods and Applications in Social and Natural Sciences / by Günter Haag.
Edition 1st ed. 2017.
Published/Produced Cham: Springer International Publishing: Imprint: Springer, 2017.
Description 1 online resource (XXV, 342 pages 109 illustrations, 19 illustrations in color.)
text file PDF
ISBN 9783319603001
Summary This book presents the theory and practical applications of the Master equation approach, which provides a powerful general framework for model building in a variety of disciplines. The aim of the book is to not only highlight different mathematical solution methods, but also reveal their potential by means of practical examples. Part I of the book, which can be used as a toolbox, introduces selected statistical fundamentals and solution methods for the Master equation. In Part II and Part III, the Master equation approach is applied to important applications in the natural and social sciences. The case studies presented mainly hail from the social sciences, including urban and regional dynamics, population dynamics, dynamic decision theory, opinion formation and traffic dynamics; however, some applications from physics and chemistry are treated as well, underlining the interdisciplinary modelling

	potential of the Master equation approach. Drawing upon the author's extensive teaching and research experience and consulting work, the book offers a valuable guide for researchers, graduate students and professionals alike.
Contents	Preface - 1 Introduction - 2 Statistical Fundamentals - 3 Derivation of the Chapman-Kolmogorov Equation and the Master Equation - 4 Solution Methods of Master Equations - 5 Some Applications in Physics and Chemistry - 6 The Master Equation in Dynamic Decision Theory - 7 Applications in Population Dynamics - 8 Migration - 9 Urban Dynamics and Transport.
LC Subjects	Chemometrics. Dynamical systems. Emigration and immigration. Game theory. Regional economics. Spatial economics. Statistical physics. Statistics.
Other Subjects	Statistics for Social Sciences, Humanities, Law. Complex Systems. Game Theory, Economics, Social and Behav. Sciences. Math. Applications in Chemistry. Migration. Regional/Spatial Science.
Additional formats	Print version: Modelling with the master equation: solution methods and applications in natural and social sciences 9783319602998 (DLC) 2017945902 Printed edition: 9783319602998 Printed edition: 9783319603018 Printed edition: 9783319868301

Multiblock data fusion in statistics and machine learning

LCCN	2022009806
Type of material	Book
Personal name	Smilde, Age K., author.
Main title	Multiblock data fusion in statistics and machine learning / Age Smilde, Tormod T Næs, Kristian H Liland.
Published/Produced	Chichester, West Sussex, UK: John Wiley & Sons, 2022.
ISBN	9781119600961 (hardback) (pdf) (epub) (ebook)
LC classification	Q175 .S614 2022
Related names	Næs, Tormod, author. Liland, Kristian H. (Kristian Hovde), author.
Summary	"Combining information from two or possibly several blocks of data is gaining increased attention and importance in several areas of science and industry. Typical examples can be found in chemistry, spectroscopy, metabolomics, genomics, systems biology and sensory science. Many methods and procedures have been proposed and used in practice. The area goes under different names: data integration, data fusion, multiblock analyses, multiset analyses and a few more. This book is an attempt to give an up-to-date treatment of the most used and important methods within an important branch of the area; namely methods based on so-called components or latent variables. These methods have already obtained an enormous attention in for instance chemometrics, bioinformatics, machine learning and sensometrics and have proved to be important both for prediction and interpretation. The book is primarily a description of methodologies, but most of the methods will be illustrated by examples from the above-mentioned areas. The book is written such that both users of the methods as well as method

developers will hopefully find sections of interest. In the end of the book there is a description of a software package developed particularly for the book. This package is freely available in R and covers many of the methods discussed. To distinguish the different type of methods from each other, the book is divided into five parts. Part I is introduction and preliminary concepts. Part II is the core of the book containing the main unsupervised and supervised methods. Part III deals with more complex structures and, finally, Part IV discusses alternative unsupervised and supervised methods. The book ends with Part V discussing the available software. Our recommendation for reading the book are as follows. A minimum read of the book would involve chapters 1, 2, 3, 5 and 7. Chapters 4, 6 and 8 are more specialized and chapters 9 and 10 contain methods we think are more advanced or less obvious to use. We feel privileged to have so many friendly colleagues who were willing to spend their time on helping us to improve the book by reading separate chapters. We would like to express our thanks to: Rasmus Bro, Margriet Hendriks, Ulf Indahl, Henk Kiers, Ingrid MaÌ‹ge, Federico Marini, AÌ‹smund Rinnan, Rosaria Romano, Lars Erik Solberg, Marieke Timmerman, Oliver Tomic, Johan Westerhuis and Barry Wise. Of course, the correctness of the final text is fully our responsibility!"-- Provided by publisher.

LC Subjects Science--Statistical methods.
Life sciences--Statistical methods.
Statistical matching.
Machine learning.

Notes Includes bibliographical references and index.

Additional formats Online version: Smilde, Age K. Multiblock data fusion in statistics and machine learning Chichester, West Sussex, UK: John Wiley & Sons, 2022 9781119600985 (DLC) 2022009807

Quantitative Infrared Spectroscopy for Understanding of a Condensed Matter

LCCN	2019746648
Type of material	Book
Personal name	Hasegawa, Takeshi. author.
Main title	Quantitative Infrared Spectroscopy for Understanding of a Condensed Matter / by Takeshi Hasegawa.
Edition	1st ed. 2017.
Published/Produced	Tokyo: Springer Japan: Imprint: Springer, 2017.
Description	1 online resource (XI, 200 pages 115 illustrations, 55 illustrations in color.) text file PDF
ISBN	9784431564935
Summary	This book is intended to provide a course of infrared spectroscopy for quantitative analysis, covering both bulk matter and surface/interface analyses. Although the technology of Fourier transform infrared (FT-IR) spectroscopy was established many years ago, the full potential of infrared spectroscopy has not been properly recognized, and its intrinsic potential is still put aside. FT-IR has outstandingly useful characteristics, however, represented by the high sensitivity for monolayer analysis, highly reliable quantitativity, and reproducibility, which are quite suitable for surface and interface analysis. Because infrared spectroscopy provides rich chemical information-for example, hydrogen bonding, molecular conformation, orientation, aggregation, and crystallinity-FT-IR should be the first choice of chemical analysis in a laboratory. In this book, various analytical techniques and basic knowledge of infrared spectroscopy are described in a uniform manner. In particular, techniques for quantitative understanding are particularly focused for the reader's convenience.

Contents	Infrared spectroscopy as a vibrational spectroscopy - Normal modes - Light absorption by a molecule: 1. Understanding by quantum mechanical approach - Selection rule of IR spectroscopy - Light absorption by a molecule: 2. Understanding by electrodynamical approach - Fundamentals of FT-IR - Two representations of spectra: Time- and frequency-domain representations - Fourier transform relationship - Introduction of Michaelson interferometer - Representative detectors - Sampling techniques - Surface analysis using FT-IR - Boundary conditions in electrodynamics - Thin-film approximation - Surface selection rules for surface spectroscopies - Chemometrics for FT-IR - Limitation of Beer's law - Expansion of Beer's law: CLS regression - Inverse Beer's law: ILS regression - Mathematical expansion of CLS: PCA - Merge of ILS and PCA: PCR - Independent residual terms: PLS.
LC Subjects	Chemoinformatics. Materials science. Physical chemistry. Spectroscopy.
Other Subjects	Spectroscopy/Spectrometry. Characterization and Evaluation of Materials. Computer Applications in Chemistry. Physical Chemistry.
Additional formats	Print version: Quantitative infrared spectroscopy for understanding of a condensed matter. 9784431564911 (DLC) 2017935560 Printed edition: 9784431564911 Printed edition: 9784431564928 Printed edition: 9784431568025

Quantitative Methods for Food Safety and Quality in the Vegetable Industry

LCCN	2019747533
Type of material	Book
Main title	Quantitative Methods for Food Safety and Quality in the Vegetable Industry / edited by Fernando Pérez-Rodríguez, Panagiotis Skandamis, Vasilis Valdramidis.
Edition	1st ed. 2018.
Published/Produced	Cham: Springer International Publishing: Imprint: Springer, 2018.
Description	1 online resource (VIII, 303 pages 46 illustrations, 41 illustrations in color.) text file PDF
ISBN	9783319681771
Related names	Pérez-Rodríguez, Fernando. editor. Skandamis, Panagiotis. editor. Valdramidis, Vasilis. editor.
Summary	This book focuses on the food safety challenges in the vegetable industry from primary production to consumption. It describes existing and innovative quantitative methods that could be applied to the vegetable industry for food safety and quality, and suggests ways in which such methods can be applied for risk assessment. Examples of application of food safety objectives and other risk metrics for microbial risk management in the vegetable industry are presented. The work also introduces readers to new preservation and packaging methods, advanced oxidative processes (AOPs) for disinfection, product shelf-life determination methods, and rapid analytic methods for quality assessment based on chemometrics applications, thus providing a quantitative basis for the most important aspects concerning safety and quality in the vegetable sector.

Contents 1. Quantitative Methods for Food Safety and Quality in the Vegetable Industry - 2. Quality and Safety Management Systems in the Production of Vegetables - 3. Relevant Pathogenic Spoilage Microorganisms in Vegetable Products - 4. Water and Wastewater use in the Fresh Produce Industry: Food safety and Environmental Implications - 5. Advanced Oxidation Processes (AOPs) and Quantitative Analysis for Disinfection and Treatment of Water in the Vegetable Industry - 6. Quality of Vegetable Products: Assessment of Physical-chemical and Microbiological Changes in Vegetables Products by Non-destructive Methods - 7. Quantifying Human Health Risks Associated with Microbiological Contamination of Fresh Vegetables - 8. Quantitative Approaches for Microbial Risk Management in the Vegetable Industry: Case-studies of Application of Food Safety Objectives and other Risk Metrics in the Vegetable Industry - 9. Optimal Packaging Design and Innovative Packaging Technologies for Minimally Processed Fresh Produce - 10. Ensuring Fresh Produce Safety and Quality by Utilizing Predictive Growth Models and Predictive Microbiology Software Tools - 11. Quantitative Tools and Procedures for Shelf Life Determination in Minimally Processed Fruits and Vegetables - 12. Quantitative Methods for Life Cycle Assessment (LCA) Applied to the Vegetable Industry.

LC Subjects Food-Biotechnology.
Microbiology.
Public health.

Other Subjects Food Microbiology.
Food Science.
Public Health.

Additional formats	Print version: Quantitative methods for food safety and quality in the vegetable industry 9783319681757 (DLC) 2018930239 Printed edition: 9783319681757 Printed edition: 9783319681764 Printed edition: 9783319885582
Series	Practical Approaches, 2626-7578 Practical Approaches, 2626-7578

Reaction Kinetics: Exercises, Programs and Theorems: Mathematica for Deterministic and Stochastic Kinetics

LCCN	2019738069
Type of material	Book
Personal name	Tóth, János. author.
Main title	Reaction Kinetics: Exercises, Programs and Theorems: Mathematica for Deterministic and Stochastic Kinetics / by János Tóth, Attila László Nagy, Dávid Papp.
Edition	1st ed. 2018.
Published/Produced	New York, NY: Springer New York: Imprint: Springer, 2018.
Description	1 online resource (XXIV, 469 pages 81 illustrations, 80 illustrations in color.) text file PDF
ISBN	9781493986439
Related names	Nagy, Attila László. author. Papp, Dávid. author.
Summary	Fifty years ago, a new approach to reaction kinetics began to emerge: one based on mathematical models of reaction kinetics, or formal reaction kinetics. Since then, there has been a rapid and accelerated development in both deterministic and stochastic kinetics, primarily because mathematicians studying differential equations and algebraic geometry have taken an interest in the nonlinear differential equations of kinetics, which are relatively simple, yet capable of depicting complex behavior such as oscillation, chaos, and pattern formation. The development of stochastic

models was triggered by the fact that novel methods made it possible to measure molecules individually. Now it is high time to make the results of the last half-century available to a larger audience: students of chemistry, chemical engineering and biochemistry, not to mention applied mathematics. Based on recent papers, this book presents the most important concepts and results, together with a wealth of solved exercises. The book is accompanied by the authors' Mathematica package, ReactionKinetics, which helps both students and scholars in their everyday work, and which can be downloaded from http://extras.springer.com/ and also from the authors' websites. Further, the large set of unsolved problems provided may serve as a springboard for individual research.

Contents	The General Framework - The Continuous Time Continuous State Deterministic Model - The Continuous Time Discrete State Stochastic Model - Selected Addenda.
LC Subjects	Chemometrics. Mathematical physics. Systems biology. Physics. Physical chemistry.
Other Subjects	Math. Applications in Chemistry. Mathematical Applications in the Physical Sciences. Systems Biology. Applications of Graph Theory and Complex Networks. Physical Chemistry.
Additional formats	Printed edition: 9781493986415 Printed edition: 9781493986422 Printed edition: 9781493993512

Recent advances in pharmaceutical science

LCCN	2021932715
Type of material	Book
Main title	Recent advances in pharmaceutical science / Ralph Hahn.
Published/Produced	Forest Hills: Foster Academics, 2021.
ISBN	9781646460687 (hardcover)
Related names	Hahn, Ralph, editor.
Summary	"Pharmaceutical sciences are a group of interdisciplinary fields of study that is concerned with the delivery, design, action, and disposition of drugs. They use tools like physical, inorganic, analytical, and biochemical chemistry, epidemiology, statistics, chemical engineering, physics, chemometrics, etc. The pharmaceutical sciences are categorized into four main branches, namely, pharmacology, pharmaceutical chemistry, pharmaceutics, and pharmacognosy. Pharmacology studies the biochemical and physiological effects of drugs on human beings and pharmaceutical chemistry is the study of drug design. Pharmaceutics focuses on the study and design of drug formulation for best delivery, stability, and patient acceptance. Pharmacognosy refers to the study of medicines that are derived from natural sources. This book discusses the fundamentals as well as modern approaches to pharmaceutical science. It traces the progress of this field and highlights some of its key concepts and applications. It aims to equip students and experts with the advanced topics and upcoming concepts in this area"-- Provided by publisher.

Statistical and multivariate analysis in material science

LCCN	2020043491
Type of material	Book
Main title	Statistical and multivariate analysis in material science / editor, Giorgio Luciano, National Research Council of Italy, Institute for the Study of Macromolecules, Genova, Italy.
Edition	First edition.
Published/Produced	Boca Raton: CRC Press, Taylor & Francis Group, 2021.
Description	xiii, 276 pages; 25 cm
ISBN	9781138196308 (hardcover)
LC classification	TA404.3 .S73 2021
Related names	Luciano, Giorgio, 1975- editor.
Summary	"This book introduces the reader to univariate and multivariate statistics applied to material science in an easy non-mathematical approach. Contains several case studies and tutorials in order to help readers apply the techniques described in this book on their own data. The book will interest scientists and advanced students specializing in material science, corrosion science, chemometrics"--Provided by publisher.
LC Subjects	Materials science--Statistical methods. Multivariate analysis.
Notes	Includes bibliographical references and index.

Statistical and multivariate analysis in material science

LCCN	2020756695
Type of material	Book
Main title	Statistical and multivariate analysis in material science / editor, Giorgio Luciano, National Research Council of Italy, Institute for the Study of Macromolecules, Genova, Italy.
Edition	First edition.
Published/Produced	Boca Raton: CRC Press, Taylor and Francis Group, 2021.
Description	1 online resource (xiii, 276 pages)

ISBN	9781315302256 ebook (hardcover)
LC classification	TA404.3
Related names	Luciano, Giorgio, 1975- editor.
Summary	"This book introduces the reader to univariate and multivariate statistics applied to material science in an easy non-mathematical approach. Contains several case studies and tutorials in order to help readers apply the techniques described in this book on their own data. The book will interest scientists and advanced students specializing in material science, corrosion science, chemometrics"--Provided by publisher.
LC Subjects	Materials science--Statistical methods. Multivariate analysis.
Notes	Includes bibliographical references and index.
Additional formats	Print version: Statistical and multivariate analysis in material science Boca Raton: CRC Press, Taylor & Francis Group, 2021. 9781138196308 (DLC) 2020043491

Statistics and chemometrics for analytical chemistry.

LCCN	2017043378
Type of material	Book
Personal name	Miller, J. N. (James N.), 1943- author.
Main title	Statistics and chemometrics for analytical chemistry.
Edition	Seventh edition / James N. Miller, Jane C. Miller, Robert D. Miller.
Published/Produced	Harlow, United Kingdom: Pearson Education Limited, 2018.
ISBN	9781292186719 (print)
LC classification	QD75.4.C45 M54 2018
Related names	Miller, J. C. (Jane Charlotte), author. Miller, Robert D. (Chemist), author.

Contents	Introduction - Statistics of repeated measurements - Significance tests - The quality of analytical measurements - Calibration methods in instrumental analysis: regression and correlation - Non-parametric and robust methods - Experimental design and optimisation - Multivariate analysis.
LC Subjects	Chemometrics--Textbooks.

Sunflowers: cultivation, uses and ecological significance

LCCN	2019059681
Type of material	Book
Main title	Sunflowers: cultivation, uses and ecological significance / [Érico de Sá Petit Lobão].
Published/Produced	[Hauppauge], New York: Nova Science Publishers, [2019]
Description	1 online resource
ISBN	9781536171969 (adobe pdf) (paperback)
LC classification	SB413.S88
Related names	De Sá Petit Lobão, Érico, editor.
Summary	"Sunflower: Cultivation, Uses and Ecological Significance is a work that brings together renowned researchers from four different countries with experience and scientific competence in the research of Hellianthus annus L. The book consists of six chapters that present information referenced in years of research on various topics such as: a short sunflower review; an organic fertilization revision; the stages of sunflower development; the sunflower under water stress conditions; the sunflower genetic variability; and the development of composite materials. In each section of this book the authors have provided a little feedback on their extensive work from years of research devoted to the knowledge, domestication and improvement of the use of the species. There are a huge range of sunflower varieties currently being used by modern civilization for different purposes and applications. It is an extremely versatile species that can deliver

	extraordinary results in different sectors, whether as a highly healthy food in the food industry or as a strategic oilseed for the bioenergy and biofuels segment. It would be impossible to gather all the available information in a single work, however, this small work aims to contribute to the global collection about the species and guide future works that are still needed for the best use of this incredible plant that is the sunflower"-- Provided by publisher.
Contents	Short Sunflower History Review / Sergey Gontcharov - Sunflower Crop under Organic Fertilization in Brazil / Érico de Sá Petit Lobão, Albericio Pereira Andrade, Pedro Fernandes Dantas, Javob Silva Souto, José Marques Pereira and Dan Érico Lobão - Stages of Sunflower Development: Correlations in Male and Female Reproductive Structures Formation / Olga N. Voronova - Classification of Sunflower under Water Stress Conditions by Means of Spectral Reflectance and Chemometrics / Antonio José Steidle Neto and Daniela de Carvalho Lopes - Genetic Variability in Sunflower after Mutagenic Treatment of Seeds and Immature Embryos / A. I. Soroka and V. A. Lyakh - Development of Environmentally Friendly Composite Materials Based on Sunflower Oil and Alfa Fibers / Sihem Kadem, Ratiba Irinislimane and Naima Belhaneche-Bensemra.
LC Subjects	Sunflowers. Sunflowers--Utilization. Sunflowers--Brazil.
Notes	Includes bibliographical references and index.
Additional formats	Print version: Sunflowers [Hauppauge], New York: Nova Science Publishers, [2019] 9781536171952 (DLC) 2019059680
Series	Plant science research and practices Plant science research and practices.

Sunflowers: cultivation, uses and ecological significance

LCCN	2019059680
Type of material	Book
Main title	Sunflowers: cultivation, uses and ecological significance / [Érico de Sá Petit Lobão].
Published/Produced	[Hauppauge], New York: Nova Science Publishers, [2020]
Description	xi, 144 pages: illustrations; 24 cm
ISBN	9781536171952 (paperback) (adobe pdf)
LC classification	SB413.S88 S862 2020
Related names	De Sá Petit Lobão, Érico, editor.
Summary	"Sunflower: Cultivation, Uses and Ecological Significance is a work that brings together renowned researchers from four different countries with experience and scientific competence in the research of Hellianthus annus L. The book consists of six chapters that present information referenced in years of research on various topics such as: a short sunflower review; an organic fertilization revision; the stages of sunflower development; the sunflower under water stress conditions; the sunflower genetic variability; and the development of composite materials. In each section of this book the authors have provided a little feedback on their extensive work from years of research devoted to the knowledge, domestication and improvement of the use of the species. There are a huge range of sunflower varieties currently being used by modern civilization for different purposes and applications. It is an extremely versatile species that can deliver extraordinary results in different sectors, whether as a highly healthy food in the food industry or as a strategic oilseed for the bioenergy and biofuels segment. It would be impossible to gather all the available information in a single work, however, this small work aims to contribute to the global collection about the species and guide future works

	that are still needed for the best use of this incredible plant that is the sunflower"-- Provided by publisher.
Contents	Short Sunflower History Review / Sergey Gontcharov - Sunflower Crop under Organic Fertilization in Brazil / Érico de Sá Petit Lobão, Alberício Pereira Andrade, Pedro Fernandes Dantas, Javob Silva Souto, José Marques Pereira and Dan Érico Lobão - Stages of Sunflower Development: Correlations in Male and Female Reproductive Structures Formation / Olga N. Voronova - Classification of Sunflower under Water Stress Conditions by Means of Spectral Reflectance and Chemometrics / Antonio José Steidle Neto and Daniela de Carvalho Lopes - Genetic Variability in Sunflower after Mutagenic Treatment of Seeds and Immature Embryos / A. I. Soroka and V. A. Lyakh - Development of Environmentally Friendly Composite Materials Based on Sunflower Oil and Alfa Fibers / Sihem Kadem, Ratiba Irinislimane and Naima Belhaneche-Bensemra.
LC Subjects	Sunflowers. Sunflowers--Utilization. Sunflowers--Brazil.
Notes	Includes bibliographical references and index.
Additional formats	Online version: Sunflowers [Hauppauge], New York: Nova Science Publishers, [2019] 9781536171969 (DLC) 2019059681
Series	Plant science research and practices Plant science research and practices.

Surface-enhanced raman spectroscopy: methods, analysis and research

LCCN	2019018262
Type of material	Book
Main title	Surface-enhanced raman spectroscopy: methods, analysis and research / edited by Vinay Bhardwaj, PhD, Department of Biomedical Engineering, Rutgers, The State University of New Jersey,

	Piscataway, NJ, USA and Anthony J. McGoron, PhD, Professor, Department of Biomedical Engineering, Associate Dean for Academic Affairs, College of Engineering and Computing, Florida International University, Miami, FL, US.
Published/Produced	Hauppauge, New York: Nova Science Publisher's Inc., 2019.
Description	1 online resource.
ISBN	9781536153798 ()
LC classification	RS380
Related names	Bhardwaj, Vinay, author. McGoron, Anthony J., author.
Summary	"The primary focus of this textbook is to try to cover the fundamentals in SERS methods of preparing substrates and sensors, and efficient spectral analysis approaches like chemometrics for diverse applications, listed in the paragraph above. However, with a goal to give a true picture to the audience when assessing the scope of spontaneous Raman and Surface-enhanced Raman for specific applications, we have briefly introduced industrially successful examples of other types of Raman spectroscopy techniques, such as SORS, TRS, and CARS"-- Provided by publisher.
Contents	Fundamentals of surface-enhanced raman spectroscopy (SERS) / G. B. V. S. Lakshmi, PhD, Vinay Bhardwaj, PhD, and Pratima S. Solanki, PhD, Special Centre for Nanoscience, Jawaharlal Nehru University, New Delhi, India, and others - Applications of SERS for ultrasensitive detection of small molecules / Frank J. Owens, PhD, Department of Physics, Hunter College of the City University of New York, NY, US / Applications of SERS in cancer diagnosis and therapy monitoring / Supriya Srinivasan, PhD, and Alicia Fernandez-Fernandez, PhD, University of Miami, Miami-Florida, US, and others).

LC Subjects	Pharmaceutical biotechnology--Technique. Molecular biotechnology--Technique. Raman spectroscopy. Vibrational spectra. Raman effect, Surface enhanced.
Notes	Includes bibliographical references and index.
Additional formats	Print version: Surface-enhanced raman spectroscopy Hauppauge, New York: Nova Science Publisher's Inc., 2019 9781536148619 (DLC) 2019017073
Series	Nanotechnology science and technology

Surface-enhanced raman spectroscopy: methods, analysis and research

LCCN	2019017073
Type of material	Book
Main title	Surface-enhanced raman spectroscopy: methods, analysis and research / Vinay Bhardwaj and Anthony J. McGoron, editors.
Published/Produced	New York: Nova Science Publishers, 2019.
Description	xiii, 265 pages: illustrations; 26 cm.
ISBN	9781536148619 (hardcover; alk. paper) (ebook)
LC classification	RS380 .S87 2019
Related names	Bhardwaj, Vinay, author. McGoron, Anthony J., author.
Summary	"The primary focus of this textbook is to try to cover the fundamentals in SERS methods of preparing substrates and sensors, and efficient spectral analysis approaches like chemometrics for diverse applications, listed in the paragraph above. However, with a goal to give a true picture to the audience when assessing the scope of spontaneous Raman and Surface-enhanced Raman for specific applications, we have briefly introduced industrially successful examples of other types of Raman spectroscopy techniques, such as SORS, TRS, and CARS"-- Provided by publisher.

Contents	Fundamentals of surface-enhanced raman spectroscopy (SERS) / G. B. V. S. Lakshmi, PhD, Vinay Bhardwaj, PhD, and Pratima S. Solanki, PhD, Special Centre for Nanoscience, Jawaharlal Nehru University, New Delhi, India, and others - Applications of SERS for ultrasensitive detection of small molecules / Frank J. Owens, PhD, Department of Physics, Hunter College of the City University of New York, NY, US / Applications of SERS in cancer diagnosis and therapy monitoring / Supriya Srinivasan, PhD, and Alicia Fernandez-Fernandez, PhD, University of Miami, Miami-Florida, US, and others).
LC Subjects	Pharmaceutical biotechnology--Technique. Molecular biotechnology--Technique. Raman spectroscopy. Vibrational spectra. Raman effect, Surface enhanced.
Notes	Includes bibliographical references and index.
Additional formats	Online version: Surface-enhanced raman spectroscopy Hauppauge, New York: Nova Science Publisher's Inc., 2019 9781536153798 (DLC) 2019018262
Series	Nanotechnology science and technology

The Suquía River Basin (Córdoba, Argentina): An Integrated Study on its Hydrology, Pollution, Effects on Native Biota and Models to Evaluate Changes in Water Quality

LCCN	2019746456
Type of material	Book
Main title	The Suquía River Basin (Córdoba, Argentina): An Integrated Study on its Hydrology, Pollution, Effects on Native Biota and Models to Evaluate Changes in Water Quality / edited by Daniel A. Wunderlin.
Edition	1st ed. 2018.
Published/Produced	Cham: Springer International Publishing: Imprint: Springer, 2018.

Description	1 online resource (XIII, 207 pages 120 illustrations, 43 illustrations in color.) text file PDF
ISBN	9783319677576
Related names	Wunderlin, Daniel A. editor.
Summary	This book covers all relevant aspects involved in conducting water quality surveys of rivers. Focusing on the Suquía River (Argentina), it starts with a description of its hydrology and hydraulics before discussing the spatial and temporal changes in its water quality. A short chapter is dedicated to its limnology and particular characteristics due to the presence of a constructed dam and reservoir. Further chapters explore the changes in native biota inhabiting the river basin and assess various physical, chemical and microbiological parameters, as well as organic pollutants, metals and metalloids resulting from human activities. The book closes with a chapter providing an integrated approach to surveying river water quality, focusing on methods for data mining, from the simple use of water quality indexes to more formal mathematical-statistical methods (chemometrics) for evaluating changes in water quality. The book offers an invaluable reference guide for all environmental managers and scientists involved in water resources, their quality and management.
Contents	Hydrology and Hydraulics of the Suquía River Basin - Limnology of the San Roque Reservoir - Native Biota Along the Suquía River Basin - Spatial and Temporal Changes in Water Quality Along the Basin - Metals and Metalloids in Water and Sediment of the Suquía River Basin. Spatial and Temporal changes - Organic Pollutants in the Suquía River Basin - Integrating Data from Suquía River Basin. Chemometrics and other concepts.

LC Subjects	Analytical chemistry. Environmental chemistry. Hydrogeology. Hydrology. Water pollution. Water quality.
Other Subjects	Environmental Chemistry. Analytical Chemistry. Hydrogeology. Hydrology/Water Resources. Water Quality/Water Pollution.
Additional formats	Print version: The Suquia River basin (Cordoba, Argentina) 9783319677552 (DLC) 2018935113 Printed edition: 9783030098100 Printed edition: 9783319677552 Printed edition: 9783319677569
Series	The Handbook of Environmental Chemistry, 1867-979X; 62 The Handbook of Environmental Chemistry, 1867-979X; 62

Index

D

E

F